ENVIRONMENTAL PLANNING IN THE NETHERLANDS: TOO GOOD TO BE TRUE

Environmental Planning in the Netherlands: Too Good to be True

From Command-and-Control Planning to Shared Governance

GERT DE ROO
University of Groningen, The Netherlands

Routledge
Taylor & Francis Group
LONDON AND NEW YORK

First published 2003 by Ashgate Publishing

2 Park Square, Milton Park, Abingdon, Oxon OX14 4RN
711 Third Avenue, New York, NY 10017, USA

Routledge is an imprint of the Taylor & Francis Group, an informa business

First issued in paperback 2016

Translation subsidised by the Netherlands Organisation for Scientific Research (NWO). Translated by the Language Centre of the University of Groningen.

British Library Cataloguing in Publication Data
Roo, Gert de
Environmental planning in the Netherlands : too good to be true. - (Urban planning and environment)
1. Environmental policy - Netherlands 2. Environmental policy - Netherlands - Citizen participation 3. Environmental policy - Netherlands - Evaluation
I. Title
363.7'009492

Library of Congress Cataloging-in-Publication Data
Roo, Gert de.
[Planning per se, planning per saldo. English]
Environmental planning in the Netherlands : too good to be true : from command and control planning to shared governance / Gert De Roo.
p. cm. – (Urban planning and environment)
Includes bibliographical references and index.
ISBN 0-7546-3845-6 (alk. paper)
1. City planning–Environmental aspects–Netherlands. 2. Regional planning–Environmental aspects–Netherlands. 3. Urban ecology–Netherlands. 4. Environmental policy–Netherlands. I. Title. II. Series.
HT169.N4R6613 2003
307.1'2–dc21 2003052123

ISBN 978-0-7546-3845-2 (hbk)
ISBN 978-1-138-25484-8 (pbk)

Transfered to Digital Printing in 2010

Contents

List of Figures

List of Tables

Preface

This book is an analysis of environmental policy in the Netherlands, which has been the basis for environmental policy in many other countries. Dutch policy is seen as progressive and far-reaching. The latter feature in particular explains why it generates so much interest. However, current developments are completely transforming this image of environmental policy. The changes are the result of increasing criticism levelled at traditional far-reaching environmental policy by stakeholders who have experienced its consequences in the Netherlands. Dutch environmental policy appeared to be 'too good to be true' and has ceased to address the policy issues encountered in practice today. It was time to set a new course ...

This book discusses, among other things, the deregulation and decentralisation of Dutch environmental policy. Given the speed at which this process has unfolded, it appears that stakeholders have a great deal of confidence in the 'new-look' policy. But this is only partly true. Many people – among them some of the people involved in formulating the decentralised policy – have reservations about the path decentralisation is taking and its possible consequences. How can we take optimistic decisions now, only to realise at some point in the future that the environment has paid the price for those decisions? Should we loosen the policy reins in order not to stand in the way of progress – including progress with regard to environmental matters? This study reflects on the administrative aspects of this trial-and-error approach.

The rise and fall of the system of environmental standards was one of the main reasons for writing this book. Until the early 1990s, the environmental standards enforced by the Dutch government propelled environmental planning in the Netherlands to great heights. Thanks to a prescriptive system of standards, environmental planning evolved into a fully recognised policy field that achieved impressive results and was able to withstand competition from other fields of policy. Indeed, this system of rigid, quantitative standards virtually excluded all interests other than environmental interests.

Ultimately, however, environmental standards proved to be too successful for their own good and the tide turned. It was particularly surprising to see how within such a short space of time, this instrument, which had been the backbone of Dutch environmental policy for so long, became an anathema that people preferred not to discuss. By the mid-1990s, environmental standards were 'out' and the curtain appeared to have fallen. There was now very little to stand in the way of the decentralisation of environmental policy.

In the wake of this almost inexplicable and sudden aversion to standards, it briefly seemed as if the pendulum of legislation would swing to the other side, introducing such concepts as 'public consensus', 'interaction' and 'compensation' – concepts which could provide a new theoretical foothold. Participative decision-making and shared governance would be the cornerstone of a new environmental policy.

This u-turn has raised many questions and gives the impression that over-commitment to an old belief has been replaced by over-commitment to a new one. It remains to be seen whether the new belief will produce the expected results. Practical questions such as these are the basis for this analysis of decision-making and planning.

A second analysis is woven into this analysis of the development of Dutch environmental policy, namely a synthesis relating to aspects of public administration. The purpose of this synthesis is to clarify the role played by government departments in terms of planning-based action and decision-making. One of the results of this study is a pluriform model for planning-based action, which can be used to establish connections between goals (in particular environmental standards) and involvement in decision-making. The model also links the concepts of 'efficiency' and 'effectiveness' in planning and decision-making.

Reference is also made to the categorisation of environmental-planning issues according to complexity. Above all, complexity is proposed here as a factor that can bridge the gulf between apparently irreconcilable theories, visions and concepts in planning, thereby enabling us to follow and explain the developments that take place in environmental planning. The underlying purpose is to contribute to the development of new theories. In giving meaning to the concept of complexity, my aim has been to facilitate the bridge-building process between the many and diverse theoretical perspectives, visions and concepts. The result is a theoretical framework that – in my view – can further our understanding of decision-making and planning, facilitate policy choices and provide insight into the possible consequences of policies – in particular environmental policy.

The third Dutch edition of this book has already been published. The English-language edition of this book has been made possible by the Netherlands Organization for Scientific Research (NWO). I would like to express my profound gratitude to the NWO for the subsidy it provided. I am also extremely grateful to the Environmental Department of the Municipality of Amsterdam for making available the additional funding required for the translation of Chapter 7. The Language Centre of the University of Groningen was responsible for the translation of the text. Special thanks go to Yvette Mead and Julia Harvey. I am grateful to Henk Voogd, Donald Miller, Hans Verspoor and Theo Aquarius, who were always available to give advice while I was writing the first Dutch edition of this book. Their advice has also made a valuable contribution to the English-language version. I wish to express my sincere thanks to Tamara Kaspers, Miranda Jager, Martine de Jong and Floris Bruil for their valuable assistance in preparing this edition for publication. Last but not least, my thanks go to Valerie Rose of Ashgate Publishing, who added this book to her list without a moment's hesitation.

Gert de Roo *Groningen, July 2003*

Map of The Netherlands

Enschede, 13 May 2000 (E.P. van der Wal)
Vergunning (Licence)

Chapter 1

Introduction

Conflicts, Decision-Making and Complexity in Environmental Planning

> No 'invisible hand' can be relied on to produce a good arrangement of the whole from a combination of separate treatments of the parts. It is, therefore, necessary that an authority of wider reach should intervene and should tackle the collective problems of beauty, of air and light (A.C. Pigou 1920; 196).

1.1 Enschede, 13 May 2000

The nightmare began at three o'clock in the afternoon. The first signs of a drama stained the clear blue sky. The billowing white smoke, bangs and the crack of fireworks seemed totally out of place. Those people who, despite the fine weather, had been sitting indoors were drawn outside by the spectacle. Somewhere in the north of the city, something was wrong. A few minutes after half past three, two enormous explosions transformed the Roombeek neighbourhood into a disaster area. The SE Fireworks factory had been blown sky high, taking with it most of the surrounding residential area. Television reports showed the full extent of the inferno. Hard figures confirmed the tragedy: 500 homes were destroyed or seriously damaged; 1500 homes suffered moderate damage, and 500 people were injured. Twenty-one people lost their lives, but the figure would have been much higher if more people had been inside their homes, rather than outside making the most of the beautiful weather on a free Saturday afternoon, watching the strange, prolonged firework display. The area was a pre-war neighbourhood, with social housing located around industrial areas that cannot entirely conceal the decline of the textile industry. The situation would have been even worse if the explosion had affected the six thousand litres of ammonia stored at the nearby Grolsch brewery. As the smoke cleared and the gunpowder dispersed, revealing so much emptiness and suffering, many people began to ask how such a disaster could have happened in a residential area like this (De Lugt 2000).

This book describes the development of environmental planning in the Netherlands and its consequences for the 'liveability' of our day-to-day environment. The Enschede disaster holds up a mirror to environmental planning in the Netherlands, precisely at a time when all manner of proposals are being made in answer to the question whether we should adopt a more flexible approach towards environmental planning that is based to a greater extent on local circumstances. The Enschede disaster also tells us that environmental planning in the Netherlands, often seen as technically advanced and sophisticated in terms of policy, is actually '*too good to be true*'.

1.2 Reservations about Command-and-Control Planning

This book deals with the day-to-day aspects of environmental planning in the Netherlands, and these are also used as an example. The example is used to determine whether conflicts (i.e. environmental conflicts) can be categorised according to their complexity. This book also attempts to ascertain the extent to which categorisation by complexity determines the decision-making method. In order to address these questions, developments in environmental planning are considered as an empirical problem area. This book appears just as we have reached a turning point in Dutch environmental planning with regard to decision-making processes, and with regard to environmental quality issues, bottlenecks and conflicts. Traditionally, the theory and practice of environmental planning in the Netherlands have been based on universal, guideline-setting environmental standards. For three decades, the government set strict, quantitative standards for human activity, with the aim of 'protecting' the environment. This form of top-down policy works through to regional and local authorities by means of generic norms, and sets limits on other types of policy relating to the living environment. Strict generic standards are, however, increasingly coming under discussion.

Standards can provide a stimulus for human activity – for example, productivity norms – but can also halt or restrict human activity. Environmental standards were traditionally designed to do the latter. Environmental standards are limits defined through policy and are implemented in order to protect the 'environment' against harmful human activity. Setting standards is a highly arbitrary process, and therefore open to discussion. In this, environmental standards are no different from other restrictions or boundaries based on policy. On the contrary, in the 1990s, criticism of environmental standards led to a discussion on the ability of hierarchically imposed norms to solve conflicts on environmental quality and spatial planning. That discussion forms the thread of this book.

This discussion is not so much a discussion of environmental standards as an instrument or means, but rather as a definite policy goal. The application of a standard as a policy goal is a forceful means of achieving a desired level of quality. However, achieving acceptable environmental quality depends on more than environmental policy. Achieving national environmental goals also depends on developments in policy areas such as spatial planning and transport. Friction is inevitable, given that environmental goals are perceived as guideline-setting within these policy areas, and therefore repressive in terms of individual policy goals.

The repressive character of environmental standards is particularly evident in urban contexts. In the 1980s and 1990s, the bundling and concentration of activities formed the foundation for urban development, after periods in which this had been based on post-war reconstruction, 'decentralisation' and 'bundled deconcentration'. The prevailing concept for urban development is now that of the 'compact city'.

Spatial concentration and compact urban development play a more than emphatic role, partly due to the substantial house-building commitments set out in the 1993 Fourth Policy Document on Physical Planning-Plus (VINEX) of the Ministry of Housing, Spatial Planning and the Environment (VROM), and the green urban buffer zones introduced in the Fifth Policy Document (VROM 1995). The various authorities see bundling and concentration, combined with an optimal mix of activities, and improved environmental quality in urban areas as the basic principles of urban development. However, "the evaluation of the first National Environmental Policy Plan [...] revealed that policy designed to achieve dynamic, liveable towns and cities was greatly hampered by rigid environmental regulations that were not geared towards metropolitan reality" (VROM 1995; 5).

We are dealing with issues that are referred to in this book as environmental-/spatial conflicts. This type of conflict is often dealt with in a technical-functional way by converting environmental standards into distances that have to be maintained between environmentally sensitive and environmentally harmful functions, activities or areas. According to its critics, this generic, guideline-setting approach does not allow local authorities sufficient scope to make allowances for specific and often unique local circumstances when dealing with environmental/spatial conflicts. If these objections by local authorities are taken into account, it will no longer be a question only of environmental quality, and other aspects of the local living environment will also have to be considered. It will then no longer be possible to define or describe environmental/spatial conflicts in a straightforward, uniform way. Conflicts will no longer be categorised according to national environmental quality standards, but will be individually defined according to the local situation. This change will affect the technical-functional working method of the national government relating to the environment.

Generic, rigid environmental regulations are a product that – despite the Enschede firework disaster – radiates a great need for full control: command-and-control planning. Increasingly, such regulation is seen as an obstacle to urban development and quality. It creates friction between physical planning and the environment that is referred to as the 'paradox of the compact city' (TK 1993). This paradox has become a familiar battle cry in the call for change in hierarchical, guideline-setting environmental policy based on standards. Groningen City Council is not alone in claiming that "in many cases, the strict adherence to environmental regulations does not lead to optimal quality for inhabitants, or – even from an environmental point of view – to desired urban development" (Gemeente Groningen 1996; 25). The RARO (Spatial Planning Council) is not the only body to ask whether it ought to be possible for 'lower-level authorities to deviate from centrally defined standards' (RARO 1994).

In response to this, guideline-setting environmental standards are used to allow local authorities greater scope than was previously the case to pursue an integral, balanced and area-specific policy. This is also referred to as the 'external integration' of environmental policy whereby, at the various levels of government, environmental and physical planning policy in particular should grow towards each other and be mutually reinforcing. At the same time, central government makes way for local and regional authorities. The Ministry of Housing, Spatial Planning and the Environment (VROM) maintains that “responsible people behave responsibly [...]. Such an approach requires that central government keeps its distance within national aims and standards” (TK 1996; 2). In addition to this process of decentralisation, deregulation has been taking place over an even longer period of time. Deregulation should encourage greater initiative and creativity on the part of local authorities and non-governmental bodies in dealing with matters relating to the local living environment. These processes of decentralisation and deregulation herald a change in what many perceive as the autocratic approach of government towards environmental planning.

The changes are best illustrated by the developments relating to two policy instruments, which were introduced by VROM more or less simultaneously and put to the test during the 1990s. The first change is the rise and fall of integrated environmental zoning as an instrument (§ 5.4 and Chapter 6), an advanced form of national standards policy. While this standard-setting policy was maligned, concepts such as consensus-building, participation and an open planning process began to gain ground. These are key concepts that are used, among others, in the ROM regional policy (§ 5.4). ROM designated-areas policy was introduced as an innovative policy approach at more or less at the same time as integrated environmental zoning. The ROM designated-areas policy is partly based on network-like strategies for ‘complex decision-making’ (§ 4.7). In contrast to integrated environmental zoning, the ROM regional policy enjoyed a positive reception. It is a policy in which environmental interests are no longer sacred, but considered in the light of other local interests. The aim of policy is no longer environmental quality at all costs; other integral considerations relating to the world in which we live also count. This policy approach is more appropriate for the location-specific – complex – circumstances of environmental/spatial conflicts: situation-specific overall planning with a shift towards shared governance.

A number of issues were widely discussed in the 1990s; in addition to the criticism of a national environmental zoning system mentioned above, and the embracing of the ROM regional policy as a ‘socially acceptable’ policy, proposals for a standard-setting odour-emissions policy were rejected twice in the Lower House, soil decontamination based on the multi-functionality criterion and the related system of standards was found to be unaffordable and time-consuming, and the Noise Abatement Act (WGH; Wet geluidhinder) was debated and will be withdrawn (see § 5.5). Nevertheless, the consensus surrounding the national City & Environment project suggests that, at the beginning of the 21st century, standards will remain as the focus in the relationship between environmental policy and other policies relating to the physical environment. However, it is clear that the role of the standards system will change, although policy-maker and parliamentarians are hesitant about change following the Enschede disaster. The developments will not be restricted to the changing use of environmental standards; the role of the various authorities is also under discussion.

In retrospect, the Nunspeet conference of 30 November 1994 was a step in the transition from a hierarchic and guideline-setting environmental policy to one that is balanced against local circumstances. The Nunspeet conference was a policy-evaluating conference at which the various authorities[1] aired their grievances and set out their wishes relating to environmental policy. The purpose was to find a way out of the deadlock between hierarchical standards on the one hand, and the need for local circumstances to be taken into consideration on the other. "Standards relating to a particular field (e.g. external safety or noise pollution) can stand in the way of developments that are desirable based on other considerations (e.g. building in the vicinity of stations)" (VROM 1995; Appendix 1). The conclusion was a recognizable one that was supported by all those involved. The goal that the various authorities had set themselves during the conference was to reach broad conclusions about how environmental policy in the Netherlands should be directed in the future, and the best way to implement it. The conclusions were essentially based on the opinion that a 'new management philosophy for environmental policy' would be sought (VROM 1995; Appendix 1). This search characterised environmental planning during the second half of the 1990s.

There are those who argue that a policy based on consensus is preferable to policy that is optimal in the technical sense. Policy must be understood and developed on the basis of a social dialogue. It is also claimed that each environmental problem has its own specific characteristics. This means, among other things, that a local approach is increasingly preferred to a more generic approach. That is why it is considered important to allow sufficient scope at local and regional level for developing area-specific policies. There must also be sufficient scope at that level to set priorities. One solution under consideration is a form of 'self-regulation within guidelines'. It must be possible to deviate from standards, and there must be scope for public discussion of the consequences of doing so. The reasoning behind this approach is that decisions reached at the level in question are more effective than measures imposed from above. Environmental goals will have to be considered with other forms of area-related qualities, "whereby an acceptable overall solution within the environmental field can lead to the most appropriate solution for the area as a whole" (VROM 1995; 11). The government's role will therefore have to be more conducive to creating favourable conditions and facilitating than in the past (VROM 1995; Appendix 1). The formulation of these conclusions[2] brought a certain amount of consensus and calm to the sometimes-turbulent discussion surrounding environmental policy-making (§ 5.5). On the basis of these conclusions, the discussion on the hierarchical and guideline-setting nature of environmental policy was formed into a structured process in order to discover new directions. The resulting proposals have not been without effect (see Chapter 5). A situation-specific and area-oriented overall policy has become an acceptable alternative to generic command-and-control planning. However, it remains to be seen whether such a policy will make a substantial positive contribution that is more beneficial than command-and-control planning.

The above outlines the practice-related problem and describes the national discussion about setting standards in environmental policy, and their implications for other forms of policy relating to the physical environment. The question is whether environmental quality can be defined as a separate area, or whether it should be considered with overlapping issues. From an administrative point of view, it is a question of the degree of emphasis placed on hierarchical or decentralised decision-making. In the first case, problem definition will be based on a generic approach. In the latter case, each environmental/spatial conflict will be considered unique and the problem will be defined for each individual case, thereby allowing for the context in which the problem has arisen. Consequently, the conflict becomes more complex than when the problem is considered out of context.

Here, the question is whether developments and changes in environmental policy can be directed by using the term 'complexity' to elucidate the policy issue. It is a question of whether issues should be considered as straightforward or complex to a greater or lesser extent. Our reality is such that it is not possible to define all issues in a straightforward way using general formulations such as generic environmental standards. Neither are strategies for 'complex decision-making' appropriate in all cases, with popular concepts such as participation, consensus, self-regulation and communicative action. It is not likely that network-related strategies will completely replace policies that set standards. But what is likely? Will greater scope for market forces be the answer, in accordance with the spirit of the 1990s: the idea that public issues can also be developed and managed according to the principles of the market mechanism? Is a policy geared to the type of issue a feasible alternative? Or is it much more a question of placing or shifting emphasis in a pluriform policy? These are questions that fit in an academic exercise in the underpinning and reasoning of standpoints and visions regarding the question whether governmental management, planning and policy development should be more (or less) centralised, or whether it should be decentralised.

The empirical objects of this book are defined in section 1.3. These objects are partly administrative in nature, and can partly be described in physical-spatial and societal terms. It is not only a question of the environmental/spatial conflict itself. Environmental/spatial conflicts will also be considered as a subject of decision-making, the question being whether their complexity can serve as a criterion for decision-making. Decision-making, in particular that relating to the environment, is an administrative subject of study here. Section 1.4 describes how the empirical issues studied can be looked at using theoretical reflection, and how that reflection can be of empirical value. The question here is also whether complexity can be a consideration in decision-making. This question is discussed on section 1.5 as the basis for this book. The chapter ends with a description of how the book is structured.

1.3 Conflicts and Decision-Making as Subjects for Study

The relationship between the complexity of environmental/spatial conflicts and the decision-making process is set out in three related accounts. The first account deals with the concrete problems on which policy is based (see also Part A of this book), referred to as environmental/spatial conflicts. These conflicts are the result of a confrontation between environmental health and hygiene and the physical allocation of space for various functions. The problems are partly material and partly administrative. It is a question of acknowledging, on the one hand, that there is a conflict involving various elements and revealing a clear connection between human activity and its consequences and, on the other hand, of evaluating that conflict and the related policy options. The second account deals with decision-making as an administrative study object. This account is about the relationship between environmental/spatial conflicts on the one hand, and, on the other, the arguments, alternatives and development of environmental policy (see Part C). The harmonisation of environmental policy and spatial-planning policy is also discussed. This section introduces environmental/spatial conflict and related decision-making as objects for study. The third and final account is introduced in section 1.5, and is a theoretical reflection on the material and administrative study objects (see Part B). The emphasis is placed on defining a workable and consistent theoretical perspective, in order to be able to comprehend the world of making decisions and resolving conflicts.

Environmental/Spatial Conflicts as Material and Administrative Objects

The concept 'environmental/spatial conflict' was born at the end of 1994, while research was being carried out into the planning consequences of pilot projects for integrated environmental zoning (Borst et al. 1995, see also § 5.4 and Chapter 6). It refers to issues that relate not only to the spatial consequences of guideline-setting environmental policy, but also to the mutual effects of environmental and spatial planning policy for specific, concrete situations. In a publication discussing the state of flux of environmental zoning (*Milieuzones in Beweging*, Borst et al. 1995), the term 'environmental/spatial conflict' referred in particular to "a conflict situation between industry and homes. This can mean that the environmental impact of industry adversely affects environmental quality in nearby residential areas, but also that it is not possible to build homes close to an industrial area because of that impact" (Borst et al. 1995; 22). The term therefore had a relatively limited meaning, primarily intended to identify issues on which the research for pilot projects for integrated environmental zoning was based.

In this book, the term 'environmental/spatial conflict' refers to more than conflict situations between industrial and residential areas. The meaning of the term is extended to cover those issues of environmental quality and the spatial planning in an area or location that conflict with each other in some way. These conflicts or conflict situations involve clashes between value judgements relating to one or more mutually influential aspects or (f)actors experienced as such by one or more parties. In practice, this can mean that emissions of toxic or carcinogenic substances resulting from industrial, port or transport activities can result in health problems among local residents if housing is not located at a safe distance from the source of the emissions. The effects of air traffic, for example at and in the vicinity of Amsterdam Schiphol Airport, may affect environmental quality to such an extent that residential development is no longer considered wise. Such problems hinder spatial development or threaten the existing spatial constellation because environmental quality is not high enough. The opposite scenario can also occur when environmentally harmful or environmentally sensitive activities and initiatives cannot proceed because of the local spatial layout. Here, environmental/spatial conflict is treated as a physical object, whereby reality falls short of the desired situation and therefore becomes a stimulus for planning and policy. It is not only a question of problems that exist in a physical sense – offensive smells, air pollution, etc. – but also of tension between *policies*, in this case environmental policy and physical-planning policy. Confrontations between these two types of policy tend to involve the role of generic, guideline-setting environmental legislation. In such cases, the question is whether conflict situations in the common areas of environmental and spatial policy can simply be dismissed as 'environment-related'. In the past, for example, it may have been the case that residential development came so close to industrial areas that industrial activity could expand no further. In such cases it is not so much standards that are responsible for hindering new initiatives, but past developments in spatial planning.

These consequences for policy are an indication of the administrative context of environmental/spatial conflicts. From an administrative point of view, environmental-/spatial conflicts can be described as problems in the common area of environmental policy and spatial-planning policy. These problems are the result of regulations, procedures and processes *in* planning and policy, and may concern the inability to meet pre-defined criteria or a conflict of interests. The essence of the problems is often the feasibility of changing an actual situation to a desired situation.

This specific type of problem – environmental/spatial conflict – will be used to consider the merits of changes in environmental policy. This consideration will focus on the changing role of environmental standards and objectives, the implications for liveability in urban areas, the consequences of changes in environmental policy for the co-ordination and integration of environmental and spatial-planning policy, and the shift in the roles of the various authorities and other actors brought about by these changes. The discussion will focus on policy content and organisational aspects (the administrative object) in relation to the discrepancy between social and physical reality and ideal (the material object).

Environmental and Spatial Decision-Making as an Administrative Object

Discussions about the relationship between environmentally-harmful and environmentally-sensitive functions and activities in urban areas cannot be considered as a separate issue. The discussions are constantly linked to developments at policy-making level, one level of abstraction above the level at which the problem is actually experienced. This is the level at which the external integration of environmental and spatial-planning policy must be realised. There is also a policy level at which even more abstract terms are used: deregulation, decentralisation, market mechanism, open-plan processes and participation by non-governmental actors. Neither should discussions on environmental/spatial conflicts be interpreted as a direct result of the aims of decentralisation and deregulation that are evident in several areas of government planning;[3] the features of environmental/spatial conflicts are too specific to be described in these terms. Nevertheless, it is almost inevitable that proposals for decentralisation and deregulation will influence the discussion of environmental/spatial conflict, developments in environmental policy and the co-ordination of environmental and spatial-planning policy (see also § 5.5).

We can therefore identify three levels at which policy development takes place, and which are important in understanding the problems in areas where environmental policy and spatial-planning policy intersect (see also Figure 4.1). In the first place, decentralisation and deregulation, among others, are encouraged in a large number of policy areas at the highest policy-making level in the Netherlands. The option of opening up the policy-making arena to market forces is increasingly mentioned in discussions at this level. The processes of decentralisation, deregulation and introducing market forces into government planning can be interpreted at 'lower levels' as the effect of long-term policy trends. There is a further, more concrete level involving policy that is usually geared towards a specific area for which the government considers itself responsible, such as spatial planning or environmental policy. This study focuses on environmental and spatial-planning policy, in particular the way in which the two policy types can slowly but surely come closer together with a view to improved harmonisation and, where possible, integration of common elements. The most concrete level is the level at which the specific problems are experienced. This is the level at which government policy should eventually be implemented. The discussion is therefore about environmental/spatial conflicts, whereby excessive environmental impact results in restrictions in terms of spatial planning, or whereby developments in spatial planning can lead to an undesirable burden on the environment.

At the various levels of policy development there is a difference between the extent to which strategic proposals for policy development are involved and the extent to which policy is geared towards concrete goals based on material reality and derives from social and physical developments. At the level at which government planning is outlined, the emphasis is on policy itself: decentralisation and deregulation must result in greater effectiveness and efficiency. Policy effectiveness and efficiency are open to discussion at the most concrete level of government, in particular the desired result of policy measures in the actual social and physical situation. This is the level at which confrontation with environmental/spatial conflicts occurs. The policy discussion regarding environmental/spatial conflicts primarily seeks to establish a causal relationship between the conflicts on the one hand, and the generic and guideline-setting regulations of environmental policy on the other. Other developments are also noted: in the light of the subsidiarity principle, environmental problems must be solved at the level at which they occur. This discussion on decision-making in environmental policy therefore focuses on the consequences of changes in policy – whether or not they are the result of decentralisation, deregulation or subsidiarity – for concrete, local environmental issues, and how this can contribute to the harmonisation and possible integration of environment and spatial-planning policy.

1.4 Complexity as the Basis for Theoretical Reflection

Decision-making in environmental policy is susceptible to change partly because local authorities have for many decades had a much greater tendency than government to regard environmental/spatial conflicts as being closely related to other local interests – i.e. more complex. For a long time, the government has separated the environment from other interests by means of generic, guideline-setting standards, with the result that environmental interests have come to have a one-sided and peremptory – and therefore relatively straightforward – effect on local decision-making processes.

This one-sided and peremptory form of decision-making can be interpreted as a functional-rational process. Without discussing the more specific meaning of functional rationality (which is discussed in § 4.6), this type of decision-making is largely based on direct causal relationships between the relevant 'parts of the whole', in accordance with systems theory (see Chapter 4). Policy relating to environmental quality issues is developed in accordance with this cause-and-result approach on the basis of the relationship between the emission source, the conveyance of the emission over a certain distance and, finally, the undesirable negative effect of environmentally harmful emissions. It is a question of the relationship between the source and the effect of the environmental impact.

This technical-functional approach is characterised in a modernistic way "by rationality, the 'objective' appraisal of means to achieve given goals, by managerial efficiency, the application of organisational and productive techniques that produce the most for the least effort, and by a sense of optimism and faith in the ability to understand and control physical, biological, and social processes for the benefit of present and future generations" (O'Riordan 1976; 11). The approach is geared towards the elements that constitute the whole, between which a direct causal relationship is thought to exist (Kramer and De Smit 1991). In terms of causality this relates to a *causa proxima*, a relationship resulting from an immediate cause. A mechanical world is assumed in which effects are produced by an unbroken sequence of components. This is known as linearity (Gleick 1988). The sequence of components must be identified in order to arrive at a clear definition of the issue in question. The mechanical nature of this process – if it is correctly interpreted – also means that the ultimate result is largely predictable.

The ultimate result will become less and less predictable where the assumed direct causal relationship increasingly serves as an idealistic image or over-simplification of reality, or where it is increasingly exposed to external influences. Consequently, the relationship between the components will be less clear than was at first assumed. If a *causa proxima* is nevertheless accepted, this can result in the exclusion – consciously or otherwise – of external influences (Kramer and De Smit 1991; 3). However, context and surroundings usually have a considerable influence on social phenomena. As the influence of context increases, and the relationship between the components becomes less stable and direct, it becomes increasingly difficult to establish a direct causality. Here we speak of a *causa remota*, or remote/indirect cause. The supposed mechanical relationship will be increasingly replaced by one based on statistical causality (Prigogine and Stengers 1990). This is why system-based and network-based approaches have been developed as elements of the formulation of theory and concepts, whereby "probability, choice and external influences [...] [are] essential to the understanding of phenomena" (Kramer and De Smit 1991; 5). The predictability of the end result decreases in such cases. This relates to issues that will, among other things, affect the context in which they arose. Such issues therefore become substantially more complex and thus more difficult to define.

The technical-functional or functional-rational approach to environmental policy has led to criticism. As remarked in section 1.2, such approaches are geared towards the generic and quantitative application of environmental standards. The criticism also extends to the generic and quantitative interpretation of environmental policy. Because the government is supposedly not in a position to be fully aware of the possibilities of local and specific situations, it is unable to make use of those possibilities through generic and quantitative regulations. This form of environmental policy 'objectifies' physical reality through national environmental standards. Unravelling the parts of the whole (reductionism) is not the only way to objectify; nor is quantifying reality (i.e. interpreting it in abstract figures) the only way to objectify a perception of reality. O'Riordan speaks of "the myth of objectivity [...]; the appeal of quantification is an appeal to 'rational' calculation because numbers have a sometimes spurious, but undeniable, aura of respectability and credibility" (1976; 16). A number of municipalities are therefore calling for a less 'pre-programmed' environmental policy (see Chapter 7) and greater individual responsibility in balancing the various local considerations, including those relating to the environment. This means that local circumstances will play a greater role, and environmental considerations can be balanced against other local considerations. Both the 'reality' to be objectified and the subjective/intersubjective 'reality' should be seen as integral elements of phenomena and issues, and therefore must not be ignored.

In abstract terms, the criticism discussed in section 1.2 is directed at the reductionist, quantitative, object-oriented and controlled approach to environmental planning. Now that the government is taking note of this criticism, environmental interest will no longer be seen only in terms of its components, but will also have to be placed in context. It will no longer be acceptable to define environmental/spatial conflicts exclusively in terms of environmental quality standards. Environmental-/spatial conflicts will be seen in terms of their local context, and will therefore become more complex. Such conflict situations are thus accumulations of problems and frictions that combine to form a complex social reality in which it will not always be possible to identify causes. End results will therefore be less predictable. This does not mean to say that nothing more can be said about the problems: "If the real problem consists in a reality of overwhelming complexity which seems not to be simply contingent (and hence inaccessible to any intellectual ordering) but structured and even more or less apparent to the relevant actors [...]" (Kaufmann et al. 1986; 16), conceptualisation is desirable (Kooiman 1996; 41). This study discusses a concept for a planning-oriented approach (§ 4.9) whereby the degree of complexity of environmental/spatial conflicts can be rendered transparent and used as a criterion for determining the decision-making method.

1.5 Towards a Complexity-Related Theoretical Perspective for Planning

The use of generic, guideline-setting standards in environmental planning involves a functional-rational approach to environmental quality issues. This approach was described above as relatively simple, in the sense that a direct causality is assumed. In addition to a functional-rational approach, we also speak of 'complex decision-making' (see § 4.7). This relates explicitly to network management and participative decision-making, which can be interpreted as exponents of a communicative-rational behaviour (§ 4.6). The functional mechanism with a direct causal link between the relationships is set against intersubjective relationships and interactions. Both forms of decision-making distinguish between the simple and the complex.

Here, there is no longer a sharp distinction between simplicity and complexity, or between a functional-rational approach and a communicative-rational approach. On the contrary, simplicity and complexity, and functional-rational and communicative-rational behaviour, are seen as the extremes of a continuum. This results in the need for a theoretical perspective (Chapter 4) that goes further than the apparent presupposition that the issues can be categorised in one of two ways: simple or complex. In the new perspective, the *degree of complexity* of an issue can determine the decision-making process, the solution strategy and the ultimate approach.

This principle is placed in the context of current discussions on planning theory. This position is also seen in terms of the empirical problem area: developments in environmental policy and the changing approach to solving environmental/spatial conflict. Planning-theoretical discussion is used as a source of inspiration and a mirror for a theoretical framework for planning-oriented approaches, formulated for this purpose (§ 4.9). The value of the framework will be measured by its internal consistency, its points of contact with existing theory and – of course – the possibilities it offers for analysing the empirical problem area. It is a question of being able to put forward arguments relating to the efficiency and effectiveness of environmental policy, whether or not in relation to spatial-planning policy and policy relating to the physical environment. Environmental policy is thus balanced against the complexity and context of environmental/spatial conflicts. The knowledge of and insight into the complexity of environmental/spatial conflicts will be assessed in order to determine how they can contribute to policy development and decision-making in environmental policy. As an extension of this, harmonisation and integration in environmental and spatial-planning policy will be assessed. In relation to the current discussion on environmental policy, the standard-based approach – command-and-control planning – will be confronted with policy geared towards a greater whole. This policy is referred to here as tailor-made overall planning and shared governance.

1.6 Reader's Guide

This book is divided into three parts. The first, Part A deals with *environmental-/spatial conflicts in a changing context*. The aspects of environmental/spatial conflict are discussed in detail and placed in a dynamic – and therefore continually changing environmental quality and functional-spatial context. This distinction between aspects, the whole and the context forms the basis for the second part of the book: we assume that the way in which environmental/spatial conflict is perceived has implications for the way in which it is defined and for its complexity, and therefore determines the form of decision-making. Part B deals with *the complexity of policy issues and pluriformity in decision-making.* A framework is proposed for linking decision-making and complexity. This is an exercise whereby – based on a systems-theory approach – insights into planning, decision-making and management are related to the concept of complexity. The concept of pluriformity is inextricably linked to this. In the third part of the book, the relationship between decision-making and complexity is assessed in terms of policy practice. Part C deals with *changing goals and interaction in environmental policy,* with reference to policy in the Netherlands. In particular, the complexity of environmental/spatial conflicts is linked to the efficiency and effectiveness of regional environmental policy. The purpose of this structure of policy practice set against theoretical reflection is to clarify the relationship between decision-making and the degree of complexity accorded to an issue, and to make that relationship manageable.

Chapters 2 and 3 deal with environmental/spatial conflict, which is seen here as the link between decision-making (about what?) and complexity (of what?). Chapter 2 discusses those aspects of environmental/spatial conflict that relate to environmental quality. Environmental quality is discussed on the basis of an *environmental problem chain.* Chapter 3 discusses the functional-spatial context. Environmental/spatial conflict is set against the background of the *compact city*, an urban concept that many spatial planners believed to be solution for achieving sustainable and environmentally-friendly development. This elaboration of environmental/spatial conflict in terms of environmental quality and spatial-functional aspects introduces the conceptual framework that is essential for referring to the physical, social and policy aspects of a conflict, among other things. The main aim of Chapters 2 and 3 is to identify the various aspects of environmental/spatial conflict. The elements of a conflict are set out, and the relationships between them are then discussed, as well as the forces that influence a conflict from the physical, social and administrative context. Chapter 2 therefore examines the technical-functional cause-and-result relationship between the source of environmental impact and its ultimate observed effects. Chapter 3 highlights various, often incomprehensible forces that determine the demand for space. The realisation that environmental/spatial conflicts can be influenced by many and various forces points to the possible relevance of the context in which a particular conflict arises. Because the context of a conflict may exert a clearly perceptible influence, the question arises – following on from the critique in section 1.2 – whether maintaining a physical distance between environmentally harmful and environmentally sensitive activities is the most logical approach. Here, the formulation of a solution strategy must be based on more than a direct causal relationship between source and effect. In the light of the Enschede tragedy (§ 1.1), the question is: what precisely is meant by 'more than'?

Chapter 4 fleshes out the approach to the possibilities and limitations of the technical-functional and other possible approaches. Developments in discussions on planning-theory and decision-making are therefore followed. Theoretical visions and concepts relating to planning and decision-making are analysed from three perspectives: the *decision-oriented perspective,* the *goal-oriented perspective* and the *institution-oriented perspective.* It is noticeable how, within the frameworks of the three perspectives, we can highlight the way in which relevant thinking in the field of planning theory has developed over the years. This development is partly the result of an increasing awareness of the complexity of the subjects, and of awareness of the importance of the intersubjective relationships on which planning-oriented action is based. Insights from systems theory are used to structure the assumed relationship between issue, decision-making and complexity. With reference, among other things, to the nascent theoretical constructs that are gaining ground primarily in the science disciplines as 'complexity theory', an attempt is made to establish a double connection between complexity and decision-making. In short, it is a question of using planning-theory discussions to analyse the relationships between complexity and decision-making. These relationships are set out in a framework for planning-oriented action designed for that purpose.

This awareness of complexity can have consequences for the type of decision-making and planning followed in day-to-day policy. Chapter 5 focuses on analysing planning-oriented approaches in environmental policy. The analysis is based on the behaviour perspectives identified in Chapter 4 and the theoretical framework for planning-oriented approaches. Chapter 5 discusses the process of *integration and shifting of responsibilities* that can be observed in environmental policy. The main aim of this is to find new ways to deal effectively and efficiently with environmental/spatial conflict, possibly in combination with other issues that are current to the local physical environment. Such shifts mean that environmental policy is particularly appropriate as an empirical problem area for testing the theoretical framework for planning-oriented approaches developed in this book. Chapter 6 sets out ideas about complexity and decision-making in a more concrete form. The usefulness of *integrated environmental zoning as an instrument* is assessed according to three categories of environmental-/spatial conflict, each of different complexity. This approach should clarify to what extent abstract ideas can also make a practical contribution to decision-making, policy formulation and related policy instruments. Chapter 7 focuses on a case study: the environmental/spatial conflicts concerning the development of the Houthavens (former timber docks) area in Amsterdam (Chapter 7). In this example, the emphasis of environmental policy has shifted from a top-down approach, based on sectoral environmental standards and zoning, to a broader, more local approach. This represents a shift towards what is referred to in this study as shared governance. The case study is used to examine the possibilities and consequences of an approach that takes greater account of local considerations: a tailor-made, overall policy. This discussion ends in Chapter 8 with a number of concluding observations.

Notes

1 The conference was attended by administrative delegations from the Directorate General of Environmental Management (DGM) of the Ministry of VROM, the Association of Water Boards (UvW), the Interprovincial Platform (IPO) and the Association of Netherlands Municipalities (VNG), all of which are members of the joint consultative platform DUIV.

2 These conclusions were officially recorded at the DUIV meeting of 15 December 1994 (DUIV: see footnote 1).

3 Prime Minister Kok's first 'purple' cabinet aimed to create a closer relationship between policy development and social dynamics by implementing its 'MDW' (Market Mechanism, Deregulation and Legislation Quality) project. The project was based on the principle that regulation must not impede economic dynamics, local responsibility or creativity.

PART A
ENVIRONMENTAL/SPATIAL CONFLICTS IN A CHANGING CONTEXT

Chapter 2

Externalities and the 'Grey' Environment

Environmental/Spatial Conflict in the Context of Environmental Encroachment

> The history of life on earth has been a history of interaction between living things and their surroundings. To a large extent, the physical form and the habits of the earth's vegetation and its animal life have been moulded by the environment. Considering the whole span of earthly time, the opposite effect, in which life actually modifies its surroundings, has been relatively slight. Only within the moment of time represented by the present century has one species – man – acquired significant power to alter the nature of the world (Rachel Carson 1962; 23).

2.1 Introduction

"For almost as long as there have been cities they have been polluted, and before there were cities there were polluted huts and houses". This pronouncement by Brimblecome and Nicholas (1995; 285) points to the eternal dilemma of human activity: man's tendency to 'foul his own nest' – the almost inevitable side-effect of development and 'progress'. This is also known as environmental health, environmental hygiene and environmental encroachment. In this chapter, environmental health and hygiene and the spillover of pollutants are referred to as 'grey' in order to distinguish them from other aspects of the environment, including the natural 'green' environment, the 'spatial' environment and the 'day-to-day' environment.

The grey environment is affected most by urban areas with a higher than average concentration of people and human activity. The relationship between a function or activity and its 'unpriced effect' is usually distance-related. These effects are also known as 'externalities'[1] (Marshall 1924, Mishan 1972, Pinch 1985). Here we are concerned with externalities that have a characteristic spatial impact. Pinch (1985) refers to 'tapering effects'. Given the consequences for the spatial environment, it is logical that sufficient distance is maintained from the source of an externality that is experienced as negative. This approach has become commonplace in seeking to achieve an acceptable environmental quality in areas surrounding sources of environmental impact. It is a technical-functional approach that translates environmental health and hygiene standards into spatial contours. These contours separate areas with excessive environmental impact from areas in which that impact has fallen to an acceptable level. This distinction can have implications for the functions and activities located in these areas. The fact that environmental impact can have spatial consequences is fundamental to what is referred to in this study as 'environmental/spatial conflict'.

Given the objections (§ 1.2) to maintaining sufficient distance between functions by means of generic and guideline-setting environmental health and hygiene standards, what are the alternatives for achieving an environment of an acceptable level of quality and liveability? This question is addressed in the present chapter from the point of view of environmental health and hygiene, in particular the functional-technical and social characteristics of environmental impact. These characteristics are less straightforward than past policies suggest. Here, we are discussing the characteristics of environmental health and hygiene that determine the quality of the urban environment, are relevant to spatial developments, and produce responses from society and in government policy. At the same time, we are discussing characteristics that are ridden with uncertainty and obscurity, and are subject to differing interpretations, perceptions and recognition. On reflection, environmental health and hygiene proves to be an extremely complex subject. This is probably the most important observation in this chapter.

The concept 'environmental/spatial conflict' and the concept 'environment' therefore both require further elaboration in this chapter. In section 2.2, the concept 'environment' is defined more closely for the purposes of this study. The definition is based on the assumption that 'environment' encompasses more than physical characteristics. The term also refers to the relationship between those characteristics and how they are perceived by the individual, by society and in government policy. In section 2.3 the relationship is represented as a 'cause-and-effect chain', which expresses as far as possible the causal link between human activity, its impact on environmental health and hygiene and the eventual perception of that impact. The individual segments of the chain are discussed in more detail in the sections that follow. Section 2.4 identifies the characteristics of environmental impact that are relevant to urban environmental/spatial conflicts, in particular environmental health and hygiene issues that, as 'externalities', lay claim to urban space. This claim leads to tensions, and consequently policy has come to focus more on the consequences of environmental health and hygiene externalities. The aim is to create a more cohesive and effective environmental policy, among other things on the basis of a number of environmental themes. The most important of these themes are discussed in section 2.5. There is a significant relationship between these themes and the environmental health and hygiene standards under discussion (see § 1.2). Although the system of standards is largely based on the physical characteristics of cause-and-effect relationships inherent in environmental impact, that impact is only evaluated to a limited extent on the basis of objective reasoning. This realisation – namely that conflicts always involve several conflicting interests – is discussed in section 2.6.

2.2 The 'Grey' Environment as a Coherent Concept

Broadly speaking, the concept 'environment' can have two meanings. First, it refers to the social surroundings in which people live, work and generally spend their lives. This is the 'intersubjective' environment in which subjects interact. Second, the term 'environment' also refers to the physical environment.[2] In this study, environmental/spatial conflicts are primarily seen as social issues that are partly due to phenomena (spillover of pollutants affecting health and hygiene) occurring in the *physical environment.*

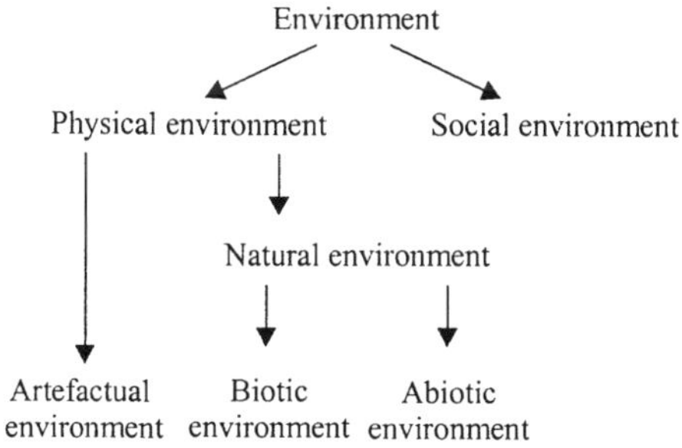

Figure 2.1 Structure of the concept 'environment'

In functional-technical terms, the physical environment can be separated into the biotic, the abiotic and the artefactual (see Figure 2.1). In this classification, the environment is seen as a combination of three physical components. Hoeflaak and Zinger (1992), however, distinguish between three types of environment for policy purposes: *green, blue* and *grey.*[3] This classification emphasises the policy context rather than the physical components of the environment. This is a simple approach which is in line with 'municipal day-to-day practice' (Hoeflaak and Zinger 1992; 11).

Policy relating to the *green environment* focuses on nature and the landscape. It is geared toward preserving species and types of landscape. According to the definitions employed by Hoeflaak and Zinger, the *blue environment* refers to aspects of the environment that contribute to 'sustainability' (see § 2.6). Their effects are usually felt on a relatively wide geographical scale and only become apparent in the long term. The *grey environment* is that part which relates to the hygiene of the physical environment and partly determines the quality of our day-to-day local and regional surroundings. Hoeflaak and Zinger refer to aspects that "obviously relate to environmental hygiene: pollution of ground and surface water, air and soil, and the limitation of hazards, noise and odour nuisance" (1992; 12). It is mainly these types of environmental impact that can have spatial implications. Environmental/spatial conflicts therefore relate primarily to the grey environment.

Discussions about the perception of the grey environment are as old as environmental issues themselves. However, it was not until relatively recently that such issues began to be considered in a social context. Initially they were considered from separate points of view such as public health (§ 2.4) and the economy. In 1920, Pigou introduced the concept 'negative external effect' in order to denote the grey environment as an economic issue. According to Pigou, 'negative external effects' – referred to by Marshall (1924) as externalities (§ 2.1) – are the result of "divergences between private and social net products that come about through the existence of uncompensated services and uncharged disservices" (1920; 191). In 1967, Mishan emphasised weaknesses in neoclassical economic theory, in which private production and consumption resulting from the prevailing market mechanism are emphasised more strongly than public goods. Hardin makes similar observations in his much-discussed work *Tragedy of the Commons*[4] (1968), and concludes among other things that the environment is undervalued as a public good. Effects other than economic effects are emphasised, such as population growth (Ehrlich and Ehrlich 1969, Hardin 1968, Malthus 1817) and biological and chemical effects (Carson 1962).

From the 1960s onwards, increasing emphasis was placed on the relationships between various aspects (e.g. Meadows et al. 1972, World Commission on Environment and Development 1987). Aesthetic and ethical values were also considered (see also Nelissen, Van der Straaten and Klinkers 1997). The shift towards cohesion and correlation was also part of the discussions focussing on sustainability (see also § 2.6) that became popular at the end of the 1980s (see McDonald 1996). According to Healey and Shaw (1993; 771) the discussion centres on the "concept of sustainable development or ecological modernization, which offers the prospect of a beneficial relationship of environmental resource conservation and economic development". Here, Healey and Shaw emphasise the relationship between the economy and the environment, with 'sustainability ' as a central concept (see § 2.6). The concept of sustainability was put on the international political agenda by the World Commission on Environment and Development[5] (1987). The Commission used the term to denote, among other things, the connection between environmental questions and other socially relevant issues, and it speaks of the "various global 'crises' that have seized public concern, particularly over the past decade" (1987; 4). According to the Commission, the main issues are energy, development and the environment. These issues must be understood collectively, not separately, but it was a long time before this was acknowledged.

The following sections discuss the relationship between the functional-technical, policy-related and social characteristics of environmental health and hygiene issues. An environmental cause-and-effect chain has been constructed in the following section in order to facilitate the visualisation of that relationship. It is important to note that the chain is a somewhat simplified version of reality. It is not true to say that all effects are predictable. Further, the links in the chain suggest certainties that do not always exist. The cause-and-effect chain is intended as a simple tool to aid understanding of the complex reality it represents.

2.3 A Problem Chain for the 'Grey' Environment

Recognising and acknowledging problems is an indication of the growing individual and social awareness of a phenomenon – or its possible effects – which are experienced as negative and unwanted. This growing awareness of human encroachment on the physical environment received an additional impulse in the 1950s and 1960s following a number of significant events, such as 'the Great London Fog' in 1952, and writings such as Rachel Carson's *Silent Spring*,[6] published in 1962. This period saw the 'rise of environmentalism' (Reade 1987; 162) and a growing awareness of the adverse *consequences* of human activity. Later, attention shifted to the often-complex *causes* underlying environmental issues.

It is now common practice to think of environmentally harmful activities not only in terms of consequences or effects, but also in terms of the underlying mechanism. This mechanism is referred to as the environmental effect chain (Udo de Haes 1991, Ragas et al. 1994) or *environmental problem chain* (Bouwer and Leroy 1995; 26). Bouwer and Leroy prefer the term 'environmental problem chain' because it expresses the interplay of social forces that influence environmental issues. Environmental issues are not only functional-technical in nature, in the sense of the cause and effect of the encroachment/impact relationship, they are also a consequence of social developments and activities, which are not so easy to explain in terms of cause and effect. In such cases, an environmental problem chain should not be seen as an accurate linear representation of reality; it is a highly simplified representation of a much more complex, non-linear social and physical reality (see § 2.4-2.7).

There is no such thing as a standard, universal problem/effect chain. There are a number of variants that differ in detail and purpose.[7] A chain of this type represents a series of links, usually beginning with the link that represents the motive for human activity – the developments and needs of the individual and society. These developments and needs initiate the next link in the chain: actual individual and societal activity. This activity in turn affects the physical environment and, depending on its duration, scale, intensity and nature, may result in excessive pressure on the environment (see § 2.4). In the short or long term, that pressure may lead to undesirable negative effects on the physical environment, and consequently to a reaction from society and intervention through policy (see § 2.5 and 2.6). The final link in the environmental problem chain relates to the total perception of all the effects and encroachments relating to the physical environment (§ 2.6). This concerns not only the negative consequences of human activity, but also the recognition of intervention and corrective measures designed to rectify the damage.

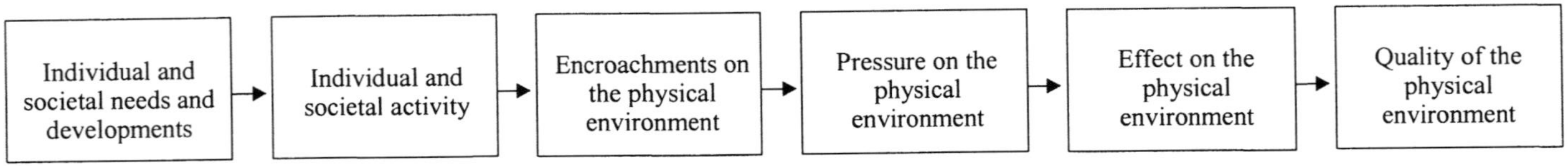

Figure 2.2 An environmental health and hygiene cause-and-effect chain

An environmental problem chain or environmental cause-and-effect chain such as the one in Fig. 2.2 shows how environmental issues are anchored in the interplay of social forces. The social dilemma inherent in environmental issues is characterised on the one hand by conflicting developments and needs, and on the other hand by the undesirable effects they cause. The fact that this dilemma can be expressed as a chain would seem to confirm the causal relationship between the various aspects. However, the links are not always self-evident. The questions surrounding the possible effects of CO_2 emissions are a good example of this. The relationship has also been represented as a cycle (Adriaanse, Jeltes and Reiling 1989) and as a network (see VROM 1984a). However, the typology is relatively simple and its purpose is to identify the aspects of the often-complex cause-and-effect relationships inherent in environmental impact. Below, individual and societal needs and developments, as well as the individual and societal developments resulting from them, will be regarded as given. By contrast, the technical-functional, policy-related and social characteristics of interventions in the physical environment and their consequences will each be examined in turn. The physical characteristics of environmental impact will be discussed first.

2.4 Environmental Health and Hygiene in an Urban Context

Urban environments are characterised by their concentration and diversity of functions and activities. These functions and activities guarantee a dynamic and multifunctional urban space, but some of those functions are mutually incompatible from the point of view of protecting the environment. This is a question of 'externalities', not all of which can be avoided. This chapter will first discuss the physical characteristics of environmental issues that can hinder spatial development in the urban environment.

Pollution, Depletion and Damage

Udo de Haes (1991; 26) divides issues concerning the physical environmental impact that are the result of human activity[8] into the following categories: physical/chemical/biological pollution, depletion and damage. *Pollution* is the excessive addition of chemical substances, physical phenomena and biological organisms to one or more of the components of which the environment is constructed, and as a result of which an equilibrium is temporarily disturbed or irreparably damaged, either locally or in a more widespread sense. *Depletion* refers to large-scale extraction by man of abiotic and biotic materials from the environment, whereby the local equilibrium is temporarily or permanently disturbed. *Damage* refers to human intervention such that the situation cannot be restored, or only with great difficulty. By definition, pollution and depletion result in environmental damage, but do not comprise all forms of impact. Land levelling and the streamlining of brooks and rivers have little to do with pollution or depletion, but nevertheless involve interventions in the physical environment that almost always have irreversible results. Categorisation on the basis of pollution, depletion and damage therefore encompasses the whole spectrum of environmental phenomena that results from human activity.

Environmental pollution, depletion and damage can also have spatial consequences. For example, traffic and industry cause pollution, which means that environmentally sensitive activities cannot be located in their vicinity. Contaminated soil can affect spatial development in or on the site in question. The depletion of raw materials can have spatial consequences at the site of extraction, and the resulting spatial conditions will determine to a large extent how the location can be used in the future (Ike 1998, Van der Moolen 1995, Van der Moolen, Richardson and Voogd, 1998). The spatial consequences of impact on landscape or stream systems are obvious. For each of the three categories of environmental damage, it is possible to establish a relationship between an intervention in the physical environments and the spatial consequences. In urban areas, on which this study concentrates, the interplay between the condition of the physical environment and spatial development is largely restricted to the category *Pollution*.[9]

Cities have always had an adverse impact on environmental health. In Roman times, the courts had to deal with cases in which smoke from workplaces was a cause of nuisance to those living nearby (Brimblecombe and Nicholas 1995). It is even argued that the fall of the Roman Empire was partly due to lead poisoning (Copius Peereboom and Reijnders 1989). Various authors (Van Ast and Geerlings 1993; 15, Van Zon 1991; 94) have pointed to the damaging results of the mediaeval practice of using sulphurous coal for heating. An anecdote from 1257 tells of the excessive smoke and smell from the town of Nottingham, which forced the royal family to cut short its stay at Nottingham Castle. During the industrial revolution, urban pollution not only increased, but also became more diverse. At the same time, populations in industrial areas were growing very rapidly. In the Netherlands this occurred particularly after 1870 (Van der Cammen and De Klerk 1986; 37). At the beginning of the 20th century, not only industry, but also urban transport – the horse, and later the automobile – left their unmistakeable odorous and noisy mark to remind us of the incidental excesses of human progress. Smog and stinking canals required government intervention. According to Van der Cammen and De Klerk (1986) the cholera epidemics at the end of the 19th century, which were the result of poor sewage systems, opened people's eyes to the poor living conditions in the industrial towns and cities of the western world. The cities spawned by the industrial revolution were the site of many more wealth-creating activities than before, but those activities had a visible adverse impact on people's day-to-day surroundings.

In the recent past, too, there are a number of interesting but dramatic cases of harmful substances released into the environment that had a heavy impact on liveability on the local scale. On 16 July 1976, an explosion at the ICMEA factory in Seveso near Milan, released TCDD (tetrachlorodibenzo-p-dioxin) into the air. Dioxins are considered to be among the world's most toxic substances. TCDD is a particularly aggressive dioxin. There were human casualties, but also serious consequences for the local ecosystem. In December 1984, a leak developed in a container of methyl isocyanate at the Union Carbide factory in Bhopal, India. A poisonous gas cloud formed as a result of the leak, and more than 2,500 were killed. The cloud affected a further quarter of a million people. In 1992, the world was horrified by a chemical accident near Guadalajara, Mexico's second-largest city. These examples (Copius

Peereboom and Reijnders 1989; 86, Cutter 1993, Ellis 1989; 167) are notable because of their scale. The Netherlands also has a number of environmental scars. The firework disaster in Enschede in the spring of 2000 is still etched in everyone's memory – largely thanks to remarkable and gripping live reports by the local television station *TV Oost* (§ 1.1). Every decade has seen a large-scale environmental disaster in and around urban areas, resulting in enormous numbers of casualties. The point is not only the contamination these incidents have caused, but also its far-reaching consequences. The incidents also show us that there is a *probability* or *risk* (TK 1989) of pollution, depletion and damage, and very great uncertainty about the nature and extent of their consequences.

The Effects of Pollution on Humans

The effects of exposure to pollution on humans in the local physical environment can be classified according to the risk to human health. Broadly speaking, there are three categories: nuisance, physical symptoms and an increased risk of illness and mortality (Borst et al. 1995; 46, De Hollander 1993). Exposure to, for example, noise, odour and vibration can be experienced as a considerable nuisance, but does not necessarily result in physical symptoms. Nuisance is largely a matter of individual perception. That is why group averages are used for formulating general criteria such as environmental standards designed to restrict emissions, whereby the legislator arbitrarily designates a certain nuisance level as unacceptable (Ettema 1992). Immissions that are perceived as a nuisance can eventually lead to physical symptoms as the duration or degree of exposure increases.[10] Physical symptoms can also result from exposure to immissions that are not directly experienced or observed. Symptoms may include loss of concentration, stress symptoms, and respiratory or skin damage, but can also lead to cancer and other complex symptoms. Exposure to pollutants does not always produce symptoms immediately. Sometimes the physical effects do not manifest themselves until many years later (De Hollander 1993). It is therefore not always easy to establish a link between certain forms of air pollution and the health of the local population, despite indications that air pollution resulting from industrial activity can seriously affect human health.[11]

In addition to the categories relating to nuisance, physical symptoms, illness and mortality, pollution is also classified according to 'nuisance and risk'. 'Risk' refers primarily to the probability that phenomena such as physical symptoms, illness and mortality will result from a certain level of exposure to environmental pollution, and/or the probability that a certain amount of pollution will be released into the environment, with all the consequences ... (§ 5.2). From the point of view of policy development, this classification is preferable because levels of acceptability can be defined on the basis of probabilities.

Classification using nuisance and risk categories is also based on a clearer distinction between what can and cannot be directly experienced. Nuisance results when our senses register exposure to environmentally harmful immissions. Perception is also an important factor. The absolute level of the environmental impact is not the only factor that determines the nuisance level. The 'background level' of environmental impact also determines to what extent pollution is perceived as nuisance. In other words: in order to be noticed, a source of noise in Piccadilly Circus would have to produce considerably more sound than it would on the Yorkshire Moors. Nuisance is therefore relative. 'Risk' indicates the probability that exposure to a certain level of pollution for a certain length of time will harm human health. The actual effects of exposure to high-risk pollution are considered separately from how it is experienced/perceived. Therefore, background levels of pollution are not considered when determining the level of risk. This could lead to the expectation that regulations relating to high-risk immissions would be stricter and more extensive than for nuisance-causing immissions. Inhabitants who experience nuisance can also take their own measures in this regard, but this is not possible in the event of high-risk environmental impact.[12] However, such a link cannot be established, not least because setting risk-related standards involves a relatively large number of uncertain factors[13] (see, for example, Copius Peereboom and Reijnders 1989, De Hollander 1993).

Characteristics of Environmental Pollution

What constitutes environmental pollution? In the urban environment, pollution can result from the emission of waste and harmful substances. In the Netherlands, however, urban waste – with the exception of litter – has been largely reduced to a logistics issue.[14] Urban environmental hygiene therefore depends primarily on the level of harmful emissions, particularly emissions into the soil and the air. In the Netherlands, almost all harmful emissions into the component 'water' are collected and purified, since most urban wastewater is discharged into closed drainage systems.[15]

Given that almost all forms of environmentally harmful emission into the soil are prohibited in the western world, one would expect that spatial restrictions due to soil contamination are slowly but surely becoming a thing of the past (Page 1997). However, in many cases, concentrated 'historic' soil contamination exceeds the accepted background level for urban areas to such an extent that this serious, widespread inheritance leads not only to adverse, undesirable effects in the soil, but also to serious stagnation in spatial development (see § 5.2 and 5.5). Furthermore, it is unlikely that gradual and diffuse soil contamination resulting from urban activity can ever be completely eliminated.[16]

The biggest challenge is above-ground emissions that spread beyond their source. These emissions are the most difficult to combat through regulations or technical measures. In contrast to soil problems – although policy relating to this area is also changing radically (§ 5.5) – the use of environmental policy for above-ground pollution is far less oriented to the almost total elimination of emissions, although it does aim to minimise and undo the negative, undesirable effects of emissions and immissions. This can be done by means of measures at-source, by ensuring that emissions take place at the highest possible point in the atmosphere, through measures at the transmission stage, the use of insulation where environmentally-harmful immission is undesirable, or through spatial measures.

The interaction between environmental health situations and spatial developments can sometimes be of a consecutive nature. This is the case, for example, with issues relating to soil contamination where that contamination has consequences for the development of the location. Concurrent and more dynamic relationships also exist between the environmental hygiene situation and spatial developments. Changes in the production process of an industrial plant, or increased traffic volumes on a ring road can affect the quality of the local environment, and consequently the opportunities for spatial development in the area.

There is another fundamental difference between above-ground and underground contamination. Underground contamination is almost always chemical. Above-ground contamination is less easy to categorise; it may be chemical in nature, but a further distinction is made between toxic and carcinogenic substances. Radioactive substances are classified as a form of physical pollution, due to the ionic radiation they release (Pruppers et al. 1993). Electromagnetic radiation from high-voltage power lines and radio/TV masts are also considered to be sources of physical pollution, and a specified distance is recommended between these sources and housing developments (Van den Berg 1994). Vibrations (e.g. from passing trains), excessive, nuisance-causing odour (Cavalini 1992), and excessive, nuisance-causing noise are physical manifestations that can be classified as sources of environmental pollution. This also applies to dust particles. Particles smaller than four micrometers can produce symptoms of irritation and lead to illness. By contrast, larger dust particles are primarily classified as a source of visual pollution (Zeedijk 1995), as are structures that obscure the view, such as acoustic baffles. Excessive light (e.g. from horticultural greenhouses and industrial plants), and the shadow and wind effects around buildings can also be designated as nuisance. Biological pollution can have an adverse and undesirable impact on the physical environment, but the extent to which it restricts local spatial development is minimal or non-existent.[17] By extension, although they fall to some extent outside the scope of the definition of environmental pollution, inner-city climate changes can also affect the liveability of the urban environment (Hough 1989, Miura 1997). This also applies to the possible synergy and cumulative effects caused by the interaction between various types of environmental impact. These effects are also outside the scope of this study.

The forms of pollution that are within the scope of this study can, broadly speaking, be said to involve a 'tapering effect'. The level of concentration and therefore the effect of the pollution fall in proportion to the distance from the source.[18] As the level of concentration falls, so does the extent to which it is perceived or classified as impact.[19] This relationship to distance exists until the point at which the emission is diluted to a level at or below the background level.[20]

It is not so that all forms of environmental pollution adversely affect human health, or have ecological or spatial consequences. This obviously depends on the nature and scale of the pollution, but the way in which pollutants enter the environment is also important. Over the past century we have seen that man has tried to prevent as far as possible the undesirable effects of these emissions by deflecting them to a higher geographic level, or a longer time scale (see also § 5.3). As emissions are deflected, the relationship between environmental health and spatial development will become less direct and will diminish. It will also be more difficult to trace the relationship between the source of emissions and the effects of immisions to individual incidents of 'cause-and-effect'. In such cases, undesirable and detrimental effects on the physical environment will be spread over a large area, and will only become apparent in the long term. The interaction between environmental health and hygiene and spatial developments with undesirable and detrimental effects will be greater than if there were a direct, identifiable relationship – that can also be located in spatial terms – between an individual source and a separate local effect.

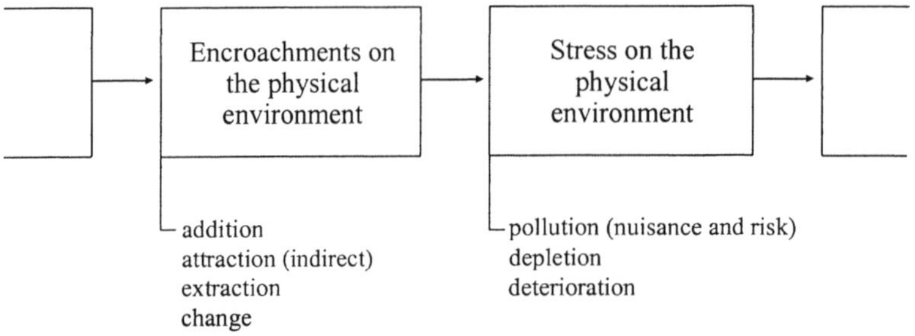

Figure 2.3 Categorisation of the physical characteristics of environmental impact

Note: The impact is set in the context of the cause-and-effect chain that influences environmental health and hygiene (see Fig. 2.2)

In conclusion, it can be said that urban environmental/spatial conflicts are partly due to environmental pollution that can lead to undesirable and detrimental effects in the local area. In the first instance, this is a matter of a direct and demonstrable interplay between the environmental hygiene situation, ecological and health effects, and spatial developments. Contaminants that can lead to excessive impact in the air and soil have a particularly high risk for man and ecosystems. When it is impossible, or possible only with extreme effort, to reduce, deflect or dilute these emissions, to take measures at the transmission stage or insulate the affected party, a spatial claim can arise owing to environmental pollution. The spatial claim is usually established by applying environmental health and hygiene standards.

2.5 Local Environmental Conflict as a Policy Theme

This discussion of urban environmental/spatial conflict focuses to a large extent on policy-based responses and action (see Fig. 2.4) relating to negative and undesirable physical impact on local environments (see the 'action' arrow in Fig. 2.4). The origin of these phenomena, which can eventually have a negative effect on the quality of the local environment, must be sought in individual and societal needs and developments.

If this negative effect on quality leads to a policy-based response, the aim of that response will usually be to remove the negative and undesirable effects or reduce them to an acceptable level. The question is, then, how the policy will evaluate the negative and undesirable effects on environmental quality, how it will be implemented, or enforced. Chapter 5 is entirely dedicated to this question, in particular the extent to which the source/effect relationship of environmental impact can provide insight and serve as a guideline for formulating policy based on standards. Standard-based environmental policy, introduced as a problem area in section 1.2, rests to a significant extent on this cause-and-effect relationship. At the very least this suggests that, in all cases, setting environmental standards will provide clarity as to how issues relating to environmental quality can be resolved in a satisfactory and 'sustainable' way. That clarity, however, is open to discussion.

The structuring and regulation of environmental policy dates back to the 1970s when it was formulated in a technical-functional way, with the cause-and-effect relationship of environmental impact as the leitmotiv. Initially, policy was geared towards 'compartmentalised' decontamination: removing the negative and undesirable effects that were visible in the soil, the air and the water. This approach ultimately proved ineffective and too limited (see Chapter 5).

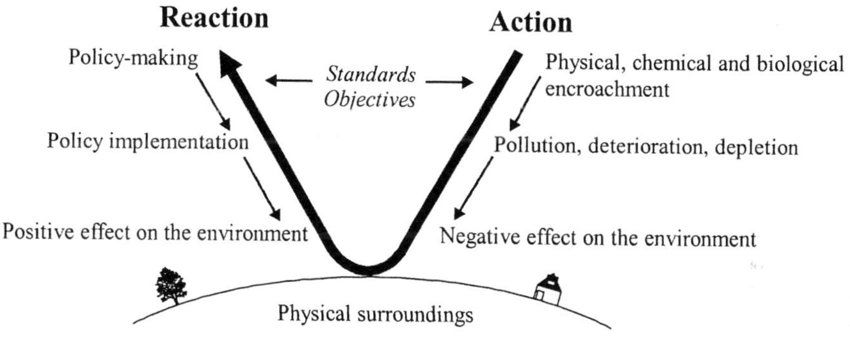

Figure 2.4 A policy-based response to undesirable effects on the physical environment resulting from individual and social developments

In the 1980s, source-oriented policy was thought to be the answer, and prevailed over effect-oriented policy. Policy-makers believed it would be better to turn off the tap than continually mop up the water. The assumption was that, after turning off the tap, only minimal low-priority, effect-oriented measures would be required (§ 5.4). This argument is probably workable from a functional point of view, but on economic and social grounds it is almost impossible to eliminate harmful emissions at source. A noise-free car – even if it were feasible – would be unaffordable and therefore economically unattractive and socially undesirable. Harmful emissions will continue, and as a consequence there will be a constant need to consider the various interests involved and the level of cohesion between source-oriented and effect-oriented environmental policy.

In the mid-1980s, the Dutch Ministry of Housing, Spatial Planning and the environment (VROM) developed a number of themes as foundations for environmental policy, in order to give it a cohesive structure (see § 5.3). These themes refer to a number of relevant environmental issues, such as climate change, acidification, eutrophication and disturbance.[21]

The aim of this is not so much to approach phenomena from the point of view of their individual aspects and express these in a continuous cause-and-effect chain, but to see them as a coherent whole composed of characteristics, factors, links and interests. The priority is not to establish practicability, but to understand how the themes are connected. The added value of this thematic approach lies primarily in its structuring effect (Chapter 5). One of the results of this thematic structure of environmental policy is that relevant environmental issues have come to be seen as a sum of constituent parts. For local environmental hygiene issues with possible spatial consequences, the themes of disturbance, dispersion and acidification are particularly relevant.

Disturbance

Table 2.1 shows how Hoeflaak and Zinger (1992) link the environmental policy themes, which were introduced in the Indicative Long-Range Programme 1985-1989 (VROM 1984b) and developed in the first National Environmental Policy Plan (TK 1989), to the physical environment categories they identify. Disturbance affects "the realisation and preservation of a liveable environment in terms of the environmental aspects noise, odour, vibration, external safety and local air pollution" (TK 1989; 150). These aspects are primarily the result of industrial activity, agriculture, or motorised traffic. The aim of the theme 'disturbance' is to reduce the detrimental and unwanted effects of chemical immissions and physical phenomena that are (or can be) brought about by human activity. This theme focuses on our immediate surroundings. The importance of the quality of our local environment is expressed in the theme document on disturbance, *Verstoring* (VROM 1994), on the basis of three aspects. The first aspect is the importance for human health and well-being, described above as 'risk and nuisance'. The second aspect relates to the influence of the quality of the local environment on the emotional value we attach to the world we live in. The third aspect is how the quality of the local environment affects spatial planning in a country with limited space, due to the fact that the mutual effects of spatial activity have to be considered. This last aspect expresses the essence of environmental/spatial conflict.

The theme document *Verstoring* recognises three forms of environmental pollution, of which at least one must be present if national policy is to be applied (VROM 1994; 3):

- Forms of environmental pollution of which the impact is perceived as nuisance and/or is a danger to public health. Such impact is directly noticeable, tangible or visible.
- Forms of environmental pollution of which the intensity decreases as the distance from the source or source area increases. The area within which the pollution occurs can be indicated by means of zoning.
- Forms of environmental pollution that occur mainly on a local or regional scale.

Table 2.1 A structure for environmental policy themes and environmental interests

Category	Environmental themes[a]	Interest
Blue environment	Climate change, acidification, dispersion, removal, eutrophication[b]	Future generations (sustainability)
Grey environment	Disturbance, dispersion, acidification[c]	Public health / nuisance to today's generations (liveability)
Green environment	dehydration, fragmentation, disturbance	Nature, landscape, present and future generations

a Environmental themes referred to in the NMPs. 'Fragmentation' is an additional theme that was originally a theme in the Nature Policy Plan (LNV 1990).

b Eutrophication is not included in Hoeflaak and Zinger's classification.

c Here, acidification has been added to Hoeflaak and Zinger's classification (see explanation in text).

Source: Based on Hoeflaak and Zinger 1992

The theme of disturbance only refers in part to environmental pollution that has an impact at local level. More particularly, it refers to above-ground emissions and immissions. A general characteristic of above-ground emissions is that "the pollution decreases to the background level soon after the source of the pollution has been removed" (Van Velze and Maas 1991; 397). These immissions mainly affect urban areas. By introducing spatial zones it is possible to achieve an acceptable – and possibly 'sustainable' – separation of environmentally sensitive functions and areas and environmentally harmful functions and activities. This sustainable separation is realised primarily through standards that express the desirable level of environmental quality. Furthermore, the standards exert a direct influence on local spatial development. The instrument of integral environmental zoning was also developed in the context of the Disturbance theme, and is an example of generic standard-based policy put forward by the government as a solution to environmental/spatial conflicts between complex industries and the local residential environment (§ 5.4, TK 1989; 152). Zoning is therefore the policy instrument that functions as a bridge between environmental policy and spatial planning.

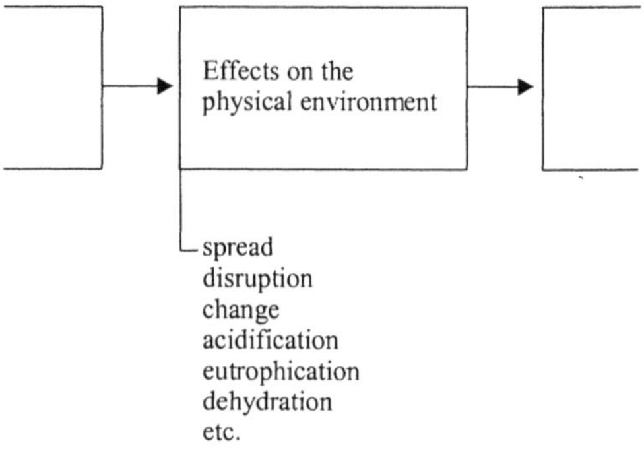

Figure 2.5 Environmental policy themes geared towards preventing adverse environmental effects, or ensuring they remain at an acceptable level

Note: See also Fig. 2.2

Dispersion

This theme refers to a form of environmental contamination (i.e. soil contamination) that affects the local environment *and* can restrict spatial development[22] (see also § 5.2 and 5.5). In this context, policy has been developed that should lead to effective measures for dealing with environmentally harmful emissions resulting from disasters. In general terms, the theme 'dispersion' refers to "a very large number of harmful substances, including [...] pesticides, heavy metals, radioactive substances and genetically modified organisms that, individually or together, constitute a risk to human health or the environment" (TK 1989; 140). Policy that can be categorised as 'dispersion policy' is heavily 'substance-oriented', and is therefore much less 'region-oriented' than disturbance policy. In the context of 'dispersion', a great deal of emphasis is placed on reducing emissions. The KWS 2000 project (VROM 1989) is an example of policy that is designed to encourage the target group 'industry' to introduce source measures. Quality standards have also been developed for soil and water, whereby limits are set for immissions of harmful substances (TK 1991). In summary, the theme of dispersion is geared towards the emission, subsequent transmission and eventual immission of environmentally hazardous substances. Dispersion-related policy can have consequences for spatial planning, but not necessarily in all cases. The spatial dimension is most explicit in soil-decontamination policy. Soil contamination has spatial consequences because, in principle, regulations prohibit property development on contaminated soil (§ 5.5). Soil contamination influences the possibilities for urban development to such an extent – the second National Environmental Policy Plan (NMP-2) refers to "a risk of stagnation of social processes due to the increasing number of cases of soil contamination" (TK 1993; 90) – that soil decontamination and protection policy are seen by many not as a direct part of dispersion policy, but as a separate, recognised category of policy development. In the NMP-3 (VROM et al. 1998) this is expressed by the fact that 'soil decontamination' is referred to in addition to the theme of 'dispersion'.

Acidification

Hoeflaak and Zinger (1992) identify disturbance and dispersion as policy themes for the grey environment, but this does not include the problem of ammonia emissions at national level. Ammonia is an acidifying substance that, in terms of policy, comes under the theme 'acidification'. Acidification is caused principally by the emission of sulphur dioxide, nitrogen monoxide and ammonia. Indirectly, acidification is increased by the emission of volatile organic substances, which contribute to the high concentrations of ozone and thus increase the acidifying effects of sulphur dioxide and nitrogen monoxide. Because most acidifying substances are either released high in the atmosphere (SO_2) or quickly disperse to higher atmospheric strata (NOx), their immediate local effect is limited. The effects only become apparent in the long term, following decades of emission leading to excessive concentrations at supra-regional level. Hoeflaak and Zinger therefore place acidification in the 'blue' environment category, but this line of reasoning does not apply to ammonia. Ammonia does not pass into the higher atmosphere strata and therefore does not disperse over a wider area. Moreover, ammonia sources – primarily livestock manure – are such that it cannot be released high into the atmosphere. Ammonia remains in the region in which it was released and produces negative and undesirable effects. Ammonia can therefore be compared to issues relating to disturbance, and may give cause to implement spatial measures. The negative and undesirable impact of ammonia immissions only affects the urban environment to a limited extent, or not at all, but their effects are felt primarily by rural flora and fauna. The problem of ammonia therefore has consequences for the green environment as well as the grey environment.

Based on these themes relating to the grey environment, three forms of pollution can be identified that can have a direct, reciprocal effect on the local environment, and for which specific policy has been and is being drawn up in the Netherlands:

- above-ground contamination, particularly in urban areas, categorised under the theme of 'disturbance'. Generally speaking, policy is geared towards maintaining a distance between the source and environmentally sensitive functions or areas, by means of zoning or other methods;
- underground contamination, categorised under the theme of 'dispersion'. Area-oriented policy is not geared towards maintaining a specified distance, but towards decontamination and the decision to allow or prohibit particular functions at or near a contaminated site;
- ammonia contamination, which mainly affects rural areas. Policy is geared towards maintaining distance between livestock farms and environmentally sensitive flora/fauna.

Because this study focuses on urban environmental/spatial conflicts, only the first two forms of pollution are relevant.

Table 2.2 The main 'environmental bottlenecks' reported by municipalities for the City & Environment COBBER project

Procedural

i **Noise:**
Current practice in noise regulation (exemptions, prosecutions, etc.) can lead to inefficient use of manpower and resources.

ii **Soil:**
Residential development delayed by soil decontamination procedures.

Financial

iii **Soil:**
Cost of soil decontamination leads to restrictions in use-designation: if building is not permitted in the inner city, a peripheral location is required.

Policy-specific

iv **Noise:**
Urban expansion is restricted by the standards set out in the Noise Abatement Act (WGH).

v **Soil:**
Compulsory soil decontamination hinders optimum resource deployment.

vi **Odour:**
Standards are not in line with the public perception of noise nuisance.

vii **External safety:**
Group risk and individual risk standards (emplacements and individual companies) restrict building in a compact city.

viii **Noise:**
WGH standards are not in line with the local perception of noise nuisance.

ix **All environmental aspects:**
In some cases (e.g. in inner-city areas), limit values for odour, traffic noise, and industrial noise cannot be realised. Target values would only appear to be feasible if urban development is not compact.

x **Mobility:**
The reduction of traffic noise is hindered by high costs. Autonomous growth in traffic volumes produces new problems. The situation is also influenced by traffic policy that is also desirable from an environmental point of view, namely the 'bundling' of road traffic on main traffic routes. This has consequences for air quality, which, from an environmental point of view too, must not deteriorate.

Design/technology

xi **Design/feasibility of new technology:**
Measures are required at structure-plan level before new technology/design can be implemented at the level of land-use planning.

Source: VROM 1996e; 31

Table 2.2 is an overview of the main problems reported by large and medium-sized municipalities. They are the outcome of an analysis carried out for the City & Environment project COBBER.[23] The analysis is interesting because it is divided into four categories: procedural, financial, policy-specific and technical. This structure renders transparent the environmental/spatial issues experienced by the municipalities.

It will also be clear that the 'standards discussion' introduced in section 1.2 is not an abstraction, but is perceived as an actual problem at local level. Financial and procedural problems are also experienced to the same extent. The following are seen as the causes of stagnation in spatial development and the inefficient deployment of manpower and resources: inadequate co-ordination with spatial planning; delays in housing development due to procedural obligations relating to environmental policy; restrictive and unrealistic environmental standards, and the high cost of meeting environmental quality standards. A striking conclusion is that local councils attribute the cause of environmental/spatial conflicts not so much to the pollution itself, but primarily to guideline-setting, generic standards and procedures.

The results of this analysis support the conclusion of Bartelds and De Roo (1995; 65) that "the dilemmas of the compact city are largely policy-related". Such problems can characteristically be defined as local or regional, and translated into spatial 'claims', with the obvious aim of separating environmentally sensitive functions from environmental contamination. In urban areas, this mainly relates to the risk of underground and above-ground contamination that can be categorised under the policy themes of disturbance and dispersion. Such contamination determines the quality of the local environment and can influence local spatial development through guideline-setting, generic regulation.

2.6 Evaluating the Grey Environment

Appreciation of the grey environment is measured not only by interpreting phenomena – which are as objective as possible – observed in the physical environment. Intersubjective evaluation is also important. The level of appreciation will also depend on the interplay between personal perception, social dialogue and policy visions. This interplay inevitably influences policy relating to the grey environment. After all, the degree of acknowledgement of 'grey' issues and the way in which they are subsequently dealt with depends on how they are appraised.

Sustainability

Environmental appreciation is also related to the concept of sustainability, i.e. the necessity of exercising due care in human activity in order to preserve the quality of the physical environment. Although the concept 'sustainable development' was introduced as early as 1980 in the IUCN publication[24] *World Conservation Strategy*,[25] it became fashionable worldwide following the publication in 1987 of the UN report *Our Common Future* by the World Commission on Environment and Development. The term sustainability was used in the report by the World Commission on

Environment and Development to indicate the need to ensure that future generations have the same access to natural resources as today's generations. Today's activities and interventions must not burden future generations with a legacy that could restrict their scope for development. In other words: sustainability relates to the development of welfare and prosperity, allowing for the fact that society must content itself with meeting "the needs of the present without compromising the ability of future generations to meet their own needs" (World Commission on Environment and Development 1987; 8). In this sense, the definition of sustainability is heavily anthropocentric. In the report *Our Common Future*, such sustainable development is thought to be feasible provided that present generations are willing to make the necessary efforts[26] and continue technological progress (see also RIVM 1988; 11).

For Opschoor and Van der Ploeg, sustainability means "the continued existence of components of the ecosystem that are functional for human society"[27] (1990; 102). Here, sustainability is more or less synonymous with the support base[28] (TK 1989; 92) or capacity[29] (Healey and Shaw 1993) of the ecosystem and the ability to substitute within the physical environment for the purpose of social development (Opschoor and Van der Ploeg 1990; 99). Pearce et al. (1990) define sustainable development as the wish to preserve "a set of aspirations of a group in society" in the long term. Here, they are not referring specifically to the physical environment, but to the aspect of time. However, this definition also expresses a general feeling of concern and awareness of the necessity of protecting aspects and functions that – directly or indirectly – contribute to economic development, are links within and between ecological functions, and merit acknowledgement by individuals and society. In *Zorgen voor Morgen* (*Caring for Tomorrow*) the RIVM points out that "the time factor also plays a role in the development of environmental problems" (1988; XV). It claims that "even after emissions have ceased, [...] it takes a long time for the original situation to restore itself, if this is possible at all" (1988; XV). The NMP-1 (TK 1989; 43), following the report *Caring for Tomorrow* (RIVM 1988), speaks of an increasing frictional time[30] and delayed effect.[31] Apart from the time aspect, the increasing scale on which environmental issues are experienced is also a factor. "In the past, the main environmental issues were relatively local" (Winsemius 1986; 27), while "today, environmental problems are often regional or even global" (1986; 27). As environmental issues occur on an increasingly larger scale, it takes longer to identify them. These issues are not experienced in the local environment until much later, and they are much more difficult to deal with than issues on a local scale. Therefore, sustainability refers not only to future generations (deferral in time), but also to the higher scales on which environmental issues can occur (spatial deferral) (see also § 5.2 and De Roo 1996; 20-23).

Although striving for sustainable development primarily involves caring for future generations and preventing or undoing undesirable and detrimental effects occurring elsewhere, far from our own local environment, choices made here and now can also contribute to sustainable development. Obviously, this does not relieve local administrators and policy-makers of their responsibilities. In the opinion of Breheny: "local initiatives must be seen as contributions to the achievement of sustainability at larger, national and global scales. Hence the exhortation to 'think globally, act locally’”’[32] (1992; 280).

Opschoor and Van der Ploeg (1990; 81) point out, however, that there are several problems inherent in the concept of sustainable development. They claim that the concept is difficult to define and arises out of a compromise designed to achieve consensus between various parties, including the environmental movement and economic interest groups. Opschoor and Van der Ploeg thus ask “whether the term sustainable development encompasses all aspects of environmental policy” (1990; 81). They believe it does not: “it is a matter of optimum liveability and the integrity of natural systems”[33] (1990; 81) (see also Fig. 2.6).

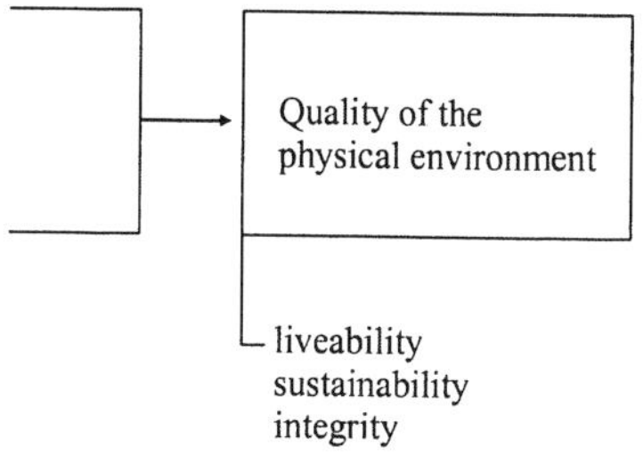

Figure 2.6 Possible distinctions in thc quality of the physical environment
Source: Based on Opschoor and Van der Ploeg 1990 (from Fig. 2.2)

Liveability

It is not always possible to defer negative and undesirable environmental effects resulting from human activity to distant regions or future generations. Ammonia released from manure remains a local and regional problem that has consequences for the agriculture sector and for rural areas close to intensive farming areas. Noise emissions by traffic and industrial activity are primarily local and heard in the vicinity of these activities. Odour emissions, vibration and dust are most highly concentrated at their source. Also the effects of traffic accidents, whereby hazardous substances are released, and industrial accidents will be most heavily felt at the site of the accident. The deflection of pollution into the soil, which was common practice until relatively recently, also leads to present-day restrictions in land use. In all these cases in which human intervention harms the environment on a local scale that is almost directly perceptible, the quality of the local living environment is harmed in a way that constitutes a threat to liveability for man and society.

Consequently, the concept ‘liveability’ also acquires a meaning that, in the same way as the concept of ‘sustainability’, is geared towards “guaranteeing functionality for society” (Opschoor and Van der Ploeg 1990; 103). However, in contrast to sustainability, it refers not so much to matters relating to effects that will be felt at a distance and in the long term, but to the situation here and now.[34]

Here, liveability is taken to mean the quality humans experience in their daily world: a quality that is experienced in terms of hygiene, safety and (multi-)functionality. This is not only a matter of the functionality of that environment; liveability also depends on how we perceive the environment. Liveability relates to the 'here and now'. It is therefore a quality concept that also encompasses the grey environment and local spatial/environmental conflicts. Definitions of liveability are largely subjective, particularly with regard to safety in the community and the need for green areas. But it is also possible to use empirical indicators in order to objectify our day-to-day surroundings, particularly when measurable health aspects are involved, for example. Liveability has not proved to be a straightforward means of categorising the quality of the local environment: Van den Bergh et al. (1994; 11) conclude that "neither policy-makers nor the users (of space) have a uniform understanding [of liveability]". All too often, liveability is confused or identified with sustainability. The English-language literature makes little or no distinction between sustainability and liv(e)ability with regard to the environment and environmental health and hygiene. According to Beatley: "terms such as *livability* and *quality of life* [...] beg for definition and description" (1995; 387) with regard to "creating and supporting humane living environments, liveable places, and communities that offer a high quality of life" (1995; 387). Neither does Dutch-language literature always make the distinction between sustainability and liveability employed by Opschoor and Van der Ploeg, as well as Hoeflaak and Zinger. The RIVM considers the number of households experiencing excessive noise nuisance and "the number of streets that are 'unhealthy' due to air pollution in inner cities" (RIVM 1988; 340) as sustainability issues. The first and second National Environmental Policy Plans also refer mainly to sustainability, and this is partly because these plans are strategic and aimed at the long term. The Amsterdam city council, however, focuses strongly on the difference between sustainability and liveability: "Liveability, in contrast to sustainability, does not refer to the future but to the current state of the environment, which must offer an acceptable quality of life"[35] (Gemeente Amsterdam 1995; 9).

Liveability relates not only to environmental quality, but also to all other components of the physical environment. On the one hand, liveability is partly determined by environmental health, an aspect that can already be quantified to a certain extent using environmental quality standards developed in the Netherlands. On the other hand, liveability is determined by the social, spatial and economic qualities of a local environment,[36] which in are in turn determined partly by 'hard' criteria, and partly by perception and experience.[37] Liveability is therefore an aspect of quality that can be defined in many different ways for our day-to-day surroundings, and the quality of the grey environment is part of it.

Environmental/spatial conflicts that arise as a mutual and usually direct impact between environmentally harmful activities and environmentally sensitive areas require a 'sustainable' approach in order to ensure that the solution does not quickly become outdated. Nevertheless, 'liveability' will continue to be the most important, location-based quality aspect of the grey environment because that quality is largely determined by the extent to which human activity has a direct impact on the local environment.

Appraisal, Responsibility and Management/Control

In the Netherlands, the government has attempted to define acceptable surroundings primarily by setting standards for emissions, immissions and environmental quality (see § 5.2). This is an arbitral exercise that focuses on the functional-technical mechanism underlying the cause-and-effect relationship of pollution (see § 2.4). The government's aim has been to implement these standards on a national scale. In theory, therefore, the same restrictions on environmentally harmful activities and exposure to pollution apply to all situations and locations. This method of appraising the grey environment makes insufficient allowance for local circumstances. In response to this, certain areas of responsibility have been decentralised (see also § 1.2). A shift is also being promoted from a 'technically oriented policy' to a policy that is based on public consensus (see § 5.4 and 5.5).

According to Voogd, creating a support base in the community is "primarily a matter of information and communication" (1996; 38), which is increasingly seen as an integral part of the planning process. In this sense the planning process is "a political and administrative process, within which aspects such as consultation, opinion-forming, consensus-building and political decision-making play an important role" (Voogd 1995; 51), and whereby the basic principle is to achieve consensus between the objectives, resources and interests of different social groups and institutions (Voogd 1995; 31). As mentioned in section 2.3, technical-functional standards policy is not without uncertainties. This also applies to building public consensus. This is not so much due to object-related or functional-technical uncertainties as to the assessment and applicability of intersubjective relationships and the way in which they can be used in decision-making. The relationship between the appraisal of 'grey' issues and the decision-making method and government control partly depends on *who* has defined the quality goals. In the light of the developments outlined in section 1.3 it would seem that, in the near future, policy-makers formulating policy principles for standards will wish to base these less on an 'objective' and quantitative interpretation of the quality of the grey environment (see Chapter 5). The 'social' appraisal of the grey environment, i.e. the perception of those directly involved, is becoming increasingly important.

This also leads to the question *for whom* quality is defined. This question is particularly relevant when determining to what extent deterioration in environmental quality is or must be accepted in order to achieve a general improvement in environmental quality at the supra-local level. Public interest in environmental issues is sometimes due to an increasing sense of dissatisfaction at local level with activities that serve the general or supra-local interest, but lead to adverse, unwanted effects locally. Individual citizens and action committees are increasingly led by their own perception of environmental/spatial conflicts and developments that have an impact on the quality of local life. In this context, the terms 'Lulu'[38] and 'Nimby'[39] have acquired a special meaning. These concepts are so emotionally charged that, according to Bakker (1995; 183), they barely leave room for comprehension or qualification. Although Lulu and Nimby are not necessarily restricted to environmental/spatial conflicts (see e.g. Ashworth and Ennen 1995), both 'syndromes' are generally associated with a local perception of unwanted environmental nuisance. The 'Nimby' syndrome generally arises when a relatively small group of individuals experience deterioration in the quality of their day-to-day surroundings due to pollution caused by an activity that benefits a relatively larger group.

Nimby and Lulu refer explicitly to a locally perceived imbalance between local interests and supra-local or general interests. This involves externalities, as referred to in section 2.1, which decrease (taper) as the distance from the source increases. Both terms implicitly refer to differences in the scale on which the benefits and disadvantages of encroachment are experienced. Such 'externalities' are experienced as negative on a local scale, and are caused by activities that, for other reasons and on a wider scale, are perceived as positive (see also § 3.4).

Pinch argues that "the extent to which an externality is positive or negative will largely depend upon the perception of the individual" (1985; 92). Although the appraisal of externalities is a largely individual matter, the Nimby syndrome is characterised by the sense of community it evokes. Nimby sentiments have little to do with social status: "environmental problems are now affecting rich and poor alike" (Blowers 1990; 96).

It is not only individuals who group together as a result of shared dissatisfaction with local environmental nuisance. Authorities, too, are increasingly acting on their sense of duty towards the local community. Local resistance to the expansion of Schiphol Airport (Linders 1995) moved the Leiden city council to purchase trees for the *Bulderbos*, the woods planted by Friends of the Earth Netherlands to obstruct the building of the airport's fifth runway. Local authorities have also threatened to withdraw their support from other major projects, including the Betuwe Line (the rail route for freight through the green river delta of the Netherlands). Nimby sentiments arise primarily out of dissatisfaction with policy choices that can create local environmental pollution or cause it to increase.

Summarising, we can say that the quality of the grey environment has more to do with liveability than sustainability. Traditionally, the government has used quantitative standards to express the 'liveability' of the grey environment. However, appraisal of the grey environment is not only a matter for governments. The interplay of social forces produces, on the one hand, reactions to a locally perceived imbalance between negative and unwanted externalities resulting from nearby functions or activities and, on the other hand, reactions to the supra-local or general interest of these functions or activities. Environmental/spatial conflicts are therefore more than simply policy-based issues that relate only to the attribution of environmental and spatial claims (§ 2.5). Neither can they be unravelled or understood purely in terms of their technical-functional characteristics (§ 2.4). Environmental issues must also be placed in their social context. If current trends prevail (§ 5.5), individuals, social groupings and authorities will have to jointly determine which tensions between environmental and spatial claims are acceptable, and how the quality of the local environment will be realised.

2.7 Environmental/Spatial Conflict and Environmental Health and Hygiene

Problems that arise because the spatial constellation in its present form is threatened by poor environmental quality, or because the spatial layout obstructs the consequences of harmful immissions, have been introduced above (§ 1.3) as environmental/spatial conflicts. Previously, the term environmental/spatial conflict was only used in studies of integrated environmental zoning projects, in which industrial activities were seen as environmentally harmful and a cause of conflict with housing development, which itself was seen as an environmentally sensitive function. In a general sense, environmental/spatial conflicts relate to contamination of the environment that can lead to unwanted negative effects on a local scale, and in which there is a large degree of direct and demonstrable interplay between the environmental health and hygiene situation – described here as the 'grey' environment – and spatial developments. There is, to a greater or lesser extent, an undesirable overlap of externalities between environmentally harmful functions and activities, and environmentally sensitive functions and areas. This involves the environmental health and hygiene externalities of functions or activities that, depending on their nature and scale, can be characterised as:

- negative, undesirable and unsolicited;
- mainly chemical and physical contamination and affecting the natural environment;
- forms of pollution that, at least in the Dutch situation, occur mainly in the air and soil;
- demonstrable or high-risk effects: nuisance, physical symptoms and/or the risk of disease and mortality;
- according to their spatial manifestation, which decreases (tapers) as the distance between the source and the environmentally sensitive area increases.

It is a question of the externalities resulting from emissions that pollute the air, and immissions and soil contamination. Restrictions can be placed on spatial development in order to limit the impact of these externalities. Because these externalities are distance-related, environmental standards can be translated into zones, which in turn lead to spatial claims that may conflict with other spatial claims. Environmental quality standards and the consequent spatial claims both derive primarily from the cause-and-effect relationship of environmental pollution.

The environmental cause-and-effect chain in section 2.3 establishes a model-based connection between individual and societal needs and developments, environmentally harmful activities, the harmful effects of these activities and policy-related response to and social evaluation of the effects. In the following sections, the cause-and-effect chain is discussed in conjunction with the concepts, categories and themes that relate to environmental/spatial issues. The discussion is summarized in Figure 2.7. The area highlighted in grey contains the concepts that relate to the environmental health and hygiene aspects of environmental/spatial conflicts.

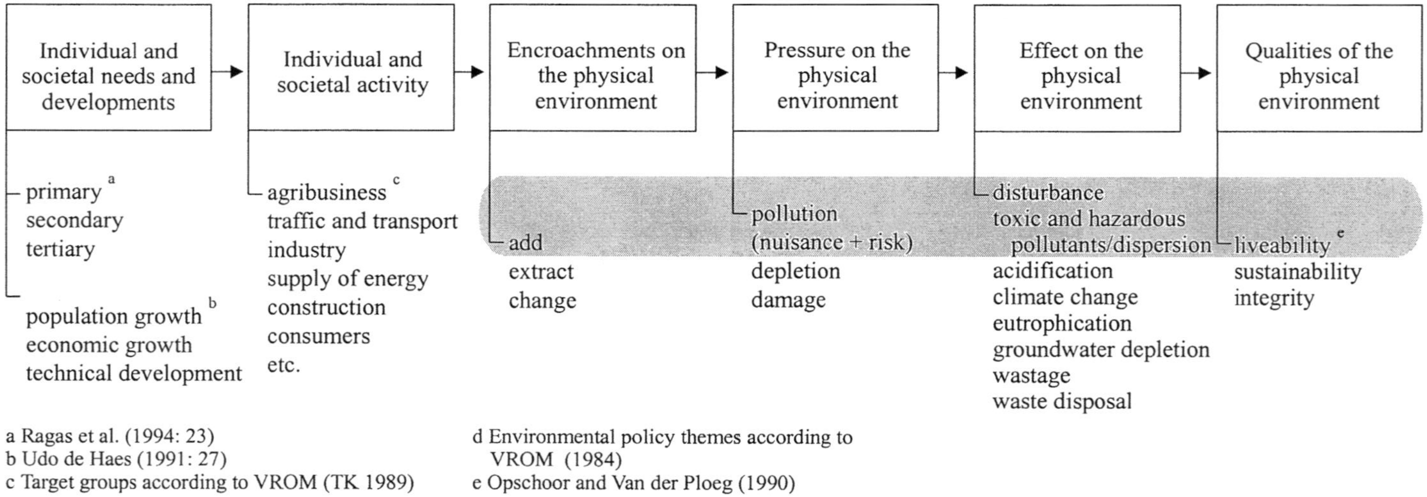

Figure 2.7 **The links in the environmental cause-and-effect chain**

Note: Concepts relating to environmental/spatial conflict are highlighted in grey

The elaboration of the cause-effect chain shows that externalities relating to environmental health can only be partly understood in terms of a direct relationship between cause and effect. Pressure on the environment in urban areas is more than a physical problem. When the functional elements of the problem have been unravelled on the basis of physical characteristics, its social context must also be considered. The recognition and evaluation of a problem, and the weighing-up of claims and interests are no less important than the observable state of the local physical environment. Such issues involve the balancing of interests, and phenomena such as the Nimby and Lulu syndromes can be associated with them. Bouwer and Leroy (1995; 26) thus refer to the 'sociogenesis' of the cause-and-effect chain. The chain's components, which give a global indication of the origins of environmental problems, have a high analytical value. But at the same time it must be remembered that "the connection between the 'objective' characteristics of environmental impact and the social appraisal thereof as an environmental problem [...] is in no way causal and only correlative to a certain extent" (Bouwer and Leroy 1995; 38). The aim of this chapter is, then, to offer only partial insight into the problems and the related policy-based discussions on the discrepancy between wish and reality in environmental/spatial conflicts. One equally important aspect must not be forgotten: the development of an awareness of the *complexity* of themes. That complexity is not apparent in the cause-and-effect chain, but is contained within the underlying situation.

Up to this point, we have discussed environmental/spatial conflict largely in terms of environmental health and hygiene. However, the concept 'environmental-/spatial conflict' also refers explicitly to the relevance of environmental issues for the spatial environment. The conflict between environmentally harmful and environmentally sensitive functions and areas is most tangible in the spatial context of the compact city. In the next chapter, the 'compact city' is introduced as an environment in which the spatial characteristics of environmental/spatial conflicts manifest themselves in the most complex circumstances.

Notes

1 Pinch (1985; 89-93) has summarised a number of generalisations that occur in the literature and relate to activities or functions which result in 'externalities':
- the nature and scale of the externality depend on the nature and scale of the activity;
- the nature of the activity determines the distance decay curve for the undesired effect;
- the activity can result in different types of externality;
- the activity can have positive as well as negative (spatial) effects;
- the extent to which these effects are experienced depends on the perception of the individual.

2 The physical environment can be defined as the whole that encompasses living and non-living elements: flora, fauna, soil, water, air and artefacts, both separately and taken together.

3 In addition to the green, grey and blue environments, Van der Wal and Witsen also distinguish between *blue networks* (water flows), *green networks* (ecological relationships) and *black networks* (traffic flows) (1995; 4-5).

4 In *Tragedy of the Commons*, Hardin warns us of the possibility that valuable social resources and communal values could be lost in a world that gives free rein to the pursuit of individual interests and local goals that are unilateral and separate. Nijkamp calls this 'society's snare': 'In a democratic society it is highly probable that the rational behaviour of individuals based on enlightened self-interest will not lead to optimal or consistent choices at the collective level' (1996; 133).

5 This UN commission became known under the name of its president Gro Harlem Brundtland, then Prime Minister of Norway.

6 *Silent Spring* was an important catalyst in initiating the study of the ecological consequences of human activity and the development of regulations (De Koning 1994; 22). In *Silent Spring*, Carson describes how excessive use of pesticides damages the ecosystem. Carson claimed that intensive use of pesticides did not increase their effectiveness – as was widely believed at the time – but rather led to immunity in the organisms they were designed to destroy, and to the poisoning of the ecosystem – man included.

7 For a comparison see Bouwer and Leroy 1995; 22-39, Ragas et al. 1994; 23-24, Udo de Haes 1991; 26-27.

8 According to Johnson, such interventions are seen as "'interference' with the environment, rather than just activity within it; it is sometimes considered outside the 'natural' aspect of an environment, with the implication that human 'interference' is 'unnatural'". He rightly argues, however, that "the distinction is really one of scale; all human activities are natural, but have far greater potential impact – vide the nuclear weapon – than those of any other species" (Johnson 1989; 34).

9 This must not lead to the conclusion that urban activity does not lead to depletion and damage. On the contrary: the needs of urban populations result in flows of materials and fuel, which usually come from another location where they are extracted from the environment, leading to depletion and environmental impact in that location too. At the local urban level, choices can be made that can lead to reductions in these flows. This means that the depletion of sources elsewhere will also decrease, something that can be seen as a positive development in terms of sustainability (see § 2.6). The policy of avoiding using tropical hardwoods in housebuilding projects is a good example. In terms of cause and result, urban planning choices can be made to remove or restrict the causes of depletion (expressed in terms of economics: decreasing demand will influence the supply side). However, the effects of such choices on the physical environment will not be felt in the urban area itself, but at the remote locations.

10 The standpoint of the Health Council (in line with that of the World Health Organisation) is that exposure to nuisance-causing immissions must be considered as a infringement of human health, and therefore as unacceptable (Gezondheidsraad 1994).

11 In addition, the likelihood of illness and death caused by exposure to toxic and carcinogenic substances depends to a large extent how susceptible individuals are to disease and the effects of pollution.

12 However, there is 'indirect' perception because the general public has certain feelings with regard to high-risk activities. "The acceptance of risk activities by the general public often depends strongly on social-psychological aspects, such as their (general)

relevance, the degree of trust in government and management, familiarity with or dependence on high-risk activities, the assumed (un)manageability of activities and their negative consequences, the ratio of pluses to minuses, and voluntary exposure" (De Hollander 1993; 607). Consequently, in a certain sense, scientific estimates of risk have only a limited significance, which means that risk standards are not always the most appropriate instrument for safeguarding public health (Vlek 1990, Midden 1993).

13 In the Netherlands, nuisance standards relating to noise have a particularly strong support base, and the planning consequences of noise impact have been worked out in great detail. With regard to high-risk immissions, the translation into spatial zones and the related spatial consequences are less clearly underpinned, or placed in a less clear legal framework. This contrast can be described as a 'nuisance/risk paradox' (see also De Roo 1996; 111). For nuisance caused by substances and vibration, too, regulations in the Netherlands were either sober or completely lacking. In the past, therefore, we looked to standards in other countries – especially Germany – for guidance in a number of cases.

14 The waste-collection sector in the Netherlands is now highly professional. Council or private waste services remove domestic and industrial waste at regular times from the consumer's premises or from collection points. The waste may be separated into various categories. If the waste cannot be reused, it is burned or dumped (Niekerk 1995). The location of refuse dumps and waste-incineration plants has been a subject of heated debate between local authorities and people living in the vicinity because of the effect they have on the area's image and possible emissions from these sites (Niekerk 1995). Waste does not therefore become an 'externality' until it reaches the refuse tip or incineration plant.

15 In this situation, environmentally harmful and domestic discharges can in principle be rendered harmless, although the siting of a water-treatment plant evokes similar reactions to the siting of a refuse tip or incineration plant (StoWa 1996). However, not all industrial and domestic emission points are connected to the sewer system. In such cases, the decomposition of environmentally harmful substances into harmless components must take place by means of dilution and natural decomposition. Where this process is unsatisfactory, rural and urban canals can become a source of odour nuisance.

16 The deposition in the soil of pollutants generated by normal daily activity is inherent in urban life. In many cities, for example, there are increased concentrations of lead in the soil due to the prolonged emission of lead from petrol engines (Copius Peereboom and Reijnders 1989). These emissions have decreased substantially following the introduction of lead-free petrol. Despite this decrease, diverse pollutants of this kind will continue to encroach on 'urban' soil. So long as they do not constitute an immediate risk to public health, such pollutants will possibly be considered as an inevitable background level that is difficult to eliminate.

17 The Dutch government considers it as its task to limit the risks of biological pollution using regulations and guidelines (Ragas et al. 1994; 172). In the first version of the VNG publication *Bedrijven en milieuzonering* (*Companies and Environmental Zoning*) from 1986, vermin were also named as harmful to the environment. In this type of environmental impact "distance can play a role, but is usually not a determining factor" (VNG 1986; 20). In later versions of the document, 'vermin' is no longer a zoning category. Other forms of biological pollution, such as genetically modified organisms

and disease organisms, can produce negative and undesirable effects on the environment with human intervention. In a number of cases, such as the cultivation of genetically modified soya beans, genetically modified organisms are deliberately introduced into the environment. Here, the government's main concern is the lack of clarity regarding the possible effects of this (Bulletin of Acts, Orders and Decrees 53; 1990). There is no systematic procedure for quantitative risk analysis, and therefore no procedure for formulating environmental quality criteria to apply to the introduction of such organisms into the environment (Ragas et al. 1994; 173). The accidental or deliberate introduction of primarily aggressive organisms – genetically modified or not – into the environment can also have serious consequences. The deliberate introduction of bass populations into Lake Victoria has completely disrupted the ecosystem, and is considered to be the one of the biggest ecological disasters resulting from well-intentioned development aid (Wanink 1998). In such cases, we can no longer talk of the 'tapering effects' of environmental pollution. Medical geography shows that the distribution pattern of some forms of biological pollution is consistent with that of epidemics (Haggett 1979). Instead of showing a decrease in impact in proportion to the distance from the source emission, the distribution pattern of biological pollution be completely different and much less predictable. This means that biological pollution can be very different from chemical and physical pollution.

18 Pollution can also be caused indirectly, e.g. litter at shopping centres, vermin at fish-processing sites, and large concentrations of seagulls at refuse tips.

19 This tapering effect depends to a certain extent on the definition of the source. If the source is considered to be the factory as a whole, rather than its chimney, the concentration in the vicinity of the factory may be lower than the concentration further away.

20 There are, however, forms of pollution in which there is tapering that produces no more than a minimal 'effect', e.g. pollution by substances such as CO_2, O_3 and SO_2, which lead to negative and undesirable effects because continual long-term emission increases the global or continental background level of these substances, thus causing damage to the physical environment on a wider scale than a local or regional scale.

21 These themes came to be known as the '*VER* themes' of environmental policy. This was because, in the National Environmental Policy Plans (NMPs), the Dutch nouns for the eight themes all began with the inseparable prefix '*ver*': *verzuring* (acidification), *vermesting* (eutrophication), *verspreiding* (dispersion), *verstoring* (disturbance), *verdroging* (dehydration), and *verspilling* (wastage). Other Dutch strategic plans were also based on '*VER* themes'. The 1990 Nature Policy Plan (LNV 1990) referred to the themes *versnippering* (fragmentation), *verzoeting* (sweetening of salt water) and *verwaarlozing* (neglect). In addition to these national '*VER* themes', there are regional '*VER* themes' such as *verzilting* (silting) of the IJsselmeer and *verstuiving* (drifting sand) on reclaimed peatlands (Province of Groningen 1990).

22 In the first National Environmental Policy Plan (TK 1989), soil decontamination policy was placed within the context of removal (*verwijdering*). In the second National Environmental Policy Plan (TK 1993) it was placed within the context of the 'dispersion' theme (*verspreiding*). In the third plan (VROM et al. 1998) the decision was taken to make soil policy an independent component of the 'dispersion' theme under the description 'soil contamination' ('*verontreiniging van de bodem*').

23 COBBER is the Dutch acronym for 'Creative Solutions within Existing Regulations'.

The project, which began in 1995 as a joint project between several authorities, and in 1996 produced the City & Environment publication *Binnen regels naar kwaliteit*, was primarily geared towards "areas of tension between existing regulations and the realisation of the compact city" (VROM 1996e; 14, see also § 5.5).

24 IUCN = International Union for Conservation of Nature and Natural Resources.

25 In particular the relationship between conservation and sustainability is emphasized (see also Thibodeau and Field 1984). The IUCN report for the Netherlands was compiled by Dankelman, Nijhoff and Westermann (1981), and adapted for the Dutch situation by the Netherlands National Committee for IUCN/Steering Group World Conservation Strategy (1988).

26 As early as 1972, the Club of Rome were discussing the need for "a fundamental change in values and aims at individual, national and global level" (Meadows et al. 1972; 187).

27 Opschoor and Van der Ploeg define sustainable development as "development such that the resulting pressure on the environment is 'ecologically acceptable'" (1990; 101).

28 The main aim of environmental management is "to preserve the carrying capacity of the environment in order to allow sustainable development" (TK 1989; 92). According to the Ministry of VROM, the carrying capacity of the environment can be said to be deteriorating if "within one generation, irreversible effects occur such as disease or death among humans, serious nuisance and decreasing well-being, extinction of plant and animal species, destruction of ecosystems, damage to water supplies, soil fertility or cultural heritage, and the restriction of spatial and economic development" (TK 1989; 92).

29 'Carrying capacity' is "the notion that a given ecosystem or environment can sustain a certain animal population and that beyond that level overpopulation and species collapse will occur" (Beatley 1995; 339).

30 "Due to the higher scale at which problems arise, and the large number of sources, a long period of time elapses between recognising the problem and taking measures at source" (TK 1989; 9).

31 A long period of time elapses "between changes at source and the ceasing of negative effects, and sometimes the negative effects cannot be undone" (TK 1989; 9).

32 'Think globally, act locally' is the motto of Agenda 21, set up following the United Nations Conference on Environment and Development, held in Rio de Janeiro, Brazil, from 3 to 14 June 1992 (United Nations 1992). This conference prompted many local authorities and cities worldwide to implement a Local Agenda 21, proposing measures that can be taken at local level to contribute to sustainable development on a global scale.

33 Opschoor and Van der Ploeg assume that the protection of "specific forms of biological wealth and the diversity of species and ecosystems" (1990; 103) does not necessarily follow from the desire for social development, and can therefore only be categorised in a limited sense under liveability and sustainability. They discuss integrity, which refers partly to the care for and protection of the physical environment driven by ethical and aesthetic needs.

34 Following this line of reasoning, the predicate 'compact city' used by a number of municipalities should refer primarily to decisions taken locally to prevent negative and unwanted effects on the physical environment that would manifest themselves elsewhere in the longer term. It could also apply to the sustainable nature of decisions

aiming to improve environmental quality, in the sense that local decision-making guarantees liveability for longer than one generation. Sustainability at the local level is, however, sometimes confused with policy initiatives designed to realise or preserve the liveability of the local physical environment (see Bartelds and De Roo 1995; 42-43).

35 It is argued that liveability encompasses "environmental harm (emissions, effects) caused by the function, as well as the environmental quality of the function. The emphasis differs from function to function. With regard to the functions 'work' and 'traffic', the influence of work-related and traffic-related emissions on total environmental quality is central. With regard to the functions 'living', 'public spaces' and 'nature', greater emphasis is placed on the environmental quality and liveability of the function itself" (Gemeente Amsterdam 1995; 9).

36 Van der Bergh et al. (1994) identify no less than 15 "facets of 'liveability in the compact city'" (1994; 53-70): level of enjoyment derived from a property, nuisance from neighbours, traffic intensity, safety of slow traffic, availability of public transport, accessibility of shops for basic necessities, presence of open spaces, presence and quality of recreational, aesthetically pleasing and ecological green areas (identified as three separate facets), air pollution, nuisance caused by noxious odours, noise nuisance, hygiene and unsafe places. In its report on liveability in residential neighbourhoods, 'Beyond the front door', the National Housing Council uses a similar set of indicators that are adapted specifically for residential neighbourhoods (Camstra et al. 1996; 26-27). In its report *Thuis*, the Ministry of VROM (1996) specified seven criteria: identity, comfort and security, innovative capacity, value of the 'grey' environment, value of the 'green' environment, economic value and cohesive spatial development. The RIVM has attempted to objectify these criteria by expressing them as a living-environment equation (1998).

37 In Dutch spatial planning, three values are used as the basis for enhancing the quality of our day-to-day environment (TK 1988). The first value is the *future value*, which relates to the appraisal of spatial functions over time (see also Voogd 1987) and is in many ways similar to the definition of sustainability. The second value is *use value*, and relates to the current functional use and allocation of space. Spatial planning is concerned not only with the functional use of space, but also with its *amenity value* for users, the third criteria. The use and amenity values are to a certain extent synonymous with the concept of liveability (see also Bartelds and De Roo 1995; 68-74).

38 Locally Unwanted Land-Use.

39 Not In My Back Yard.

Chapter 3

The Compact City

A Concept of Overexpectation

> Great cities can rise out of cruelty, deviousness, and a refusal to be bounded. Liveable cities can only be sustained out of humility, compassion, and acceptance of the concept of 'enough' (Donella Meadows 1994; 138).

3.1 Introduction

When sustainable development appeared on the global agenda, many spatial planners – both inside and outside the Netherlands – believed that they already had a 'sustainable' concept: the compact city. Beatley is not alone with his remark that "sustainable communities are [...] places that exhibit a compact urban form" (1995, 384). The compact city is the spatial concept that has served as the guiding principle for urban development for some time. Notions such as clustering and concentration have become important in the attempts to achieve compact, multi-functional and 'sustainable' spatial planning. Compact urban development, however, can also mean that functions and activities that are not really compatible for environmental reasons are located close together, possibly resulting in a loss of urban quality.

In such cases, compact urban development is the proverbial spanner in the works when dealing with urban environmental conflicts that have spatial consequences. Here, the approach is in fact based primarily on maintaining sufficient distance between environmentally sensitive and environmentally intrusive functions, activities and areas, as a result of the translation of environmental standards into spatial zones. In the attempts to achieve compact urban development, it may be hard to maintain sufficient distance and use this as a measure against excess environmental encroachment. This dilemma is sometimes referred to as the "paradox of the compact city" (TK 1993; 203).

The compact city is the ideal context for mapping the friction between spatial development and the quality of the grey environment. It also offers possibilities for examining a variety of environmental conflicts against the background of a dynamic and complex urban landscape. Within the compact city, space is a relatively scarce resource, while the number of claims on that space is relatively high and tends to fluctuate. The question is, therefore, how spatial claims resulting from environmental encroachment relate to other spatial claims in the context of the compact city. It is also a question of whether alternatives are available when maintaining a sufficient distance is not an adequate measure. Roseland (1992; 22) believes there are alternatives: "Cities provide enormous, untapped opportunities to solve environmental challenges". If that is the case, the question is: how do we recognise such 'opportunities'?

Chapter 2 focused on the environmental perspective of environmental-/spatial conflict. This chapter will emphasize the spatial aspect. The relation between the environment and space in a compact urban setting will also be discussed here. Section 3.2 will deal with the 'breakthrough' of the compact city as a spatial concept. In the late 1980s to early 1990s, the belief grew that the compact city can have a positive effect on the environment. Indeed, the compact city was soon hailed as a concept for sustainable development. Section 3.3 discusses the developments during the 1990s that affected the demand for urban space. These include demographic developments and expected developments in housing construction. The developments motivated central government to embrace the concept of the compact city. However, the developments also reveal some objections to the idea of the compact city. Section 3.4 will discuss some of these objections, describing, for example, the 'quality of life dilemma' that threatens to ensue from the radical concentration of functions. This 'dilemma of the compact city' relates to the tension between spatial development and environmental quality, one of the effects of lack of space and spatial pressure. As mentioned above, this dilemma makes it difficult to separate functions that are harmful to the environment from those that are environmentally sensitive. This exerts pressure on a relatively simple and obvious strategy for dealing with environmental/spatial conflicts: keeping a safe distance.

3.2 The Concept of the Compact City

Decades of reconstruction, decentralisation and clustered de-concentration – during which the population of the Netherlands increased strongly – resulted in the rapid urbanisation of rural areas and a growing need to travel. Urbanisation of rural areas and increasing mobility threatened to put a heavy claim on the highly limited space available in the Netherlands. The period of rapid population increase was reflected in a shift in household size. As a result, the demand for housing in the Netherlands – and consequently the pressure on urban space – continued to rise. Lower economic growth forecasts and the demographic developments in the 1970s became a motivating factor for dealing more carefully with what was already available and required an adjustment of 'de-urbanisation' policy (Frieling 1995; 7).

The Compact City as a Spatial Concept

An answer was found in the spatial concept[1] of the compact city. This concept was expected to facilitate "the control of mobility and economical use of space" (VROM 1993; 203).[2] In the course of the 1970s, larger cities – led by Amsterdam and Rotterdam – began actively to reverse de-urbanisation in favour of developing compact cities (Borchert, Egbers and De Smidt 1983). In the report *De compacte stad gewogen* (*The Compact City Evaluated*, RPD 1985) the National Spatial Planning Agency (*Rijksplanologische Dienst*; RPD) lists the leading principles on which the various cities based their urban concentration policies. The main principles are:

- To maintain and, if possible, increase the city's population, limit the increase in the use of urban space, and increase the civic foundation of the city. [...]
- To halt and reverse the unfavourable undermining of the city's economic, social and cultural functions. [...]
- To adapt the city to the growth in motorised traffic and promote public transport and low traffic speeds in the city. [...]
- To strengthen spatial and functional cohesion in the city. [...]
- To utilise investments already made, particularly with regard to facilities (RPD 1985; 30, summarised in Bartelds and De Roo 1995; 38).

The concept of the compact city is seen as a solution for various problems in urban areas, their immediate surroundings, and in rural areas. It therefore involves more than spatial and environmental benefits; indeed the first stimuli for concentration policies arose primarily out of economic and social considerations (see RPD 1985). In the 1970s, the erosion of the city as a basis for economic and social development became a problem in the Netherlands (Zonneveld 1991). Outside the Netherlands, too, self-containment was a reason to promote "intensification of the use of space in the city" (Elkin et al. 1991; 16). The development of ghettos and depopulation of the inner cities, phenomena that characterised cities in the United States[3] (see Jacobs 1961, Kunstler 1993, Smyth 1996), require attention in Europe, too.[4] "'Spread City' seems the rule of the day," laments Beatley (1995; 384). It is caused by what he calls "the evil of urban sprawl" (1995; 384). The result is an unbalanced population structure dominated by the less affluent population within and directly around the inner cities. In the Netherlands, too, this development is leading to a fear of deterioration and lack of an economic base. Not only the decrease in economic functions in inner cities, but also their disequilibrium is seen as an undesirable development. These factors make cities less attractive places and have a negative effect on public safety in the inner cities, especially in the evenings and at night. For that reason, not only the intensified use of space, but also diversity and multi-functionality are seen as essential characteristics for preserving the urban base.[5] So the main issues here are not compactness or concentration as such, but the function and quality of urban space. That is why concentration rather than compactness is now the principle with which urban planning must conform.

In the 1970s, policies still concentrated on channelling urbanisation through designated centres of urban growth. As the urban problems resulting from decentralisation became increasingly tangible, so too did the criticism of the uninteresting and un-dynamic urban growth centres. Initially, the RPD was criticised for its lack of interest in the concept of the compact city (Korthals Altes

1995; 97). However, concentration and compactness gradually became accepted as the basic principles for national spatial-planning policies.

For that reason, the Policy Document on Urbanisation (*Verstedelijkingsnota*, VRO 1979) not only concentrated on the policies regarding urban growth centres and decentralisation, but also on improving the quality of cities. The document contained the first initiatives for the concept of the compact city that was to be developed later. The economical use of space was to be promoted, and priority given to building within the city.[6] Partly at the insistence of the Dutch parliament (Second Chamber; now Lower House of Representatives), the report "emphasized clustered urbanisation, both nationally and in urban areas" (VRO 1979; XI-2, 96). This shift from a policy geared towards the development of growth centres to a policy that focused on urban concentration was reaffirmed by the revision of the Outline Structure Plan for Urbanisation (*Structuurschets voor de Verstedelijking*,[7] VROM 1983). An urbanisation policy that concentrated on conurbations would limit the overspill from cities to suburban and rural areas (VROM 1983; XI-2, 27).

In 1985, the National Spatial Planning Agency (RPD) report *De compacte stad gewogen*[8] (*The Compact City Evaluated*) heralded the final breakthrough for the concept of the compact city in the Netherlands. In this report, the RPD describes the key element of the compact city as "increased emphasis on the concentration of functions (living, working, services and facilities) in the city" (1985; 30). The compact city eventually became a popular spatial concept as a reaction to spatial policy based on clustered deconcentration.

Many cities in the Netherlands have always been relatively compact.[9] This applies mainly to inner cities where the mediaeval structure is still dominant.[10] However, not all parts of a city are equally compact in their structure. In particular, post-war developments relating to decentralisation and dispersion of urban space seem to conflict with the careful utilisation of the scarce space.[11] According to Kok and Van Wijk (1989) it is a matter of "preventing the ever-increasing fanning-out of urban elements across the region as well as the strict division of functions in the city itself" (1986; 16).

Le Clercq and Hoogendoorn (1983; 161, see also De Jong and Mentzel 1984) have outlined the compact city in functional-spatial terms on the basis of the following principles:

- emphasise city and landscape, build by adding to the existing structure;
- combine functions at an urban district level;
- distribute facilities in order to limit necessary traffic and improve accessibility for inhabitants;
- high-density construction;
- emphasise public transport.

These principles emphasise an expected change in traffic patterns as a result of compact building. This change in pattern should lead to a reduction in mobility. The faith in the healing effect of the compact city is not limited to Le Clercq and Hoogendoorn[12] alone. Expectations are generally high regarding travel-reducing opportunities resulting from compact-city policy. This belief in the reduction of mobility is only a small step away from the notion that the compact city can contribute to sustainable development.[13]

The Compact City as a Concept for Sustainable Development

The principle of compact urban development was followed not only in the Netherlands, but also at international level. The previously quoted report by the World Commission on Environment and Development, *Our Common Future* (1987), plays a major part in the discussions on this subject (see § 2.2 and 2.6). Many national governments have adopted the sustainable development advocated in this report as a major element in their policies. They are also looking for 'sustainable' concepts for urban development, and have a notable preference for the compact city.[14]

In Europe, the European Commission is a strong advocate of sustainable urban development. In its *Green Paper on the Urban Environment*, the European Commission (CEC 1990) lists a number of causes of urban problems that hinder sustainable development. In particular, it points out that "dependency on motorised vehicles in general and private transportation in particular has increased" (1990; 48). According to the Commission, urban pollution is partly the result of congestion, which leads to high levels of air and noise pollution. The Commission chooses several policy approaches:

- combining functions in urban areas and promoting inner-city living is preferable to a strict zoning approach;
- urban problems should not be solved by extending cities into their peripheries, but should be solved within the existing city boundaries (CEC 1990; 45).

With these policy guidelines, the European Commission is embracing the concept of the compact city (see also Breheny and Rookwood 1993; 155, and Hall, Hebbert and Lusser 1993; 23). It also underscores the view 'preserve what exists'. The Commission encourages the rejuvenation of existing areas and the renovation of derelict urban areas, and promotes diversity (CEC 1990; 60).

The UN Conference on Environment and Development held in Rio de Janeiro in June 1992 was a major impulse for 'sustainable' urban development (United Nations 1992). The UN considers changes in traffic and transport patterns as an essential step towards 'sustainable' cities. Efficient and ecologically sound public-transport systems and non-motorised methods of transport are considered important. However, the Commission believes that practical details need to be determined at local and regional levels.[15] Local governments are therefore expected to adapt this programme of action to their local situation: a form of Local Agenda 21. Once again, the belief is that the concept of the compact city should provide the scope for this adaptation process (see also Jenks, Burton and Williams 1996).

Elkin et al. (1991; 12) believe that a sustainable city must "be of a form and scale appropriate to walking, cycling and efficient public transport, and with a compactness that encourages social interaction". Thus, the idea of the compact city from the early 1980s seems to be the ready-made answer to the call for sustainable urban development. Jenks, Burton and Williams (1996) subsequently conclude that "recently, much attention has focused on the relationship between urban form and sustainability, the suggestion being that the shape and density of cities can have implications for their future. From this debate, strong arguments are emerging that the compact city is the most sustainable urban form".

Density versus Sustainability

The belief in the benefits of the compact city is widespread. Thomas and Cousins (1996) have summarised the benefits acclaimed by many (see, for example, Engwicht 1992, McLaren 1992, Newman and Kenworthy 1989, Sherlock 1991): "Less car dependency, low emissions, reduced energy consumption, better public transport services, increased overall accessibility, the re-use of infrastructure and previously developed land, the rejuvenation of existing urban areas and urban vitality, a high quality of life, the preservation of green space and a milieu for enhanced business and trading activities" (1996; 56). The belief in reduced mobility is particularly strong. Thomas and Cousins, however, point out that these claims "are at the very least romantic and dangerous, and do not reflect the hard reality of economic demands, environmental sustainability and social expectations" (1996; 56). In Dutch policy, too, the positive expectations of compact urban development have been tempered slightly. A substantial reduction in automotive mobility has not been achieved and the giant building projects resulting from VINEX (Fourth Policy Document on Physical Planning-Plus) and the *Vijfde Nota* (Fifth Policy Document on Physical Planning) have resulted in the further urbanisation of rural areas (see § 3.3).

Welbank (1996) concludes that the pursuit of sustainable urban development founded on the principle of compactness is primarily based on beliefs rather than on rational arguments (1996). Breheny, too, has doubts about the objectivity with which the idea of the compact city is pursued as a sustainable concept; 'suburban development' is dismissed as 'sprawl' by the European Commission. It reasons that the compact city will ensure "a superior quality of life for its residents" (Breheny and Rookwood 1993; 155). Jenks, Burton and Williams point especially to the weak relationship between the formulation of the compact-city concept, its foundations, and its assessment by means of research (1996; 7).

English-language authors, in particular, attack the European Commission's view that equates the 'sustainable' city with the compact city. In the Netherlands, however, the 'compact city' was never intended to be a spatial blueprint for sustainability. Here, it was a directive that underlined the desire to increase the emphasis on intensifying use of existing urban spaces with the necessary care. In this sense, the concept of the compact city was a reaction to developments that were considered politically as well as socially undesirable. This reaction predicted a number of benefits that, in retrospect, were based on wishful thinking and perhaps on naïve partiality. However, we should separate this notion from the wish to reverse the ebb of urban decentralisation.

The notion that compact cities are also sustainable cities has proved to be a step too far, and is based on a belief in a simplicity that does not exist. Cities are complex systems that are hardly influenced by predictions about aspects such as mobility. As a reaction to the waning belief in compact sustainable cities, Breheny (1996), Scoffham and Vale (1996), and Thomas and Cousins (1996) are seeking to open-up the discussion on density levels and the new desirability of decentralisation. Above all, this discussion will be a critical consideration of the one-sided and somewhat unsubtle standpoint expressed in European policy on sustainability.

This study will deal only in part with the discussion between Breheny and others about the desirability and the degree of compactness because the scope of this study is limited to a specific 'contradiction': spatial compactness versus acceptable environmental quality. Here, it is more important to be able to conclude that there is an essential difference between the spatial concept of compactness and the environmental, sustainable concept of compactness. Compactness as a sustainability concept seems to be too heavily based on beliefs, while compactness as a spatial concept is a reaction to 'an easy way out' that, for various reasons, is considered socially and politically undesirable.[16] The choice in favour of compactness and concentration must therefore be seen primarily as a concern for what already exists and for the quality of life in rural and urban spaces.

3.3 Clustering, Growth and Contour Planning

Despite the growing awareness that the idea of the compact city could result in fewer positive effects than was hoped, there was still focus on concentrated urban development in the 1990s. Growth of mobility will have to be reduced further, however difficult this may be. Houses, workplaces and facilities will have to be located so that the distances between them allow easier access by bicycle and public transport. This is one of the reasons that a distinction is made between urban and rural areas in terms of policy. Urbanisation of rural areas must be limited, while – in accordance with current policy views – support for the city must not diminish (VROM 1993; 6, VROM 2000; 7-20).

In the course of the 1990s, the demand for urban space increased even more, partly as a result of demographic, economic and social developments. On the one hand, this growing demand is an additional reason for controlling the development of urban areas. On the other hand, however, these developments impede the control of urban growth. In fact, the pressure on existing urban areas is constantly increasing. The question is therefore whether the preference for the compact city and for planning by contours, as introduced in the *Vijfde Nota* (VROM 2000; 5-36), can be more than a declaration of intent to aim, as far as possible, for compact spatial planning in urban areas.

Clustering by Consensus

In the 1990s, spatial policy was dominated by VINEX (the Fourth Policy Document on Physical Planning-Plus; VROM 1993). VINEX was preceded by VINO, which set the tone for VINEX: a selective approach based on participation. In contrast to spatial plans preceding these two documents, VINO was an elaboration of agreements between the various participants in the spatial planning process. This meant a shift from centralised control at national government level to control based on agreements and shared responsibilities[17] (see also § 4.7). The plan was designed to serve as a framework for the various government participants (Korthals Altes 1995; 151, see also Brussaard and Edwards 1988). A change in spatial planning was proposed, as well as a clarification of responsibilities and empowerment. Social developments became the basis for spatial planning. General solutions were not sought; policies would be developed only for the most important and spatially relevant social developments.[18] The intention was not to pursue an integral policy, but to seek selective responses to current or expected developments (Korthals Altes 1996; 155).[19] Motions and policy changes were accepted on the basis of consensus and further developed under joint responsibility (VROM 1988).

Based on consensus among all participants, strategic visions on the future were proposed for specific projects in order to realise a broadly supported approach, a plan of action, priorities,[20] and financing. These include key projects, the ROM area policy (see § 5.4), urban hubs and liveability projects. These projects served not only to control and promote spatial planning in the area where the project was realised, but they also served as examples and as a means of communication.

Once VINEX was realised, this policy was reinforced. "What was new, [...] was the co-ordinated position of the national government in negotiations with other levels of government and the desire to reach joint decisions and set them out in formal agreements" (Zwanikken et al. 1995; 19). In early 1991, the national government entered into negotiations with the provinces, the four major cities and the anticipated city regions about the execution of VINEX building tasks (see Table 3.2). It was hoped that these negotiations would lead relatively quickly to joint implementation agreements.[21] The idea was that "urban districts (or provinces) would also be risk-bearers within the provisions of the agreements. With the implementation agreements, the principle of decentralisation [is] being actively pursued",[22] according to the Ministry of Housing, Spatial Planning and the Environment (VROM) (1996; 5). By signing the agreements, local and regional administrators undertake to make every effort to realise the building projects designated to the area for which they are responsible.

At first sight, VINEX appears to be a blueprint for urban development. However, this impression is qualified by the emphasis on negotiation, and on agreements concerning how construction projects are to be realised and where responsibilities lie, which also result in a broad support base for VINEX.[23] The Spatial Planning Act (*Wet op de Ruimtelijke Ordening*) and the Spatial Planning Decree (*Besluit op de Ruimtelijke Ordening*) set out the legal instruments for policy implementation, but the following instruments are now also used in spatial planning: communication, consensus, establishing priorities, a tailored approach and shared responsibility (Zoete 1997; 111). VINEX has thus become a communication and negotiation plan.

In the preamble to the *Vijfde Nota*, this line is pursued further (VROM 1999, VROM et al. 1999). In line with the slogans of the "purple" coalition cabinets (social democrats and liberal-conservatives) of the 1990s, the term 'green polder model' is even used.[24] The *Vijfde Nota* can be drawn up in "very close collaboration with other social organisations" (VROM 1999; 14, see also WRR 1998).[25]

However, VINEX is not just a communication and negotiation plan that serves as a framework for making decisions. VINEX also elaborates what is seen as the most desirable urbanisation pattern, and describes how this can be achieved (Zwanikken et al. 1995; 18). Concentrated urban development is called *cluster policy* in VINEX. This means that "in areas with a large population, the increasing demand for housing, jobs and facilities (with a support base that transcends the local level) is met by building residential and commercial buildings and providing facilities in conurbations" (VROM 1993; 5).[26] Efforts are focused on building a concentration of residential and commercial buildings within the planning period. The building should take place in a limited number of large locations in and adjacent to existing cities and conurbations. Suburban building must be kept to a minimum. At the same time, VINEX finally moves away from the policy of the past: large-scale single-function residential areas (growth centres). Instead, the aim is to cluster the functions 'living', 'working', 'recreation' and 'nature'. This aim, however, did not prevent criticism of VINEX locations. In general, these locations are considered uninteresting, monotonous and not very imaginative – the same criticisms that were also levelled at 'growth cities' in the 1980s.

Five criteria were formulated for cluster policy. The criterion *proximity* gives an order of priority for siting buildings whereby inner-city locations are considered first, and in particular the possibilities for preserving buildings, rejuvenating derelict sites, and filling up empty spaces. When the VINEX proposals were developed in detail, *proximity* proved to be the most important criterion (Zwanikken et al. 1995; 19). The *access* criterion is in line with the aim of developing proper public transport in the cities and conurbations, and also promotes low traffic speeds. Other criteria are *cohesion* (i.e. between functions such as living, working, recreation and green spaces), and *keeping open spaces open*. The latter refers specifically to the open spaces between cities.[27] The final criterion is policy *feasibility*, which relates not only to financial and economic aspects, but also to social and environmental considerations. These location criteria enable provinces and municipalities to translate their obligations under VINEX into new residential and work locations ('VINEX' locations) that are appropriate for the local area in terms of scale and site. However, spatial-functional control by means of clustering and cluster-oriented location criteria is applied not only in the (compact) city, but also at a higher geographical level: that of the Randstad (the area covered by Amsterdam, the Hague, Rotterdam and Utrecht) and other conurbations. Clustering and proximity are geared towards the compact-city concept of the unfragmented city on the one hand, and conurbation strategies for cohesive urban centres on the other.

Planning by Contours

This desire to cluster urban functions at various levels was followed-up in terms of administration policy in the *Vijfde Nota*. Several new concepts were introduced, including 'intensification and combination' and 'transformation' of town and country, and 'red' and 'green' contours. These are designed to reinforce the clustering policy and refine compact-city policy. They will also encompass, affirm and – if possible – ensure that established, more or less autonomous developments, or developments that have proved difficult to control, become acceptable for new policies.

Even before the *Vijfde Nota* was published, for example during the process of reviewing and updating VINEX (VROM 1996d), nuances were sought in the conceptual framework. By now, the concept of *the complete city* was being used. This concept expresses the need to integrate spatial policy in a much broader 'large city' policy aimed at increasing the *vitality* of several aspects of the city (Van de Vijver 1998). At the same time, the concept considerably refines the policy of concentrated urbanisation. The 'complete' city underlines differences between inner cities, particularly differences in density and the great importance of space within the city (VROM 1998).

The coalition agreement for the second term of the Kok cabinet (1998-2002) outlines a mobile urban society which – following the recommendations of the Council for Housing, Spatial Planning and the Environment (*VROM-Raad*) scenario, City Land-Plus (*Stedenland-plus,* VROM 1988) – requires further elaboration in terms of the compact city approach, together with the controlled development of regional *corridors* (also see VROM et al. 1999 and VROM-Raad 1999). Despite the fact that the corridor concept is carefully avoided in the *Vijfde Nota* and the National Traffic and Transport Plan (*NVVP; Nationaal Verkeers- en Vervoersplan,* V&W 2000), the importance of corridors as transport axes that provide opportunities for economic and spatial development is beginning to become evident in practice. This involves 'unplanned growth' that could provide opportunities for "optimising the various functions using a structured approach" (VROM 1999; 31).

Compact, complete cities that are situated on corridors and other regional and supra-regional transport axes will develop under a regime of controlled (i.e. structured) growth. Policy-makers expect this development to lead to the development of regional clusters of *network cities* (VROM et al. 1999, VROM 2000, V&W 2000). The concept of network cities must constitute a policy-induced acknowledgement of the existence of relevant relationships that transcend conurbations and, at the same time, are determining factors in the development of cities and conurbations.

However, the publication of the *Vijfde Nota* and the wish to decentralise responsibilities could not suppress a deep-seated need, namely to control spatial development as far as possible. Subjects such as 'controlling the demand for space' were raised and, according to the Ministry, required "enhanced concentration and intensification of urban development in zones based on clusters of main roads" (VROM 2000; 7-19). This also involves an enhanced contrast of the city versus the countryside "hung on the contradiction between urban concentration zones that are carried by the main infrastructure on the one side and large natural landscapes on the other" (VROM 2000; 7-20). This contrast is also achieved by the introduction of 'contour policy'.

By introducing contour planning, the national government aims to reduce, to a certain extent, its involvement in local spatial planning. Provinces and municipalities should be given and should take greater responsibility in regional and local spatial developments. There is one condition, however: red and green contours must be included in their strategic spatial plans. The purpose of this is severely to restrict externally oriented urban development that adversely affects rural areas and nature reserves. *Red contours* must be drawn around all built-up areas and necessary urban expansion areas in the Netherlands. The *Vijfde Nota* specifies that the functions 'living' and 'working' should be developed within those contours (VROM 2000; 5-36). Areas that are of value in terms of landscape, culture or history will be enclosed within *green contours*, which are designed to "put a stop to new expansions of towns and villages, fragmentation caused by infrastructure, and intensive forms of agriculture and recreation" (VROM 2000; 5-37). All the above must all be set out in ten-year spatial planning programmes.

This eruption of new concepts and proposals is symptomatic of the search for new ways to escape the static – and constraining – principles of the compact-city concept. The desire for clustered growth clashes with the dynamics of urban development. At the same time, however, there is reluctance to accept the full consequences of too much flexibility and freedom. The deep-seated need to control is apparently difficult to overcome – however impossible this may be, given the dynamics of everyday urban reality.

This criticism of contour planning is not entirely unfounded, however. Particularly in environmental planning, many battles were fought in the 1990s to translate the pros and cons of rigid, framework contours as effectively as possible into policy – in this case to determine the level at which environmental encroachment in terms of spatial planning can still be considered acceptable (§ 5.4 and 5.5). Framework-setting contours and environmental zoning policy in environmental planning must now make way for more refined and flexible instruments for dealing with everyday dynamics. Herein lies an important lesson that will dampen the excessively high expectations relating to contour planning (see Chapter 5).

A Distribution Formula for Growth

VINEX contains the distribution formula for a "substantial increase in the housing stock"[28] (VROM 1993; 11) up to the year 2015, with a view to controlling urban development and growth. This 'quota method' – a quantitative distribution formula by which housing quotas are allocated across the Netherlands – is omitted from the subsequent policy document, the *Vijfde Nota* (Fifth Policy Document), in order to devolve responsibility to regional and local governments more explicitly than before. The *Vijfde Nota*, however, also focuses on the high level of demand for space for future urban development, although no longer in terms of the number of homes to be built. The growth in the national demand for housing is given in hectares (VROM 2000; 4-6). Along with clustering, growth will become a key element in spatial planning decision-making in the Netherlands.

The aim is to control the spatial processes of clustering and growth by means of a three-step approach, based on the density criterion. The approach is geared towards "locating new residential, work and recreational areas and public facilities in and as close as possible to large and medium-sized cities". The following three steps determine the order of preference: "the first priority is to utilise the possibilities offered by urban areas, then to move to the city periphery, and only then to use locations that are further away but connected to existing centres" (VROM 1993; 6-7). Under the *Vijfde Nota*, local and regional governments must endorse the urbanisation process by establishing red contours and 'fixing' them for the long term.

Both VINEX and the *Vijfde Nota* reaffirm the faith in the concept of the compact city. VINEX is based on the notion that space in the urban environment should be utilised as intensively as possible, "making use of opportunities for preserving or enhancing spatial quality whenever possible" (VROM 1993; 13). The *Vijfde Nota* suggests that compact cities should develop as networks, but without losing their individual character or filling the spaces between the cities within those networks (VROM 2000; 5-41). Priority should be given to using open spaces *inside* the boundaries of the cities, and redeveloping disused industrial sites, military barracks and railway yards. In order to support this concentration policy, it is made explicit that "concentration policy must be hindered as little as possible when regulations are applied (including the Noise Abatement Act) and consultation takes place [...]," (VROM 1993; 15). VINEX is very clear about the importance attached by the national government to a more concentrated development of urban space to deal with the expected extensive growth of spatial functions, notwithstanding environmental preconditions.

The agreements set out in VINEX and the demand for space specified in the *Vijfde Nota* can be seen as attempts to anticipate largely autonomous developments. The most important aspect is the ability to meet the future demand for housing, but the need for infrastructure and business parks is also an issue. Depending on the prognosis,[29] approximately one million homes will have to be built between 1995 and 2015 to meet demand (see Table 3.1). If the expansion of housing stock and the development of other spatial functions in urban areas are linked to the proximity criterion, this will inevitably increase the pressure on existing urban space.

Table 3.1 Estimates (1988 and 1999) of the need for expansion[30] of the housing stock in the Netherlands (1995-2015)[31]

		numbers x 1,000				
		1995-1999	2000-2004	2005-2009	2010-2015	total 1990-2015
Trend report 1988	2% variant	52		38		
Trend report 1992	Median scenario	332	169	149	150	800
Trend report 1995	Median variant	353	264	229	225	1,071
Wotab 1999 (VROM 1999)		243	299			

The growing need for housing is primarily due to the sharp increase in the number of households. This increase is itself a direct result of decreasing household size and the growth of the 'adult' population,[32] which is expected to grow from 11.7 million to 13.0 million by 2015 (CBS 1997; 64).[33]

Developments relating to household size also explain the expected increase in the number of households in the coming years. Household size will decrease further in the next few years. As a result, the percentage of persons living alone will increase in the near future, although the actual rate of increase will slow (see VROM 1995; 43). A low estimate predicts an increase from 2.1 million in 1995 to 2.7 million in 2015 (VROM 1995; 48). A reduction in household size and an increasing adult population means that the number of households is estimated to increase: a low estimate predicts an increase from 6.5 million in 1995 to 7.5 million in 2015 (VROM 1995; 48, see also CBS 1996). The nature and the level of the demand for space cannot be derived solely from demographic indicators. It is also influenced by changes in behavioural patterns, social and cultural trends, economic developments and government policy (see De Beer and Roodenburg 1997; 4). These largely autonomous factors can only be influenced to a limited degree, and they limit the extent to which clustering, concentration and contour planning are possible.

Studies of VINEX 'housing consumers' show "a general preference for larger houses, preferably single-family houses" (Kempen 1994; 191).[34] People living alone are also increasingly looking for larger homes. Consequently, there is a discrepancy between the needs of the VINEX 'housing consumer' – who is interested in larger, suburban houses – and attempts by the national government to restrict suburban living to a minimum.[35] The ability to predict the claim on space on the basis of these 'soft' indicators is decreasing substantially because of their inherent uncertainty. Nevertheless, it is a fact that the adult population is increasing and that the attitude to housing is changing. The result is an increasing claim on space.

Now that clustering and planning by contours are considered to be the structuring principles for dealing with the growing demand for housing, suitable building locations are being sought, mainly in and close to cities. Taking all the VINEX agreements together, we can see that 71 per cent (455,000 houses) of the total VINEX housing target (638,000 homes) will be built at VINEX locations (RPD 1997; 106). One-third of the houses will be built in existing urban areas, and two-thirds will be realised at the city periphery (see Table 3.2, Kolpron 1996, VROM 1996).[36]

It is striking that the locations in urbanised areas of the Netherlands (such as the Randstad conurbation), which showed little or no growth in the years prior to the publication of VINEX, are allotted an above-average housing construction target. Equally striking are the considerable differences between the various urban districts with regard to the ratio of intraurban and exurban construction. The city of Amersfoort will expand only to a limited extent; most of its housing construction target will be realised in peripheral locations. In the Rotterdam region, on the other hand, locations within the city are to be found for more than half (28,000) of all the homes to be built. However, this region also has peripheral expansion locations, including Noordrand II and III. These are located so far outside the existing city boundary that they are not connected to existing urban centres (step 2 of the proximity criterion).

In the latter case, it is more a question of locations that "belong strongly to a polynuclear metropolitan constellation within which communities are interwoven rather than wishing to cling to the nearest historic city" (Van der Poll 1996; 33). Rotterdam is not alone in this respect. Planned residential areas such as Driel Oost in the municipality of Arnhem, Eschmarke near Enschede, Delftlanden near Emmen, and Haagse Beemden around Breda lack a connection with existing urban areas. Almost all these locations are situated far from city centres (see VROM 1996). Based on the outcome of his study of some twenty structural concepts for VINEX locations, Van der Poll speaks of "isolated locations with predictably ordinary living environments" (Van der Poll 1996; 32). Kolpron (1996a) measured the distance from a large number of VINEX locations to the centre of the main city of a conurbation. For intraurban locations, the average distance was 2.9 kilometres. The corresponding distance for peripheral locations was 9.2 kilometres. Amsterdam was at the top of the list with no less than 15.5 kilometres.

Table 3.2 Overview of construction targets for 1995-2005 per conurbation, according to VINEX agreements

	intraurban	exurban	total	
Amsterdam conurbation	34,500	65,600	100,100*	
Rotterdam region	28,000	25,000	53,000	
Haaglanden conurbation	9,000	33,500	42,500	
Utrecht region	5,600	26,000	31,600	
Dordrecht conurbation	3,450	10,250	13,700	
Haarlem conurbation	3,100	3,700	6,800	
Hilversum conurbation	3,350	1,150	4,500	
Leiden conurbation	8,000	3,840	11,840	
Proportion for Randstad area	*95,000* *36.0%*	*169,040* *64.0%*	*264,040*	*58.0%*
Eindhoven region	12,130	16,270	28,400	
Breda conurbation	approx. 5,500	10,400	15,900	
's-Hertogenbosch conurbation	approx. 6,000	6,000	12,000	
Tilburg conurbation	approx. 5,500	10,000	15,500	
Amersfoort	1,600	11,100	12,700	
Arnhem-Nijmegen development hub	7,011	17,109	24,120	
Urban interconnecting area	*37,741* *34.7%*	*70,879* *65.3%*	*108,620*	*23.9%*
Other conurbations	38,032	44,654	82,686	
Other conurbations	*38,032* *46.0%*	*44,654* *54.0%*	*82,686*	*18.1%*
Total	170,773 37.5%	284,573 62.5%	455,346	

* including 27,000 homes in Almere

Source: Adapted from VROM 1996

The autonomous growth in the adult population, the changing composition and size of households in the Netherlands, and changing attitudes to housing have led to a sharp increase in the demand for living space, and this increase does not even account for growing industrial activity and increasing traffic volumes. Space is in great demand. Areas outside the urban environment, both close to it and farther away, are also being claimed in order to meet the increasing demand for space. Despite the good intentions of compact-city policy, this has inevitable consequences for rural areas and mobility. Clustering and contour planning as planning principles thus have an effect not only at neighbourhood and city level, but also at the level of the conurbation when the objective is cohesion in urban spatial planning. It is not surprising, therefore, that the *Vijfde Nota* refers to regional city networks (VROM et al. 1999, VROM 2000; 5-41). Clustering and red contours also involve co-ordination between the conurbation level and the levels below. Each level develops in its own way and this affects the other levels. This certainly applies to the intraurban environment, too. It is an environment that, in a sense, is already compact in itself but is nevertheless subject to change, has its own dynamic and may provide opportunities for urban development.

The Changing Intraurban Environment[37]

Most municipalities will only be able to realise intraurban housing developments by using locations that have not previously had a residential use. This could mean that the blending of housing with other existing functions in urban areas could increase multi-functionality, diversity and spatial cohesion between functions. However, as Van der Poll (1996; 34) concludes, few strategies have been developed for mixing residential and, for example, workplace functions. Combining housing and other spatial functions within a single location has proved less successful than combining social housing and homes built for the market. So far, the realisation of intraurban housing developments have resulted in rather monotonous sites based on variants of the ‘living quietly at the edge of town’ theme (Van der Poll 1996; 37). The restructuring or transformation of disused and run-down areas into multi-functional or residential areas may perhaps help to break through the monotonous results of the VINEX housing programme.

VINEX points in particular to the possibilities offered by disused industrial sites, barracks, stations and shunting yards for possible housing development. Insofar as these areas are not already empty and deserted (for example because of serious soil contamination), there is usually a need for a change of function or designated use. Such areas are sometimes referred to as transition areas (Bartelds and De Roo 1995), transformation areas (Brouwer et al. 1997, Heidemij 1996) or change-of-function locations (Kolpron 1996a). Brouwer et al. (1997) point out the many opportunities that urban dynamics create for housing development: “Whereas, until now, it was a matter of transforming the 19th century and early 20th century city – including harbour areas, hospitals, warehouses, slaughterhouses, gasworks, water board sites, breweries, etc.; the priority during the coming decades will be urban renewal in the cities of the welfare state” (Brouwer et al. 1997; 16).

What the authors observe above all, in addition to the progressing urban renewal that will also focus on post-war residential areas in the near future, are the strong dynamics and the functional shift in urban facilities. According to Brouwer et al., the strong trend towards merger in education alone will create "an entirely different need for space, while thousands of badly equipped buildings become empty" (1997; 16). The developments in healthcare, in sports accommodation and around shopping centres can be compared to this, enabling a future change in the use of space.

Functional transitions are already well under way at outdated or disused industrial locations and harbour sites. This development will continue in the near future. Pellenbarg (in RPD 1997) estimates that the Netherlands has over 60,000 hectares of industrial sites, 10,000 of which are probably outdated (Brink 1996; 30, EZ and Heidemij 1996). This means that approximately 15 to 20 per cent of the total area of industrial estates in the Netherlands are outdated and qualify for major maintenance, restructuring, rejuvenation or transformation (Koekebakker 1997; 4). When a site is restructured, the type of activity for which it is used often changes. Obsolescent industrial sites in and near inner cities are particularly interesting in terms of functional change – transformation – along with rejuvenation. The transformation may be from industrial to residential, or from single-function to multi-functional (§ 3.4).

The above appears to describe a situation in which outdated industrial sites, social, cultural and (health)care facilities, and obsolete or disused harbour and railway sites can be used for intraurban housing construction. This is partly true, but there is a somewhat one-sided emphasis on the VINEX housing programme. Although the recent wave of mergers in the education sector may well lead to concentration and a resulting change in the demand for space (Brouwer et al. 1997), this function will not disappear. On the contrary, new locations will be sought within or adjacent to the cities. This applies even more to industrial sites and office buildings. Some estimates predict that the demand for office space will lead to quantitative shortages in many regions, in addition to the shortages resulting from an insufficiently segmented supply (EZ 1994; 61-63). Such shortages will occur primarily in inner-city office locations (EZ 1994), despite the fact that, in general terms, a 'centrifugal movement' (Kassenaar 1997; 21) and 'economic suburbanisation' (Oosterhaven and Pellenbarg 1994) have been observed. Companies became much more mobile in the first half of the 1990s, and this development is attributed primarily to the lack of space resulting from company growth and worsening accessibility. "For a long time, business-to-business service industries remained strongly bound to inner cities – information-sensitive as they are. With mass motorisation and telecommunication, they now see the urban periphery as an attractive alternative, too" (Hemel 1996; 17). In fact, most relocations involve short distances and are outward movements towards the periphery (Oosterhaven and Pellenbarg 1994).

VINEX focuses on a number of spearhead activities (mainports) in the Randstad conurbation, in particular, but does not go much further. In the updated VINEX, the demand for space for business and industry has been quantified, and agreements with provinces and regions have been drawn up on the number of hectares to be realised for industrial sites in the period 2005-2010 (VROM 1997; 2). This involves a total of more than 3,000 hectares, most of which will have to be realised in the Randstad conurbation and the provinces of Brabant and Gelderland (see VROM 1996d). These agreements are adapted to local situations in a similar way to the housing agreements; the State determines the main quantitative requirements, while local governments implement the agreements and ensure that qualitative requirements are met. Compact urbanisation is the basic principle for this form of development, too.

In spatial planning, the main function of clustering has proved to be primarily a *directional* one for spatial developments. Because of the rather intangible growth in the population and the number of households, as well as the changing attitude towards housing, the resulting urban expansion will inevitably lead to a further 'encroachment' on rural space. These social processes are mostly autonomous and difficult to direct. In addition, there is no public support base for legislation to control such processes. Nevertheless, they result in substantial spatial claims, although the possibilities for regulation remain limited. That is why neither the distribution formula for growth as specified in VINEX, supported by agreements and the like, nor the ideas set out in the *Vijfde Nota* for using contours to limit outward development at least in terms of space, provide any guarantee against large-scale 'outward' development beyond the urban fringe.

Because urban development can only be regulated to a limited extent, the emphasis of policy is expected to shift towards creating opportunities – which are plentiful. Restructuring and transition areas, for example, show that the concept of the compact city may actually provide scope for high-quality urban development. Some opportunities have a downside, however. The intensified use of urban space that results from these developments makes it difficult to maintain sufficient distance between environmentally harmful and environmentally sensitive functions in order to guarantee a high-quality environment. This may result in a temporary increase in environmental/spatial conflict. In that case, compact urban development is not only of limited value for sustainable development, but also exerts pressure on the quality of life.

3.4 Environmental Conflict in the Compact City

Obviously, the interest in urban spatial developments is not limited to the spatial planners and policymakers. Environmental policymakers, too, are very interested in the spatial effects of urban development. The first National Environmental Policy Plan (NMP-1, 1989-1993) still supports the compact city as a spatial concept (VROM 1989). The second National Environmental Policy Plan (NMP-2), however, emphasizes the adverse effects of compact urban development on the quality of the local environment: "Increasingly, diverse functions are being concentrated in the cities. This development has positive as well as negative effects, albeit on different levels. One positive effect is the management of mobility and the economical use of space. At the same time, the process is leading to a concentration of environmental problems within the urban area, and this may result in friction between the 'living' and 'working' functions" (TK 1993; 203).

The paradox of the compact city refers to the confrontation between 'environment' and 'space' resulting from compact urban development. The paradox between environment and space may lead to stagnation in urban development – and certainly in economic development, too. For that reason, compact building is no longer an appropriate solution for unwanted developments from the urban past. However, the consequences of compact-city policy extend beyond the competencies of spatial planning. It has already been noted that compact urban development only results in limited synergy between spatial and environmental values. However, when the effects of compact urban development are viewed in terms of their contribution to the 'green', 'blue' and 'grey' environments (§ 2.2), the balance is negative, particularly for the latter.

There is no unequivocal paradox here; the problems are too complex and too diverse. It is therefore logical to speak of urban *dilemmas* where 'environment' and 'space' intersect. These dilemmas are the result of choices that lead to a complex interaction of environmental and spatial aspects (see also Bartelds and De Roo 1995). Dilemmas of the compact city manifest themselves at various levels. The nuances will be described below in further detail, in order to explain those aspects that relate to urban environmental conflict.

Scale, Area, Distance and Location

Subtle distinctions can be made between the 'dilemmas of the compact city' in terms of the geographical relationship between the grey environment and compact urban development. The distinctions relate to the geographical characteristics of 'externalities' (see § 2.1 and 2.6) that can be defined on the basis of concepts of scale, area, distance and location. Our prime concern here is the spatial deflection of environmental pollution.

Urban areas have a concentration of a wide diversity of facilities and activities that serve the entire country or parts of it but which, at the same time, adversely affect the local environment in which they are situated. Only rarely are adverse effects experienced in the same area that also benefits from the activity in question. The generation of energy, for example, is one of the few activities of general (i.e. national) importance whereby environmental effects – caused by the emission of SO_2 – impact at a higher level. Where this is the case, adverse environmental effects are not specifically deflected downward to a lower geographical scale or time scale. The deflection of adverse environmental effects to another time scale or geographical scale is only possible to a limited extent, and is difficult to 'sell' on ethical grounds. One example is the 'shifting' of waste between various locations. In urban areas there are many facilities and activities whose function transcends the local level, but their adverse environmental effects are nevertheless experienced at local level. By definition, government and administrative activities are almost always located in cities. This can generate a great deal of local traffic, thereby causing increased congestion and nuisance. Cities are also intersections within the national road and railway networks. A dense traffic and transport network is therefore important both nationally and locally, but can also cause nuisance at local level. Here, we are referring to local environmental harm caused by activities and functions that benefit the population as a whole. The current discussions on the Betuwe Line (the controversial rail route designed to improve freight links with Germany, which will run through the ecologically important Betuwe region between the main rivers which cross the Netherlands from east to west) and the expansion of Amsterdam Schiphol Airport (Gijsberts 1995, Linders 1995) are focused – apart from the question of their profitability – on balancing encroachment on the local environment against the nationwide benefits (see also § 5.4). The benefits will be shared by a large group of people, and will only affect their individual lives to a limited extent. A relatively small group of individuals, however, will suffer considerably from the adverse effects of the activity or function, for example: noise pollution, increased traffic volumes and litter. The positive and negative effects of a single function can vary considerably from level to level.

'Cost-benefit' friction exists not only between the local (i.e. urban) level and supra-local level and the national level, but also between urban and rural areas. Rural areas can be protected against urbanisation to a limited extent by concentrating functions in the cities. In the 1980s, this was one of the arguments in favour of compact-city policy (§ 3.2). It was hoped that this policy would keep rural areas 'green' and 'open'. Yet the current discussion on the 'Green Heart' of 1996 has made it very clear that urban policy alone is not enough to preserve rural areas. At the very least, a restrictive policy is needed (Van Bueren 1998) but even then the autonomous growth in rural populations appears to be leading to an expansion of the housing stock in rural areas (RPD 1996). The Ministry of Housing, Spatial Planning and the Environment, too, acknowledges that urban and rural policies cannot be separated because they are strongly interwoven: "Compact housing development, for example, is only feasible if there is no competing housing capacity elsewhere" (VROM 1996b; 6). Also, making a distinction between cities and rural areas will lead to 'pressure' variations. "Urban spatial pressure is expressed, for example, in a social need for more 'spacious' living" (VROM 1996c; 2) and more spacious working environments. Larger houses in a suburban environment, the use of holiday homes as permanent homes, and the tendency of companies to establish themselves at the urban periphery increase the pressure on rural areas. At the same time, the size of the agricultural sector is decreasing in economic and spatial terms, and there are not enough suitable functions for rural areas to "fill the gap and compensate for economic decline" (VROM 1996c; 2). In environmental terms, the compact city means a heavier burden in urban areas in favour of reduced environmental encroachment and spatial pressure in rural areas. This is not so much a deflection to another level, as an *exchange between area typologies*.

Dilemmas between levels and locations can be found *within* urban areas, too. The intensified use of urban space resulting from compact-city policy, for example, exerts pressure on green open spaces in the cities. Replacing these zones with other functions can adversely affect the quality of life (Bartelds and De Roo 1995; 49). In urban areas, the focus is on maintaining specified *distances* between functions in order to ensure that the adverse effects of environmentally intrusive functions or activities remain at an acceptable level (see also the discussion on the tapering effects of externalities in Chapter 2). There are many examples of conflicts or dilemmas in urban areas (see Chapters 5 to 7) in which environmental effects are a determining factor in the decision to permit or forbid a function or activity in a particular area. The construction or extension of a road will lead to increased traffic, which will inevitably result in greater traffic nuisance. In the case of a road in a residential area that is intended to improve access for residents, it could be argued that local costs and benefits are largely in balance. However, in the case of a through route or ring road, the adverse effects experienced by local inhabitants will possibly outweigh the benefits provided by the nearby infrastructure. In such cases, spatial policy will not only determine whether to designate a particular use to a location, but must also consider its *distance* in relation to the local environment.

Distance-related environmental conflicts are about forms of above-ground pollution, such as air pollution. In cases of sub-surface pollution (i.e. soil contamination) opportunities for spatial-functional development are restricted by the *location* of environmentally sensitive functions rather than by the distance in relation to a source. Thus, environmental conflicts can be distance-related and location-related. The physical aspects of such conflicts can be dealt with by means of land-use and/or effect-reducing measures. In view of the preference for concentrated urban development, the emphasis will be on the latter. In the case of air pollution, this will involve environmental-protection measures to be taken at-source, during transmission and/or where the emissions are 'received'. In the case of soil contamination, the options are to remove it, contain it or cover it, in order to enable land to be designated for environmentally sensitive uses.

The grey environment is under pressure from a concentration of various functions, including functions that are incompatible for environmental reasons. A distinction must be made between pollution that affects the supra-local level, pollution that varies depending on the area typology, and pollution for which the source/effect relationship can be expressed in measurable distances and which involves functional-spatial relationships based on environmental protection.[38] Environmental/spatial conflicts concern the latter; pollution can be traced to its source and has a tangible effect experienced by local residents, existing spatial functions and spatial developments.

Conflict Locations

According to Bartelds and De Roo (1995), most environmental/spatial conflicts occur in particular types of location. The first type is urban industrial sites (Borst et al. 1995, SCMO-TNO 1993). In their study of the dilemmas of the compact city (1995), Bartelds and De Roo describe other characteristic locations. These include industrial sites destined for restructuring, sites with contaminated soil, railway station locations and zones adjacent to road and rail infrastructure.

It is not easy to obtain a complete picture of the nature and the size of these locations. It is possible to give indications, however, based on environmental conflicts reported by local governments in the context of VROM projects. Integrated environmental-zoning projects, for example, show the friction that exists between heavy industry and the surrounding residential areas (see § 5.4 and Chapter 6). There are also examples relating to the 'ROM area' (areas with integrated area-specific policy) policy that occur specifically in urban areas (see § 5.4).

The project with the largest range of locations, however, is the City & Environment (*Stad & Milieu*) project (see also § 5.5). The aim of the project, which was announced in NMP-2, is to find structural solutions to urban conflicts arising from an environmental/spatial clash. In 1994, municipalities were asked to submit details of conflicts for a trial project. Eighty-seven written submissions were received, and 25 of these were eventually selected as projects in 1995. They were granted the status of 'experimental *Stad & Milieu* project'.[39] If the locations or areas in these experimental projects are divided into location types, seven basic types emerge (see Fig. 3.1). In addition to the location types already referred to above, the projects relate to docklands, riverbank areas, residential areas and inner cities.[40] Notably, these are almost all restructuring locations (Kuijpers and Aquarius 1998). Taken together, the location types present a relatively comprehensive picture of the spatial constellation in which the various environmental conflicts arise (TK 1998).

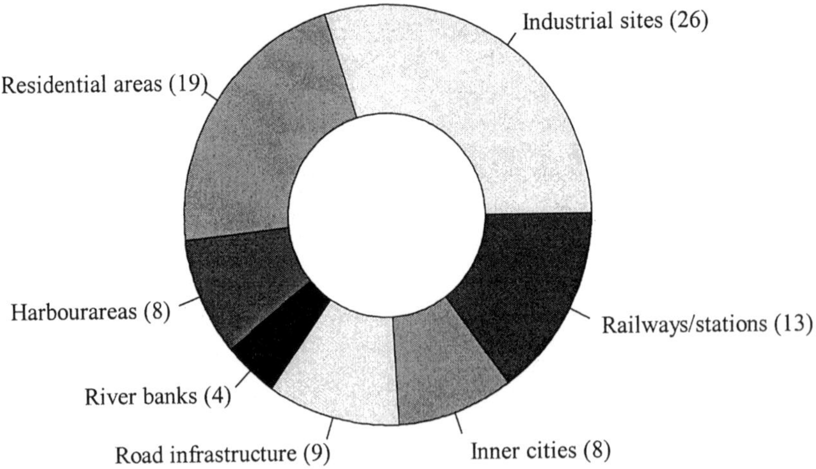

Figure 3.1 City & Environment project submissions, by location type
Source: Blanken 1997

Industrial sites In the Netherlands, industrial activities contribute substantially to the overall disruption of the quality of life in residential areas. Four per cent of the population in the Netherlands is said to experience 'serious' nuisance from pollution resulting from industrial activities (VROM 1993; 96). Other sources state that six per cent of the population in the Netherlands experiences serious noise nuisance from industrial activity (VROM et al. 1998). This amounts to a modest ten per cent (the figure includes building activities), of the total noise problem (VROM 1993).[41] The situation is quite different with regard to odour nuisance, substance emissions and external safety. Industry causes 45 per cent of all odour nuisance and 35 per cent of air-polluting emissions. Industry is also responsible for causing the majority (i.e. 80 per cent) of external safety problems (VROM 1994).

Between 1990 and 1993, VROM commissioned several studies into the extent of environmental disturbance caused by industry in the Netherlands (Akkerman 1990, Bartels and Van Swieten 1990, Hauwert and Keulen 1990, SCMO-TNO 1993, Van Swieten and Keulen 1992). It was found that, at 257 of the 1,700 sites investigated, relatively heavy industrial activity caused several types of environmental pollution (SCMO-TNO 1993, see also Chapter 6). At most (i.e. 176) of these sites, only two types of pollution arose, and the combination of noise and odour nuisance was the most common problem. One of the studies also examined the size of the environmentally sensitive area exposed to pollution, based on the predominant sectoral type of pollution. The study produced a rough estimate,[42] but that estimate presents a reasonably accurate picture of how industrial activity affects the local environment. According to the estimate, only 59 of the industrial sites producing multiple emissions affected a relatively small number of local residents (1,000 or less). By contrast, 99 of the 257 industrial sites examined were considered responsible for causing adverse environmental impact affecting 1,000 to 10,000 local residents, while 77 of the sites were considered responsible for impact affecting 10,000 to 50,000 local residents. An estimated 22 industrial sites in the Netherlands cause an excessive environmental burden affecting more than 50,000 local residents (SCMO-TNO 1993).[43]

Site redevelopment With a view to using space economically and concentrating urban functions, VINEX proposes the redevelopment of disused sites in and close to urban areas (VROM 1993; 13). Railway yards are becoming available as shunting activities are concentrated at a limited number of locations and inner-city shunting activities are transferred to outlying locations. The Amersfoort railway location is a good example of this (Hoogland and Kolvoort 1993). The reduction of armed forces as a result of the fall of the Berlin Wall, the end of the Cold War, and the abolition of compulsory military service has also led to a situation in which sites that are partially situated in or near urban areas became available. A good example is the site of the former Kromhout barracks, which was submitted as a City & Environment project[44] (§ 5.5) by the Municipality of Utrecht. The project has been halted. The site was submitted because finding a new use for the site was at odds with its environmental quality (Gemeente Utrecht 1997). The spatial opportunities for such sites are often clear in advance because, in most cases, only a single function has been or will be discontinued. The same applies to a limited number of industrial sites including former gasworks (see e.g. Gemeente Groningen 1997). These sites are more or less 'cleared' when the activity is discontinued. At these sites, the level of soil contamination and encroachment on the local environment are the factors that determine when, where and on what scale environmentally sensitive functions can be developed.

Although the vast majority of industrial sites considered for housing construction are not disused, they are nevertheless ready for redevelopment for various reasons (Gemeente Apeldoorn 1998, Gemeente Leiden 1998, Hoogland and Kolvoort 1993, Kuiper Compagnons 1996). These are sites without a predominant use and are often used for many, widely diverse purposes. Vacant premises, land that has not yet been allocated, economically weak companies, locations that are no longer prestigious, and accessibility and environmental quality are factors conducive to redevelopment. Often, such sites are no longer easily accessible because they have been partially enclosed by housing developments. Furthermore, the infrastructure and connections to main roads are no longer adequate, and transport modality has changed. In the past these sites were served by rail and harbour facilities, where possible, whereas today most transport is by road (Koekebakker 1997). Now that redevelopment of such sites is necessary to prevent them becoming further run-down, opportunities are being found to turn them into high-quality, multi-functional sites that are compatible with their immediate surroundings. Because there is a political preference for cohesive, varied and multi-functional urban spatial planning, and because compactness will result in intensified use of space, we can expect that environmental/spatial conflicts will manifest themselves at these locations.

Planners see the relocation of activities as a serious problem. Some of the activities have already moved elsewhere, some will move in the foreseeable future, some want to move but do not have the means to do so, and, for some, moving to another location is not even an option. This lack of clarity makes it difficult to develop these areas for multi-functional purposes or housing.

Built-up areas at the urban periphery (uitleglocaties) This type of VINEX location (hereafter 'peripheral location') is a specific type of 'transition location': "VINEX locations encompass new residential areas, new facilities centres, new industrial and commercial sites and modifications to existing areas" (Koekebakker 1997; 3). The idea that these sites are 'untouched' and can be used for any purpose is far from correct. Conflicting spatial claims arise because desired residential development is incompatible with the environmental situation at the sites in question. Furthermore, "by coincidence or otherwise – VINEX peripheral locations are situated at the urban periphery, where most industrial sites are also situated" (Koekebakker 1997; 3). The impact of industrial sites on the local environment is evident from emissions (licensed or otherwise), but also from the traffic to and from these sites. Neighbouring activities are therefore an important factor in choosing a site for VINEX peripheral locations. Kreileman (1996) shows that even VINEX peripheral locations are not entirely immune to environmental claims resulting from existing local functions or activities (see Fig. 3.2). These include functions that are not necessarily destined for relocation, for example: underground and above-ground energy infrastructure, transport infrastructure and existing activities (see also Kreileman and De Roo 1996).

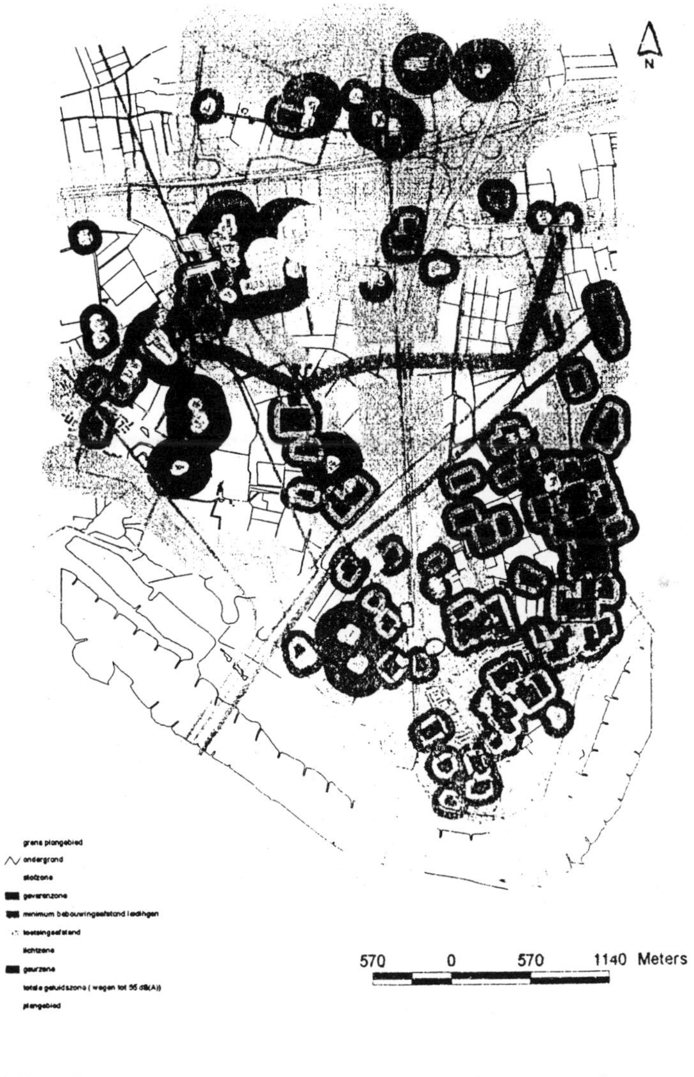

Figure 3.2 Contour map of environmental encroachment at the 'Waalsprong' peripheral location in Nijmegen

Source: Kreileman 1996

Note: This area is situated on the north bank of the river Waal

Van der Poll notes that, in many cases, noise-pollution contours determine whether or not a VINEX location is suitable for housing. He states that, as a result, the Noise Abatement Act often determines how a location will be developed. This picture of environment-driven 'contour planning' (Van der Poll 1996; 5) is supported by Kreileman's study of environmental conflicts at VINEX locations. In the conflicts Kreileman discusses, spatial development is obstructed or stagnates due to environmental contours, and problems arise sooner or later through factors such as lack of information, emphasis on only those environmental aspects that are subject to legislation, and the belated inclusion of environmental aspects in the spatial planning process (Kreileman 1996).

Locations with contaminated soil At many locations in the Netherlands, all manner of past activities have led to soil contamination. The level of contamination caused by hazardous substances is so high that the natural condition of the soil has already been affected or is under threat, and there is a potential threat to the ecosystem and human health. In most cases, these are or were industrial sites (§ 5.2). However, soil pollution has also been caused by activities other than industry. Illegal waste dumping (at Diemerzeedijk, Volgermeerpolder, etc.) and the underground storage of domestic fuel oil are well-known examples. Although many cases of soil contamination were caused in recent decades, other cases have a much longer history embracing several centuries. The soil contamination in Amsterdam's Jordaan quarter, now a mainly residential area, was caused by the many small tanneries and other activities related to shipbuilding in the sixteenth and seventeenth centuries. The pace of urban development is such that the cause of the contamination is not always immediately apparent. For that reason, historical research is often the first step in the process of locating urban soil pollution (see also § 5.2 and 5.5).

Railway station sites and railway zones The national government has elaborated several ideas regarding the development of railway station sites, and translated them into policies that do not necessarily supplement or reinforce each other. At a number of railway station sites such as Amersfoort, Dordrecht, Hengelo and Venlo, there was potential or actual stagnation in spatial development. This was mainly due to friction between the 'ABC-location policy' and the policy formulated under the theme 'disturbance' (see § 2.5). In 1998, VROM introduced the ABC policy aimed at offices and business in order to stimulate clustered urban development from the viewpoint of traffic and transport, and to facilitate mobility management in and around urban areas. The aim of the policy in terms of mobility was to 'get the right companies in the right place' (VROM 1990c). The 'ABC' policy was designed to improve accessibility in urban areas in general, with the emphasis on the workplace. At the same time, it attempted to have a positive effect on the quality of the environment by stimulating a shift in the modal split.[45] In the *Vijfde Nota* (VROM 2001) and the *Nationaal Verkeers- en Vervoersplan* (National Traffic and Transport Plan, V&W 2000), the practical experience gained with this policy induced the government to promote an 'active development policy' in addition to the specified planning arrangements. Bottlenecks were "the result of an imbalance in the objective, inflexible standards and an unsatisfactory delegation of responsibility" (V&W 2000; 63). To counteract these bottlenecks, the *Vijfde Nota* set the 'ABC location' policy in a wider context: "such an establishment of businesses and facilities that they provide a maximum contribution to the enhancement of the vitality of urban networks and the towns and villages" (VROM 2001; 5-45).[46] Responsibility is thus devolved to local government. Despite this fundamental policy change, infrastructure nodes remain crucial in policy development. The main NS (Dutch Railways) railway stations were designated 'A' locations and will remain so. These areas are therefore potential office locations.

However, functional-spatial developments around a railway station site and close to the railway infrastructure can conflict with existing rail activities. The spatial and environmental barrier effect of rail infrastructure is considered a problem for urban development (Blanken 1997, Gemeente Leiden 1998, Hoogland and Kolvoort 1993). In addition to excessive noise levels (Blanken 1997, NS Railinfrabeheer 1996), there are also problems relating to environmental risk (Bakker 1992, Bakker 1997; see also Van der Laan 1992). For example, standards for group risks[47] specify that dense housing development is not permitted close to stations where railway wagons carrying substances such as ammonia, LPG, ethylene oxide and chlorine are shunted.[48] Spatial development at 'A' locations with shunting sites is therefore being restricted and curbed.[49]

Road infrastructure There are several reasons, relating to traffic engineering, urban development and social developments, why the environmental pressure on and along the road infrastructure has a major impact on the quality of the surrounding environment. Car use is still increasing, for example, and commuting distance has decreased only slightly, in spite of compact urban development.[50] Furthermore, the dynamics and reordering process in residential and working areas are developing in different directions (Hemel 1996; 17), and there is an increasing number of double-income couples, whereby each partner works in a different location (Camstra 1995). Public facilities and institutions are being clustered and the volume of recreational traffic has increased sharply. The quality and quantity of public transport may be insufficient as pull factors; circulation measures, (re)development measures and price-policy measures meet with opposition from car owners, retailers, institutions, and inner-city businesses; and the infrastructure for cyclists is not adequate everywhere. And there are more examples. Traffic and transport problems in and around urban areas are highly complex and cannot all be solved with a single catchall policy. Autonomous processes, technical limitations, public opposition and psychological factors help to explain why the car is still a sacred cow. The fact that it has not yet been 'slaughtered' is not simply a question of disinclination. For many people, the car is still the best alternative in terms of affordability, time and convenience. In terms of traffic and transport problems, congestion leading to reduced accessibility still goes hand in hand with the harmful effects on the local environment. Some of the submissions for the City & Environment project (see also § 5.5) show that municipal authorities believe that they could make use of space adjacent to roads if legislation on noise (8 submissions) and air pollution (four submissions) were not a barrier to doing so (see Blanken 1997).

Other location types Four location types have not yet been mentioned: harbours and riverbanks, and the categories 'inner city' and 'residential areas'. The problems regarding docklands and riverbanks are roughly similar to those described above regarding the location types 'industrial sites' and 'site redevelopment'. The problems relate to the redevelopment of disused docklands and river/canal banks (see also Chapter 7) into residential and/or recreational areas (Gemeente Apeldoorn 1996, Gemeente Den Haag 1992, Gemeente Smallingerland 1997), the rejuvenation of docklands and riverbanks into multi-functional areas (also see Chapter 7), and the development of open spaces. The categories 'inner city' and 'residential areas' partially overlap. The former involves relatively small-scale projects in city centres (Gemeente Arnhem 1997, Gemeente Delft 1998, Hofstra 1996, Projectgroep Raaks 1997, Van der Veeken 1997) or projects of a more thematic character. These may involve the development of a policy for the hotel, restaurant and catering ('*Horeca*') sector, for example (see Blanken 1997). The 'residential area' category can be characterized as heterogeneous and often involves spatial developments that must contribute to social renewal in a residential area (Van Riel and Hendriksen 1996).

The previous paragraphs deal mainly with the spatial *context* within which environmental/spatial conflict arises. This paragraph discusses the spatial *characteristics* of the conflict itself. Some of the many and diverse dilemmas resulting from compact urban development can be described as 'environmental/spatial conflicts'. These are distance-related conflicts concerning above-ground environmental impact, and location-specific conflicts resulting from sub-surface pollution. From a functional-spatial perspective, these bottlenecks arise in an existing spatial constellation – in particular the functions 'industry' and 'traffic' – or are caused by changes in the existing spatial structure. In the latter case, the conflicts are mostly temporary.

3.5 Conclusion

The compact city has proved to be a concept based on over-expectation; a metaphor for good intentions that may have had – and could still have – a stimulating, activating and directional influence on urban development in the Netherlands but, at the same time, only provides a partial solution to urban dynamics and the prevailing rate of urban growth. The decision in favour of compactness, clustering and growth-constraining contours in urban development must be interpreted above all in terms of the concern for the existing physical constellation and for the quality of life in rural and urban areas. These concepts manifest themselves not only at local area and city level, but also affect spatial planning and structure at the level of the conurbation.

Urban development in the Netherlands is determined not so much by compactness and clustering, but by the *growth* in the demand for housing and commercial accommodation, and increasing traffic and transport volumes. These developments partly explain why compact urban development has only been of limited value as a concept for sustainable development. Terms such as 'compactness' and 'clustering' only serve to indicate the *direction* urban growth should take. They express the intention, which is *stronger than before*, to achieve a concentration of urban functions.

The multitude of factors and conditions, which makes up the complex range of opportunities for and limitations on urban development, results in differentiated concentration policies for urban development that are tailored to local and regional circumstances. The realisation that such policies offer little scope for control has led to new thinking in spatial planning. Policy based on a prioritising, specifying approach has replaced the comprehensive catchall approach. In addition to legal and planning instruments, other less direct instruments are considered as possible methods for achieving goals. Furthermore, shared and local governance increases as a result of increased emphasis on consultation, negotiation and agreements. According to this adapted decision-making model of shared responsibility based on joint agreements, the role of central government is primarily to monitor future spatial needs and, at the same time, to mediate in and direct the distribution and allocation of housing construction targets and location choice. In such cases, central government only sets limited preconditions. It is primarily the responsibility of local authorities to ensure that the spatial needs identified centrally are met. In doing so, they will aim to achieve compactness and clustering as far as possible. Given the complex dilemmas inherent in compact urban development, this is not an easy task. The government has made it clear that it does not intend to loosen the reins too quickly. With the publication of the *Vijfde Nota*, it checks the development of shared responsibilities by using an instrument that sets rigid preconditions for urban growth, namely planning within red and green contours. It is a 'this far and no further' approach that appears to deny the *complexity* of urban problems and the autonomous and dynamic processes within urban areas. Urban development is apparently reduced to a *simple*, controllable issue, which, in reality, it is not. At the same time, this approach largely ignores the experiences with contour planning under the environmental policy of the 1990s (see § 5.4 and 5.5).

Striving for concentration and clustering leads to several dilemmas where environmental quality and spatial development intersect. Some of these dilemmas relate to environmental/spatial conflict as defined in this study. They involve cases of underground pollution that lead to location-specific conflicts, and cases of above-ground pollution that primarily result in distance-related conflicts. Urban growth will almost inevitably lead to an increase, which may be only temporary, in the number of environmental conflicts. At the same time, it will restrict the possibilities for finding solutions based on spatial decisions. Not only is there less and less physical space available in which to realise a 'sustainable' separation of environmentally harmful and environmentally sensitive functions, but the pace of urban growth also means that there is less and less time available for finding solutions. Perhaps the most complicating aspect is the fact that environmental/spatial conflicts do not have a single underlying cause that can be described in terms of compact urban development. On the contrary, there are many different causes, ranging from 'historical' factors (soil pollution from the past, and the historical development of an area) to a lack of financial resources for remediation measures such as the 'ABC'-location policy, environmental friction resulting from function relocation *and,* as mentioned several times above, the autonomous growth of different urban functions. These drivers create an urban dynamic that can only be controlled to a limited extent, and they ensure that the nature of environmental/spatial conflict is not only location-specific, but also complex.

Environmental/spatial conflicts can be considered in various ways. They can be assessed on technical-functional grounds on the basis of their source/effect relationship. Above, it was pointed out that this is a limited view that allows the creation of a clear and unambiguous, precise, and almost systematic policy, while excluding all manner of contextual factors. Chapter 2, in fact, shows that the significance of environmental issues derives largely from the societal context in which they arise. Furthermore, we may conclude from Chapter 3 that spatial concepts only facilitate the understanding of – and the ability to control – spatial processes to a limited extent because of the many and various societal factors that affect such processes in a more or less intangible and autonomous way. One alternative is to see environmental/spatial conflicts as complex and dynamic issues in which contextual factors play a role. In this case, the approach will be less unequivocal and less predictable. Increasingly, environmental/spatial conflicts will also be seen as part of the process whereby interests are balanced. However, this will also make the issue more complex. In other words, environmental/spatial conflicts can be defined in several ways, depending on how they are viewed and – inextricably linked to this – how complex they are considered to be. This choice will shape the decision-making process. The following chapter will discuss the decision-making process when the complexity of an issue (in this case an environmental/spatial conflict) is considered.

Notes

1 Zonneveld sees a spatial planning concept as "a concise form, expressed in word and image, that indicates how an [...] actor regards the desired development of the spatial environment, as well as the nature of the interventions that are considered necessary" (1991; 21). According to Zoete, this makes it "a metaphor with a strong communicative function and it derives [... its value ...] from the possibility for combining objectives (i.e. realisation of the perspective) and measures" (1997; 14).

2 As early as 1973, Dantzig and Saaty suggested the 'compact city' as a means of reducing urban sprawl and preserving open rural areas. They took this almost to the extreme: "A quarter of a million people would live in a two-mile wide, eight-level

tapering cylinder. In a climate-controlled interior, travel distances between horizontal and vertical destinations would be very low, and energy consumption would be minimised" (1973).

3 The term 'Americanisation of the city' is sometimes used (Borchert et al. 1983; 93).

4 Sudjic, among others, observes that "rapid urban decentralisation has been a feature of most Western countries from the Second World War onwards, and earlier in the United States. The nature of this decentralisation has differed in different countries. In the United States, Canada, Japan and Australia it has tended to take the form of massive suburbanisation, creating at its extreme form The 100 Mile City" (1992).

5 Diversity and multi-functionality are not new concepts in compact-city policy in the Europe of the 1980s and 1990s. In her book *The Death And Life Of Great American Cities* (1961), Jane Jacobs was already advocating diversity and multi-functionality. Jacobs presented several conditions required "to generate exuberant diversity in a city's streets and districts". The first condition, which specified that "The district, and indeed as many of its internal parts as possible, must serve more than one primary function; preferably more than two. These must insure the presence of people who go outdoors on different schedules and are in place for different purposes, but who are able to use many facilities in common" (1961; 150) – was taken up by Dutch municipalities in order to preserve the quality of life in urban areas.

6 However, people were sceptical about preserving existing buildings. Pre-war areas of the city, in particular, were in a bad condition. Nevertheless, urban renewal became more important than urban expansion, especially in large cities.

7 The Outline Structure Plan for Urban Areas (*Structuurschets Stedelijke Gebieden*, VROM 1983) was the revised Outline Structure Plan for Urbanisation (*Structuurschets voor de Verstedelijking*), which was published as one document with the Urbanisation Policy Document (*Verstedelijkingsnota*).

8 The RPD report does not fail to mention possible adverse effects of concentration on socially weaker groups of the population and on facilities such as green areas and play areas. Furthermore, it already points out the need to prevent and restrict, as far as possible, environmental and noise nuisance caused by traffic.

9 Bartelds rightly states that, when one looks at city characteristics, one is always looking at a (high) concentration of people and activities within a limited area. "To a certain extent, the concept of the 'compact city' can be considered a pieonasm" (Bartelds and De Roo 1995; 33).

10 The fact that this situation does not apply to the spatial development that characterizes American cities is described in a fascinating way by Kunstler in *Geography of Nowhere* (1993).

11 Various studies show that, for many years, jobs (Breheny 1996) and population in Great Britain (OPCS 1992) have been moving from the city to rural areas. Consequently, Breheny and Rockwood are sceptical about the opportunities for compact urban development in Great Britain: "It requires a complete reversal of the most persistent trend in urban development in the last 50 years: that is, decentralisation" (1993; 155).

12 Buijs (1983), too, mentions several characteristics of the compact-city concept. He considers the *interspersion of functions and high-density building* to be useful methods for reducing commuting distances. This would be *a stimulus for reduced*

traffic speed and public transport. The *preservation of what already exists* is a priority. Wholesale demolition can be avoided by renovating and modifying existing structures, and this will lead to the *economical use of space.*

13 There was some criticism, however. As early as 1986, Owens was expressing her doubts about the effectiveness of compact-city policy with regard to reducing energy consumption and mobility: "The compact city idea has generally, and not surprisingly, been greeted with some scepticism [...]. Quite apart from the serious questions that could be raised about the flexibility and social implications of such form, its apparent energy advantages do not stand up to detailed scrutiny" (1986; 62). She refers to the changing 'commuting patterns', often from 'suburb to suburb', which in fact result in increased mobility. Hall (1991) is critical, too, because travel distance and 'modal splits' are not determined by urban density alone; but are also affected by the form and structure of cities, among other things.

14 Van der Wal and Witsen define a 'sustainable city' as "an urban system that does not deflect the adverse environmental effects of its functions to future generations or to other areas" (1995; 3).

15 The objectives formulated in Agenda 21 (the UN's environmental programme for the 21st century) are highly general in character. The following objective is relevant in this respect: "to improve the social, economic and environmental quality of human settlements and the living and working environments of all people, in particular the urban and rural poor" (United Nations 1992; 95).

16 Geleuken, Boeven and Verdaas add that "the resource 'space' is fundamentally different to natural gas or other minerals, to which the following maxim applies: the more economical, the better. The amount of space, however, is a constant. Space is not 'used up' in the same way as other resources" (1997; 16).

17 The diagonal planning formula (Ter Heide 1992) also originates from this period. Its purpose is to "translate government policy into visions for strategic areas and, subsequently, strategic projects within these areas. The 'diagonal' concept used here also assumes the involvement of other (horizontal) government partners and other (vertical) government authorities" (Zoete 1997; 112).

18 The Urbanization Policy Document (*Verstedelijkingsnota*), on the other hand, still described an incentives policy and spatial development policy for the Netherlands as a whole (Korthals Altes 1995).

19 With the introduction of urban hubs as a policy concept, government investment could be selectively allocated to a limited number of cities. However, this selectivity principle was largely undermined as a result of pressure from the Lower House of Representatives and from those cities that 'missed out' and were not designated 'urban hubs' (Zoete 1997, Zonneveld 1997).

20 Four investment priorities were identified: (1) to promote the competitive position of the Netherlands, (2) to harmonize working, living, public facilities and public transport, among others by realising 'A' and 'B' locations (see § 3.4), (3) to preserve, modify and rejuvenate rural areas, and (4) to pursue area-oriented spatial and environmental policies in 'ROM areas' (see § 5.4) and to improve liveability. The first priority was also an essential element of VINO, one of the reasons why Ashworth and Voogd qualified VINO as a "national plan for international competition" (1990; 5). Improving economic preconditions for business establishment is considered the primary responsibility of the national government,

while responsibility for priorities 2, 3 and 4 will be shared (Zoete 1997; 114).

21 This did not proceed as smoothly as planned, however. The first implementation agreements with the provinces were not signed until mid-1994. Utrecht was the first conurbation to follow this up by establishing its own regional government on 22 December 1994 (VROM 1996a). The last of the VINEX implementation agreements, that for the Eindhoven region, was signed on 6 October 1995. The stagnation of the VINEX programme was also due to the negotiations on finance and responsibility. National government and the municipalities did not interpret land development in the same way (Kempen 1994; 195), the nature of the government's contribution to urban renewal required clarification (Korthals Altes 1995; 184) and there was a discussion about the costs of soil remediation. Kempen suggests that these "far-reaching policy changes and the short period of time in which they must be realised, [...] place excessive demands on the administrative and official capacities of lower-level government" (1994; 220).

22 Provinces were more or less sidelined by the selection of conurbations or inter-conurbation collaborating groups as negotiating partners. However, not all neighbouring municipalities within conurbations have complete confidence in the collaborating groups, in which the central city has a dominating role. Under VINEX, however, the provinces have an explicit responsibility to act as intermediary between national government and the regions in the process of drawing up and realising the VINEX implementation agreements (see also Korthals Altes 1995; 193-194).

23 In the revised VINEX, the negotiation model is put into operation at an even earlier stage; the provinces (not the conurbations) submit recommendations for urbanisation locations and the national government then indicates its preferences (Zwanikken et al. 1995).

24 This is a direct reference to the economic and employment policies of the 1980s and 1990s that were considered successful. These policies were based on timely mutual consultation and fine-tuning of interests by the government and the 'social partners'.

25 The Minister of VROM, Jan Pronk, put forward two conditions. First: the parties must be willing to discuss matters with each other in order to present joint recommendations to the Minister. Second: they must accept that the final decision will be a political one (VROM 1999). The good intentions underlying the principle of mutual consultation cannot escape the watchful eye of central government.

26 The concept of conurbations is intended to provide an area-oriented approach based on inter-municipal collaboration as a means of structuring urban development, to counterbalance growth-centre policy and to counter suburbanisation, based on the realisation that cities can and do expand beyond municipal boundaries and have strong connections with their surroundings and neighbouring municipalities. In Part E of the Urbanization Policy Document (*Verstedelijkingsnota*) and in the preamble to the Outline Structure Plan for Urban Areas (*Structuurschets stedelijke gebieden*), the level of the conurbation is presented as the level on which clustered urbanisation should be realised (see also Borchert, Egbers and De Smit 1983, and Korthals Altes 1995).

27 In 1995, the Minister of VROM, Margreeth de Boer, opened a public discussion on the preservation of the *Groene Hart* (the 'green heart' of the Randstad conurbation) in the hope of warding off, for some time to come, calls for urban development in

the area. The outcome of the discussion is seen as an affirmation of current policy (VROM 1996e).

28 The VINEX revision (*Actualisering VINEX*, VROM 1996d) stated that the essentials of VINEX policy would certainly be applied until the end of the 20th century. It also stated that the government would decide on what was to be built and the preconditions, while the regions would formulate their urbanisation policies accordingly. In addition to the emphasis on housing development and housing locations, the VINEX revision places greater emphasis than before on work, green spaces and recreation, and on the location of these functions.

29 The estimate of the required expansion does not include replacement needs or the expansion of the housing stock. Taken together, these three factors determine the estimated total housing production.

30 In this period, various reports predicted the demand for housing. Over time, this demand showed a tendency to increase slightly (VROM 1990a, VROM 1990b, VROM 1999b).

31 Determining the demand for housing is a somewhat uncertain process (see Hooimeijer 1989). Criticism of the figures presented in the Trend Reports (RARO 1993) is aimed primarily at the methodology of the PRIMOS model on which the calculation process is based (De Graaf et al. 1994). The PRIMOS model has been used since 1983 to estimate developments in household size (Gordijn, Heida and Den Otter 1983).

32 Statistics Netherlands (CBS) classifies everyone over the age of 19 as an adult.

33 The size of the 19-and-below age group as a percentage of the total population has decreased sharply in recent decades from 36 per cent in 1970 (VROM 1995; 40) to 24 per cent in 1996 (CBS 1997; 64). This shift in age distribution will continue at an almost constant rate for some time to come. At the same time, however, the proportion of the population over the age of 19 continues to increase, although the annual rate of increase has slowed since the end of the 1970s (VROM 1995; 41) and will continue to do so (De Beer 1997). Within the adult population, the 65+ age group in particular will continue to grow until after the year 2030, from 13.3 per cent in 1996 to 17.2 per cent in 2015 (CBS 1997; 64).

34 The space required for the predicted expansion of housing stock is largely determined by the number of homes needed per region, shifts in household size, the size of home required by each category of household size. Given the fact that household size will continue to decrease in the years to come, one could conclude that the demand for smaller homes will increase. The opposite is true, however. The per-capita demand for living space is increasing. At the same time, housing density is falling. The result is, as Kassenaar describes it, a paradox of dilution and urbanisation: "The Netherlands is becoming increasing full, in the sense of 'urbanised', but urban density is decreasing" (1997; 20, also see Needham 1995).

35 The shift from stacked construction to homes with access at ground level, the shift from multi-family to single-family housing, and the shift to the more expensive price segments have also led to a reduction in the number of homes per hectare (Kolpron 1996b; 12-14). On the other side of the scale, there are heavy financial burdens on large VINEX locations that are partly due to relatively high land prices. According to Kempen (1994), this burden results in a need for denser construction, not decreasing density. This is particularly evident during the realisation of demolition

and new development plans for the pre-war areas of the four major cities, where density is increasing (Kolpron 1996c). Increased building density is also required in order to create a foundation for public transport, care facilities and the retail sector.

36 A study of the inventory of VINEX building locations shows that the average housing density at peripheral locations will be approximately 34 homes per hectare, while an average density of 65 homes was proposed in the plans for inner-city locations. A simple calculation using the data from Table 3.2 thus shows that, in order to meet building targets for the period 1995-2005, approximately 2,600 hectares (170,000 homes / 65 homes per hectare) will be needed for the inner-city areas, and 8,400 hectares (285,000 homes / 34 homes per hectare) for the peripheral locations. Housing density is defined as the 'net plan area for housing', including public green areas and neighbourhood facilities (Kolpron 1996a; 15).

37 The adaptation of the VINEX implementation agreements by the individual regions is geared towards varied residential development. This is primarily intended to prevent the progressive decline in the number of affluent inhabitants in existing neighbourhoods. Further aims are to combat the development of areas with an accumulation of problems, and to prevent social segregation in new and existing neighbourhoods (see Engelsdorp Gastelaars 1996). Homes at VINEX locations should be affordable for people with below-average incomes. The concept 'building for the neighbourhood' (i.e. focusing on the needs of a single but dominant social group) is no longer applied to restructuring and infill development in existing districts, and has been replaced by the aim to create a sound economic base, which means building homes for various income groups. To achieve this, the following distribution formula must be applied: 30 per cent social housing and 70 per cent of the homes to be built by and for the private sector (RPD 1997; 106).

38 The fact that environmental encroachment can manifest itself on different geographical scales and time scales more or less indicates that policy for dealing with it must be aimed at more than one level. For reasons such as this, the principle of dealing with problems on the level at which they occur requires refinement.

39 Eventually, 35 of the 87 pre-submissions were withdrawn. The 25 experimental projects were selected from 62 projects entered by 47 municipalities. As expected, most of these submissions were from the west of the Netherlands. The four largest cities were the most prolific with a total of 18 submissions (also see Blanken 1997, Kuijpers and Aquarius 1998).

40 The definition of location types was based on the area's main function before plans were made to restructure and rejuvenate it.

41 No less than 40 per cent of the population in the Netherlands experiences 'serious nuisance' from noise pollution (VROM et al. 1998).

42 In order to facilitate the calculation of the number of people exposed to serious noise nuisance, the relevant sectoral environmental contour was enclosed by a rectangle, whose sides correspond with the national triangulation system to the nearest 500 m. The number of people living within the rectangle was then counted (SCMO-TNO 1993).

43 For comments on this study, see: Borst et al. 1995; 58.

44 The project in Utrecht has been abandoned as a City & Environment project because the municipality was unable to purchase the land at the site.

45 This policy defined three types of location:

A - locations: areas with maximum public-transport access;
B - locations: favourably located for the public transport network and the road network;
C - locations: located close to motorway slip roads and exits.

Local authorities must balance accessibility profiles against mobility profiles for companies and activities, in order to designate locations for business and facilities. The accessibility profile of a location indicates how easily it can be reached by car and public transport. The mobility profile estimates the potential use of public transport by those working at or visiting a company or facility (VROM, V&W, EZ 1990).

46 The location policy of the Vijfde Nota sets the 'ABC location' policy in a wider perspective by using three home/work typologies: 'central environment', which refers to urban nodes such as urban centres and railway station areas, 'specific work environments' focusing on industrial sites and multimodal transport axes, and 'mixed environments' including residential and work functions (VROM 2001; 5-46).

47 The *Omgaan met risico's* report (Dealing with Risks, TK 1989) specifies the boundaries for individual risks and group risks (also see § 5.2).

48 "A spectacular example of conflicting spatial and environmental demands occurred in Dordrecht in 1992. Offices were built near the station, fully in accordance with compact-city requirements. The construction was stopped after objections were lodged by a regional environmental inspector, who claimed that the project was located within the risk contours of the shunting site, where trains carrying chlorine regularly passed through. A study group had to be set up to establish that the risk did not exceed standards. Expressing risk in figures is a complex task. Until the matter was resolved, all the people involved were almost at their wits' end" (Bakker 1997; 4, see also ROM 10/1992; 46).

49 A restrictive parking policy was enforced by implementing parking standards in these 'A' locations. Businesses consider these standards too restrictive (Martens 1996), and they may be prompted to move elsewhere. This can also lead to a deflection of parking congestion to neighbouring residential areas, where it has a negative impact on the quality of life.

50 Den Hollander, Kruythoff and Teule show that VINEX cluster policy "has generally had a limited but positive effect on commuting mobility in the Randstad conurbation" (1996; 121). The average weekly commuting distance per household has fallen by 15 km.

PART B
COMPLEXITY AND PLURIFORMITY

Chapter 4

Planning-Oriented Action in a Theoretical Perspective

Complexity and Pluriformity

> ... in the world in which man lives, for whilst it may be convenient to see pattern, to reduce to simpler issues, to classify and to label, the world of nature knows no such limitations or categorisations: it exists independently of these attempts of man to organise his own views of her and it exists as a whole (George Chadwick 1971; 33).

4.1 Introduction

Part A of this study discusses the fact that environmental/spatial conflict can be seen not only from the perspective of quantitative units based on densities and standards, but also from the perspective of qualitative and intersubjective assessments (i.e. the 'liveability') of the environment in which we live. This leads us to the conclusion that the consequences of environmental pollution can only be partially understood on the basis of a direct relationship between cause and effect (see Chapter 2). The belief that environmental/spatial conflict can be seen only in terms of technical or legal issues is becoming increasingly untenable (§ 1.2). It is understandable that environmental-/spatial conflicts are the subject of social and political discussion; they are issues with which people feel directly involved, which affect them, and which must be assessed in terms of the interplay of social forces. This observation does not make it any easier to describe and deal with environmental/spatial conflict; on the contrary.

The concept of environmental/spatial conflict, which at first sight seems a relatively straightforward concept, encompasses a world that can be extremely complex and dynamic. Chapter 3 discusses the dynamic character of the spatial environment in which environmental/spatial conflicts arise. The obvious solution appears to be to maintain safe distances, as a relatively 'simple' means of keeping apart environmentally harmful and environmentally sensitive functions in a 'sustainable' way, However, this measure cannot be applied in every situation, due to the many social and physical environmental aspects that must be considered. Here, complex and highly autonomous urbanization processes obstruct the creation of a 'makeable' society. The fact that society is only 'makeable' to a certain extent forces us to ask how we should deal with complex issues that affect us directly in a social sense.

Chapter 1 discussed the turning-point in area-oriented environmental planning theory and practice. The decision-making process for environmental policy is undergoing a structural change. The main development is the shift from a centralised policy based on standards to a more local, customised approach that seeks to build a foundation for decentralised governance. This shift in focus manifests itself explicitly in the development and acceptation of two innovative instruments: *IEZ* (Integrated Environmental Zoning) and *ROM-gebieden* (designated Spatial Planning and Environment areas). IEZ is based on central environmental standards, while ROM policy is based on shared responsibility. They represent two diametrically opposed visions that were introduced almost simultaneously by central government (§ 5.4). These instruments express a dualism – that is difficult to reconcile – between environmental standards and the building of public support.

This dualism in environmental planning between centralised standards and decentralised consensus has a clear parallel in discussions on planning theory. In these discussions, 'communicative rationality' is challenging the functional-rational approach in which direct causal relationships are the source of knowledge and insight. Communicative rationality sees the interactive relationships between actors, and intersubjective opinion-forming, as the central elements in planning. Communicative rationality is based on theoretical visions (see also § 4.6) that will be familiar to anyone who is involved with the rapid developments in environmental policy practice in the Netherlands.

This parallel between policy practice and abstract theoretical discussion is the foundation for this chapter, which deals with planning theory discourse that seeks to explain the developments in environmental planning. In an abstract sense, this is about how to deal with the issues that are experienced within a complex and dynamic society, and which arise from an increasingly complex urban context, whereby direct causal relationships are inadequate foundations for visions, decisions and predictions. In a number of cases, conscious intervention by means of strict control, whereby direct causal links are implicitly assumed, only partly achieves the desired goal of resolving a conflict. On the other hand, we must not simply assume that alternatives, such as building consensus, are sufficient to resolve planning issues.

This chapter attempts to construct a model that can accommodate issues of varying complexity. The model is based on current insights into planning-oriented approaches. In this connection, given the principle of subsidiarity, the following questions arise: which decisions can be taken at which levels in the decision-making process? What do the decisions relate to, and who will take them? Will decisions lead to a central framework-setting policy, or to a decentralised, integrated policy that considers local factors? Answering these questions takes us on a journey through a maze of theoretical studies, visions, arguments, viewpoints and doubts. The aim of the search is to find a suitable structure for making planning-oriented decisions, and a structure that can provide insight into the complex dynamics of the real world – particularly at the interface between the natural and spatial environments.

The first step towards a theoretical foundation for decision-making and planning is to answer the three questions posed in section 4.2: *what* has to be achieved, *how* can it be achieved, and *whom* will this involve? In section 4.3, these questions are placed in the context of developments that have taken place in planning-theory discourse. The questions posed in section 4.2 constantly recur in the discussion, which suggests that they are fundamental. Their significance lies in the relationship between the *object* to which planning and decision-making is oriented, the deciding *subject*, and the *intersubjective* and interactive context within which the subject, together with other involved parties, forms value judgements and reaches decisions. In section 4.4, it is suggested that this fundamental element of planning theory be developed into a pluriform vision of *planning-oriented action*. The vision encompasses three perspectives on planning-oriented action: goal-oriented action (§ 4.5), decision-oriented action (§ 4.6) and institution-oriented action (§ 4.7). This theoretical vision is discussed in more detail in section 4.8, and is linked to the theme that connects the three perspectives of planning: *complexity*. In section 4.9, the terms 'complexity' and 'pluriformity' are taken as the basis for constructing a *framework for planning-oriented action*. This framework is used in Part C, which analyses developments in area-oriented environmental policy, including the shifts in approaches to dealing with environmental/spatial conflict.

4.2 Environmental/Spatial Conflict as a Planning Object

The national discussion in the Netherlands on spatial-planning restrictions, which are the consequence of an environmental policy approach that is regulatory and framework-setting, focuses on the relationship between environmental-protection policy and spatial-planning policy. In this study, it is assumed that this discussion must be considered in conjunction with developments at national level, where the focus is on deregulation and decentralisation (referred to in Chapter 1 as trends in government planning, see also Fig. 4.1). Here, too, it is assumed that the discussion on the harmonisation of environmental and spatial needs is partly a reaction to actual difficulties experienced at local level, for example: incorporating framework-setting environmental requirements in the spatial environment. In practice, the recognition of local factors and developments at national level have resulted in a redefinition of the problem. We no longer ask 'What resources are required to provide an effective solution to environmental problems in region X?', but 'How can we reduce environmental impact and, at the same time, solve a number of social and economic problems?' (Kuijpers 1996; 62). This means that the context of local environmental issues is becoming more and more important. The latest credo appears to be: "Environmental problems should be solved on the level at which they arise" (VROM 1995, Appendix 1; 2), whereby the *context* of environmental issues determines the solution strategy to a greater extent than in the past (see Fig. 4.1).

Until recently, nationally mandated environmental standards were almost the only solution. However, this policy reorientation – which has been acknowledged by the Ministry of Housing, Spatial Planning and the Environment (VROM) and gratefully adopted by various local authorities – is a recognition of the fact that environmental issues do not occur in isolation but relate to many other aspects at local level that may influence the choice of approach (Fig. 4.1 describes the context of an object).

In the discussion, the unique nature of environmental issues and the extent of their interplay with other local factors are put forward as justifications for, at the very least, refining strict national standards and making them more flexible.

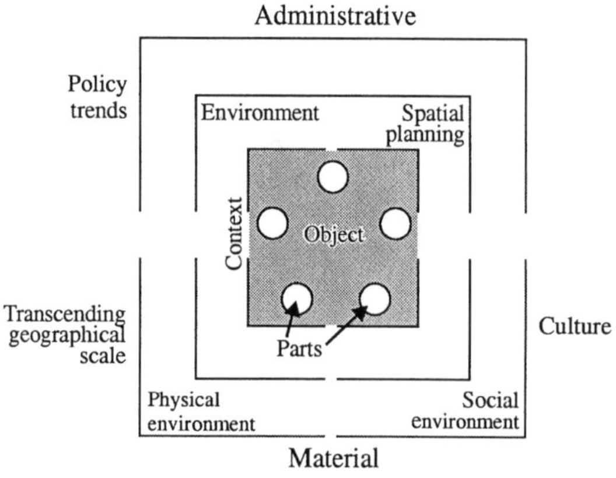

Figure 4.1 Interplay between (the parts of) the planning object and its material and administrative contexts, within the wider world

In order to reflect on such policy shifts, it is necessary to consider how conflicts or issues can be viewed in terms of policy. Three arguments are interesting in this context: (1) Policy and decision-making are ideally based on a consistent and generally acceptable perception of physical and social reality, within which intervention is based on specific grounds or a defined *goal*, in order (a) to solve a problem, or (b) to 'steer' reality towards a desired situation. (2) These interventions in the real physical and social world (as a planning object) require *choices* to be made. These choices must be reflected in the decision-making process. (3) In addition, the aim in principle will be to ensure that policy takes account of the opportunities and wishes of individuals (the 'actors'), organisations, or, at a more general level, *institutions* and institutional groupings.[1] These will be the foundation for developing and implementing policy. Policy developments can also influence how institutions act, and this can result in resistance if policy does not adequately represent the wishes and visions of the various actors involved. The above arguments give rise to three elements in decision-making and planning:[2]

- The physical and social reality that is or could be subject to purposeful intervention;
  (the material object, or content, of planning)
- The choices made in the decision-making process;
  (rationalisation of choices)
- The organisation and communication of and participation in decision-making and policy.
  (the institutional element, in terms of actors and interaction)

In principle, this is related to decision-making and planning processes that are part of the larger domain of government planning[3] (see Van Houten 1974; 109, Kickert 1986). The three elements can be developed by answering the following action-oriented questions:

- What must be achieved?
  (goal-oriented)
- How can it be achieved?
  (decision-oriented)
- Who will be involved?
  (institution-oriented)

The questions characterise the *action-oriented nature of planning,* which can be viewed from three perspectives: goal-oriented, decision-oriented and institution-oriented. Decision-making and planning usually arise out of the need for a goal-oriented approach. Lim emphasizes that "any planning activity proceeds with some notion about goals. 'Planning for what?' is probably the most crucial element in the procedural and substantive aspects of planning activities" (1986; 76). However, this inevitably leads to choices (decision-oriented), and the question 'who will be involved?' (institution-oriented)[4] (see also Fig. 4.3).

Dealing with environmental/spatial conflict relates above all to the quality of the day-to-day environment *as a policy goal* (see Part A). Here, quality is primarily understood to be the quality of the environment. Since this study focuses on a number of issues, we will also discuss the decision to pursue a single-issue goal (i.e. environmental quality standards, or an aspect of environmental quality) or a multiple-issue goal (i.e. the quality of the physical environment, which includes the quality of the spatial and natural environments). A goal is usually pursued with the expectation that policy-specific developments will be able to bridge the gap between desired and actual reality. The desired situation and the actual situation must be identified and harmonized as realistically as possible. Defining what is considered reasonable and desirable in political, administrative and social terms is a *decision-oriented exercise* that is partly based on the frameworks within which wishes and reality could or should be coordinated, on the nature and quantity of resources to be made available for this purpose, and the 'opportunity cost' of achieving the goal. The goal of policy and the decisions on how it is to be achieved are inextricably linked to choices regarding who (i.e. which actors) will be involved and the role of the policy organisation and other involved parties in achieving the defined goals. Here, then, we refer to the *institutional frameworks* within which environmental/spatial conflicts can be dealt with.

The discussion on achieving a balance between environmental and spatial needs focuses on the role of generic and framework-setting environmental standards – and therefore on the desired level of environmental quality. This is essentially a goal-oriented aspect. Defining and achieving this sectorally defined level of environmental quality, which is usually seen as separate from other goals relating to the physical daily-life environment, requires different types of decisions from those required when environmental quality is seen as an integral part of the quality of the physical daily-life environment, which is more complex. The three perspectives presented here (goal-oriented, decision-oriented and institution-oriented) are seen as the first step towards gaining insight into the relationship between decision-making and planning methods, and the complexity of environmental/spatial conflicts.

4.3 A Pluriform Approach to Planning

This study is heavily planning-oriented. It involves a scientific approach to the physical day-to-day environment, where policy measures are taken for the benefit of society. Planning is a scientific discipline with a long, rich history of discussion that is taken as the starting point of this study in order to arrive at a *pluriform vision* that can be used to understand the physical environment and the policy measures relating to it.

A Changing View of Planning: the Dutch Perspective

Since the Second World War in particular, Dutch society has developed so rapidly that the government has had to introduce, in rapid succession, more and more policy measures relating to spatial planning in the Netherlands. The emphasis was not placed on quality, but rather on the functional aspects of spatial planning. The need for greater knowledge of the spatial consequences of social change led in the 1960s to the creation of academic chairs in planning. The Dutch concept 'Planology' was introduced as early as 1929 by De Casseres in his work *Grondslagen der Planologie* (*Principles of Planology*) and evolved into an academic discipline that studied policy-based and plan-based interventions in the spatial environment. The policy field relating to 'planology' became known as 'spatial planning'.

According to Mannheim, planning should focus on "the reconstruction of an historically developed society into a unity which is regulated more and more perfectly by mankind from certain central positions" (1949; 193). Since the 1950s, planning theory has slowly but surely moved away from the belief that society is a comprehensive stable entity that can be managed in a functional-rational (see § 4.6) and hierarchical manner. A shift has occurred towards a vision that recognises the restrictions of spatial management, partly as a result of autonomous factors and developments that cannot be ignored, a rapidly changing society, the increasing complexity and dynamics of spatial developments and, last but not least, restrictions imposed by the government itself.

In the 1960s, academic planners were still focusing on the effects of policy, including the sociological implications of spatial developments. Steigenga (1964; 23) thus pointed to the society-driven nature of planning. For Van der Cammen, this meant that spatial planning as an academic discipline was "characterised by an inherent friction between science and policy effectiveness" (1979; 11), because "in daily life we [...] pay attention primarily, and rightly, to results" (1979; 11). Comprehensiveness was seen as an essential characteristic of planning (Van der Cammen 1979), in addition to realising goals and looking to the future. Comprehensiveness generally meant 'all-encompassing'. In the 1970s and 1980s, too, comprehensiveness continued to be seen as a characteristic of planning (see for examples, Kreukels 1980; 62), although the scientific focus shifted from the actual content of policy towards the policy-based processes and procedures underlying spatial developments, and their consequences. Wissink (1986; 192-193) consequently observed a one-sided emphasis on procedures, which threatened to overshadow what he saw as the essence of planning: the spatial structure and its development (see also Voogd 1986).

At the end of the 1980s, the concept of *performance* was introduced in planning (see, for example, De Lange 1995, Mastop and Faludi 1993). In practice, planning had finally shown that further investments in the spatial planning system were not greatly improving its controlling effect (Mastop and Faludi 1993; 75). The term 'performance' refers to the fact that a more or less direct relationship is assumed between the results of decision-making processes and their eventual observable effects (Mastop and Faludi 1993; 72). Performance refers to the influence of strategic policy on 'policy subjects' (Maarse 1991; 124, Herweijer et al. 1990). Strategic decisions appear to have only a limited effect, and do not necessarily achieve the desired goal. Factors such as the nature of policy content, available information, communication between actors and their individual knowledge and interpretations, and the degree of flexibility and responsibility among actors mean that although policy *performs*, it does not result in *conformance* between decisions and outcomes (Mastop and Faludi 1993; 79).

While the term 'performance' is intended to explain communication mechanisms within the organisational structure of the government, concepts such as 'building political will and consensus', 'participation' and 'network planning' refer to the influence and involvement of other actors in the management of spatial development processes. Planning, as Voogd rightly states, is always the consequence of sociological forces (1995a; 23). In the 1990s, building political will, allowing societal actors to participate, and building consensus in the community became themes in planning-oriented studies and research in the Netherlands and other countries (see, for example, Healey 1997, Innes 1996, Sager 1994 and Woltjer 1997).

Planning is a science with a spatial and administrative orientation. It deals with 'spatial problems' which, as Van der Cammen points out, "involve not only spatiality, but also matters that are a source of concern to people, which become the subject of consultation, and usually also policy measures" (1979; 17). The 'grey' environment (§ 2.2) is one such subject because the negative effects of environmental pollution are observable and give grounds for concern. They usually manifest themselves in the spatial context, too, and can impede spatial development. Of equal concern is the fact that intervention in the spatial environment can also have a positive or negative effect on the natural environment. The grey and spatial environments require management by means of policy. At least, that is the opinion of society.

The functional-spatial and administrative orientation of planning extends further than the spatial environment. Planning concerns the relationship between man and his surroundings, a relationship that is far from simple to express in patterns and to understand, let alone control. Human behaviour and sociological developments are difficult to foresee. And spatial intervention designed to support, structure or direct them are not always universally appreciated. In addition, the ability of policy measures to harmonise wishes and reality has proved limited. This is clearly illustrated in Chapter 3, which discusses how social developments and spatial dynamics have substantially eroded the foundations of the compact-city concept.

Consequently, the focus of academic discussion has shifted from object-related observations towards intersubjective interaction. This shift can also be seen in practice. Participation and shared governance are fast becoming popular concepts in policy practice. Whereas public participation had long been part of the spatial planning process in the Netherlands (§ 3.3), the importance of public support for environmental policy was not fully recognised until the government came under heavy criticism for not achieving all the goals set out in the first National Environmental Policy Plan (TK 1989) (§ 5.4 and 5.5). This led to the conclusion that the conventional 'technically oriented' policy should be replaced by 'policy based on public consensus' (§ 1.2 and 5.4). At every level of Dutch government policy there is increasing acknowledgement of the limitations of management based on direct cause-effect relationships. This has led to a shift in focus from goal-oriented policy towards the policy process itself. This, in turn, has resulted in increasing emphasis on reducing public resistance and creating a support base for policy. While developments in planning-oriented action have led to a 'reduction' in terms of specific content, they have also resulted in a broader approach, thereby doing justice to the various processes that relate to identifying and using opportunities for balancing wishes against reality by means of deliberate intervention.

A Pluriform Perspective

The observation above can be seen as confirmation of the theory that intervening in the complex, dynamic physical and social reality can serve a purpose if it is based on a pluriform vision of that reality (see also Voogd 1986; 3). Here, pluralism[5] (hereafter 'pluriformity') refers to the concept of "a *plurality* of interpretative perspectives rather than just one" (McLennan 1995; ix) and might possibly be able to give a better picture of reality, a vision that is discussed widely in the post-modern literature.[6] According to Healey, "appreciating diversity and recognising differences are key elements in this conception [...]. There is not one route to progress, but many, not one form of reasoning but many" (1992; 149-150).

This involves more than the study of the perception of reality, the discrepancy between wishes and reality, and the possibilities for harmonizing the two. This study will also discuss wishes that are not related to a problem. This has less to do with problems identified in practice (e.g. an environmental/spatial conflict) and more to do with spatial concepts – whether formulated as models and/or scenarios – on which the planning process is based. In the words of Zonneveld,[7] a *concept* is a representation of a situation or development that can arise out of "a process of collective image-forming"[8] (1991; 71). Spatial concepts (e.g. the compact-city concept and the green-heart concept) are in that case 'idealistic' sets of wishes relating to the arrangement of the spatial environment or process. These wishes may rest on a certain level of consensus, but usually arise out of past developments, a problem experienced in practice, or a problem or set of problems expected to occur in the future. Spatial concepts are therefore an expression of 'what is wanted', while spatial problems, bottlenecks and conflicts represent 'what is *not* wanted'.

What, How and Who

Developments in spatial planning, environmental planning and planning as an academic discipline are driven by many publications "that comment on policy practice, make proposals with regard to policy methodology, or describe or explain policy and policy processes" (Dekker and Needham 1989; 2). According to Voogd, this relates primarily to "future-oriented activities [...], that focus on goal-orientation [...], selecting resources and opportunities [...], taking decisions [...], and the inherent behavioural dimensions of the subject [...]" (1995a; 3). In short, planning is about *what* is wanted, *how* it can be achieved, and with *whom*. This observation is in line with previous comments (see § 4.2).

Zonneveld (1991; 16) points to three 'dimensions' that are important when agreeing on the description and scope of a problem or set of wishes: an *objective dimension*, within which the problem or set of wishes is defined 'in terms of the objective world; a *subjective dimension*, whereby decisions are based on wishes, sentiments and above all rational considerations – here we deviate from Zonneveld's description;[9] and a *social dimension*, which Zonneveld uses to refer to the interpersonal relationships between the persons involved that are of central importance in the social definition of a problem or set of wishes. Here, the position is taken that changes and/or developments can be examined within these 'dimensions', which can then be used to plan actions.

The idea that spatial issues and concepts have an objective, a subjective, and a sociological dimension (this last aspect will hereafter be referred to as the intersubjective aspect of planning-oriented action) which continually interact with each other (Fig. 4.2), is considered significant as a foundation[10] for the planning perspectives defined in section 4.2:

- The physical and social reality that could be subject to policy-based or planning-based intervention, now or in the future;
  (the object-oriented dimension; O)
- The choices made in the decision-making process;
  (the subject-based, rational dimension; S)
- The organisation and communication of, and participation in, decision-making and policy.
  (the intersubjective dimension; I)

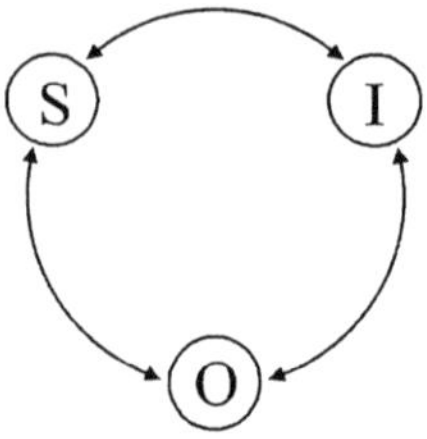

Figure 4.2 The intersubjective aspect of planning-oriented action

This study will consider planning-oriented action from a number of perspectives. This approach was chosen in order to arrive at a composite vision that provides answers to the various arguments advanced in planning-theory discussions (§ 4.5-4.7). It is assumed that this composite vision of planning-oriented action will provide coherent insight into the developments in day-to-day policy practice. Over the years, these developments have represented a shift from effect-oriented planning and results-oriented planning towards planning that focuses on procedural aspects, process and project planning, and institutional involvement (§ 4.9). There has also been a shift in visions relating to policy content; these have shifted from a comprehensive, exhaustive approach towards selectivity and prioritising. There is a growing awareness that control is only possible to a limited extent, and that complex and dynamic society is only 'makeable' to a certain extent. A study of such developments produces a colourful palette, from which we will select a number of central characteristics: goal-oriented, decision-oriented and institution-oriented developments in planning-oriented action (§ 4.2).

This idea that not one but several different perspectives of reality may be required in order to arrive at a well-founded vision will be discussed below. This relates to a comprehensive and pluriform vision or orientation that is geared towards the complex and dynamic reality of the physical environment and the societal process, the interaction between the two, and policy intervention.

4.4 A Comprehensive, Coherent Vision of Planning-Oriented Action

In order to arrive at a comprehensive and coherent vision, the three perspectives (goal-oriented, decision-oriented and institution-oriented) defined in section 4.2 are seen as pluriform, logically related and complementary *perspectives on planning-oriented action*. The most tangible relationship is that between goal-oriented action and the material planning object. The choices made in the decision-making process are central to planning-oriented action. In the case of institution-oriented action, the organisational development of the decision-making process, communication, and the commitment and involvement of social actors are central to decision-making and planning. The three aspects of planning-oriented action rest on the basic idea that a cohesive structure connects object-orientation, subject-orientation and intersubjective interaction (see Fig. 4.2). Once this idea is accepted, the next logical step is to assume a relationship between those three aspects as defined above. The only remaining questions relate to the nature of the relationship and how it corresponds with existing planning-theory arguments, visions and concepts.

There is nothing new about this proposed pluriform, comprehensive vision of planning. Visions such as that of Gillingwater[11] are above all interesting because they are a reaction to the rejection of the conventional concept of 'comprehensiveness' in planning.[12] The concept is rejected because it assumes that reality can be fully known. As Friedmann (1973) correctly points out, 'full comprehensiveness' is not possible,[13] nevertheless, 'comprehensiveness' is still used as a norm for planning-oriented action.[14]

How, then, do we approach 'comprehensiveness' when a reality has endless variations and is so difficult to grasp? We can arrive at a pluriform approach to planning by combining the essence of our own perception of reality with relevant and familiar visions or categories. Voogd makes a similar suggestion. Depending on the situation or problem, a multi-dimensional approach can be applied "whereby one or more [...] dimensions are given greater or less emphasis" (1995a; 27). In Chadwick's systems-theory description: "we introduce an order so as to constrain variety by resolving to study only certain appropriate systems..." (1971; 71). What we are discussing, then, is an 'order' which is a combination of three perspectives on planning-oriented action.

Planning-oriented action is taken to be any action performed by individuals, groups or organisations that is designed to achieve goals in a systematic way by making and implementing choices and decisions, with the help of others if necessary, and by using the required resources. In planning-based relationships, this *goal-oriented approach* embraces the entire planning 'cycle', from problem definition, through problem strategy, the involvement of actors and the use of resources, to the monitoring and evaluation of the end result of decisions and action.[15] A goal-oriented planning policy gives priority to the relationship between the *subject* that makes the decisions and takes action (possibly influenced by or prompted by an intersubjective context), and the *object* that is to be managed or changed. In certain respects, the nature of that relationship distinguishes spatial and environmental planning (as scientific disciplines) from other planning-based sciences. In spatial planning, for example, this specific subject/object relationship involves the study of policy developments geared towards the functional-spatial arrangement of parts of the Earth's surface as an interaction between, and in accordance with, existing or supposed societal needs. As such, goal-oriented action can be seen as a continuous interaction between *decision-making* relating to the physical and social environments, *implementation*[16] that affects them, and the response. Here, goal-orientation is seen as the underlying motivation for the structure of the planning system (Fig. 4.3), the most important aspects being the *decision stages* in the planning process and the eventual *effect* of planning. These aspects, however, depend on the decision-oriented and institution-oriented actions taken.

While goal-oriented action emphasises the *form* of the planning process, decision-oriented actions emphasise its *content*. Decision-oriented action relates to the way in which *choices* are made, the reasoning behind them, the rationalising of arguments and the uncertainties inherent in the foundation of the arguments (see also Faludi 1987).

Institution-oriented action is not restricted to the questions of who should be involved in the decision-making process and how the process and parties will be organised, but also considers how "other decision-making subjects [...] can intervene in the behaviour patterns of the planning subject" (Mastop 1987; 235).[17] In short, this is a matter of *interaction*.

In this supposed situation, goal-oriented action is the directing (structuring) principle underlying the planning process. Goal-oriented action is, so to speak, the third side of the triangle; without it, the two other perspectives are redundant. This argument assumes that decision-oriented, institution-oriented, and goal-directed action (which encompasses the first two), combine to form a pluriform and comprehensive decision-making structure for the planning process (see Fig. 4.3).

Up to this point, the proposals have been largely *structural*. We have discussed goal-orientation as the behavioural perspective that defines the structure of the planning process. However, if we are looking for a theoretical foundation, we can expect planning to be directed by a decision-oriented approach (i.e. the reasoning, rationalising and making of planning decisions). Decision-oriented choices are thus the foundation for goal-directed and institution-oriented actions (see Figs. 4.8 and 4.9).

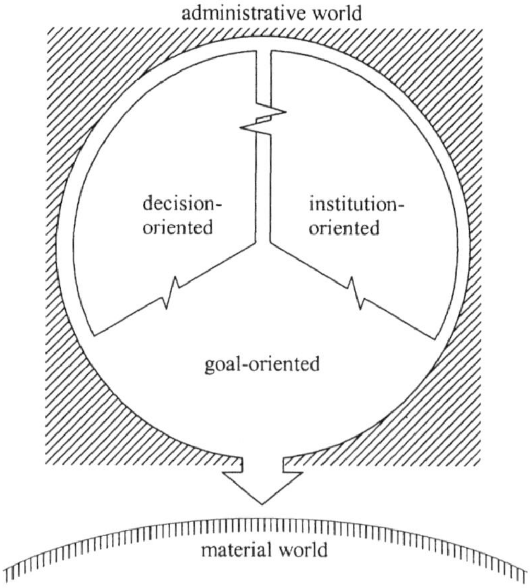

Figure 4.3 A model for planning-oriented action

This distinction between structure and content leads us to ask how the arguments in planning-theory discourse can contribute to the three behavioural perspectives. Mastop (1987) indicates the solution. He identifies four planning-related approaches to decision-making, referring to visions and concepts in planning theory:

- A system-cybernetic approach to decision-making;
- A rational approach to decision-making;
- A coordinative/communicative approach to decision-making;
- A bureaucratic approach to decision-making.

The first three approaches are discussed most often in planning literature.[18] Mastop relates the fourth approach, the bureaucratic approach, to "planning as a form of government intervention" (1987; 149). In the planning discipline, this approach is largely implicit by definition. Moreover, this bias will be robbed of its reality by the trend towards the 'communalisation' of policy (see Chapter 1). This study, however, assumes that the changing role of government in the decision-making process does not need to be seen as a separate category, but rather as a policy choice (i.e. decision-oriented) and in terms of interaction between subjects (i.e. institution-oriented).

According to Mastop, the system-cybernetic approach to decision-making is "primarily linked to the continual interaction between the subject and its context" (1987; 149). System cybernetics and, to a greater extent, systems theory in a broader sense, can be the starting point for developing a *goal-directed approach* (§ 4.5) Planned systematic action by the planning subject[19] will be repeatedly assessed in terms of its tangible effects on the functional-spatial object. The rational approach to decision-making is based on "a particular planning subject achieving a balance between goals and resources" (Mastop 1987; 149). This approach is largely based on decisions to be taken and the underlying reasoning, and is therefore in line with *decision-oriented action*. This approach refers to the arguments underlying functional-rational decisions. In planning theory, these arguments are increasingly supplemented with ideas on communicative rationality and communicative action (§ 4.6). According to Mastop, the coordinative/communicative approach to decision-making is linked to "the interaction between several subjects" (1987; 149). He shares Kreukels's conclusion that planning is hardly ever carried out or determined by a single 'actor' (person, group or organisation), but that "decision-making is based on interaction between relatively independent sub-systems that reach agreement on joint action" (Kreukels 1980; 26). Mastop refers to a pluriform planning subject (1987; 157). This is in line with *institution-oriented action*, in other words: *who* will be involved in the policy formulation process, how they will be organised and how they communicate (§ 4.7). The following three sections will discuss goal-directed, decision-oriented and institution-oriented action as aspects of planning. Developments in planning theory will therefore be discussed insofar as they have a bearing on one of the three aspects of planning-oriented action. The explanations will then be summarized and the relationships between them shown in Table 4.1.

4.5 Goal-Directed Action from a Systems-Theory Perspective

Goal-directed action is the essence of planning. It steers the systematic preparation of policy. Goal-directed action is based on a process that has a structuring effect on decision-oriented and institution-oriented action. It also establishes links between decision-making processes and the ultimate results of intervention in the real physical and social world. In effect, this is about the *effectiveness* of planning. In discussions on whether planning works, 'effectiveness' was initially taken to be a direct causal relationship. However, this belief required continual modification as time went on because reality proved to be much more complex than was supposed. This section focuses on developments in thinking on goal-directed planning.

The structuralizing aspect of goal-oriented action can best be illustrated from a systems-theory perspective. Systems theory is very appropriate for indicating structures, their constituent parts and the assumed functions of the structures. However, these are not the only arguments for applying systems theory[20] or 'the' systems approach in the discussion below. Systems theory is a valued basis for planning-theory discourse and is adapted to that context as necessary. A good example of this is the acceptance of the network approach in planning (§ 4.7, see also Voogd 1995b). This is undoubtedly due to the formal, methodical way in which systems theory seeks to link structure and content. The way in which systems theory deals with abstract concepts such as 'parts, the entity and context', certainty, uncertainty and probability, openness and predefinition, and complexity helps to provide a consistent framework for planning-oriented action. These concepts will be referred to many times in this study. They express the relationship between the way in which an issue can be considered, the degree of complexity that can be attributed to it, and the subsequent decision-making method that is applied to it.

The Systems-Theory Approach to Planning

The quest for a structured relationship between the decision-making process and the material object to which it relates could begin with the development of control and cybernetics[21] theories, collectively known as 'systems science'. The theories emphasise control and organization, in contrast to the more conventional analytical approach (see Noordzij 1977; 20). Emphasis is placed on systematic interaction between the desired situation (as defined by the subject) and the actual situation (the object) and on the 'entity' rather than its components.[22] The environment of an entity is seen as a partly determining factor (Kramer and De Smit 1991; 9).

These theories have much in common with what Kreukels (1980; 25) calls the 'system-functional concept of planning'. This concept is based on the control process that is embodied in a rational design for directing policy formulation and implementation. The system-functional approach to planning assumes that the planning process, and within it the decision-making process, can be expressed in a system design, whereby the process as a whole can be explained in terms of its associated parts. In this sense, planning can be seen as design-based and programmatic, with the design phase (policy formulation) determining the stipulation and implementation phases (policy action) (Kreukels 1980; 25). The desired goal and observed effects are then expressed as far as possible in a regulating process, which must include corrective actions, usually in an iterative process. This description suggests a mechanism that relies heavily on a direct cause-and-effect relationship, yet this is not necessarily the case. Below we will see how the mechanical step-by-step method that emphasises cause and effect is shifting towards an approach that is more dynamic, interactive and choice-related. This approach explains the administrative and material object of this study without departing from systems theory.

In the post-war years, experience of systems theory, mostly gained during military operations, was applied in industry and government organisations. These 'first-order' system-cybernetic models relied heavily on the concept of a 'makeable' society and "linear, mechanical regulating processes in relatively closed systems with fixed goals" (Kreukels 1980; 26). Key principles included *optimising* the structure and behaviour of systems, system regulation and preservation.

In the 1950s and 60s, the assumption grew that all human behaviour is *goal-directed* (Simon 1960). Goal orientation, formulated at that time in terms of objectives, was consequently seen as the element that distinguished human or social systems from physical systems (Checkland 1991; 67), and became a dominant paradigm in hard system thinking.[23] However, it has now been accepted in systems theory that "it is the case that in real life the goals change all the time" (Checkland 1991; 63). The concept of a 'physical' direct cause-and-effect relationship is no longer the only or leading paradigm in systems theory.

Here we are drawn to make a comparison with planning, in particular spatial planning, which has also moved away from linear processes and fixed goals. 'Blueprint planning', a form of planning based on pre-defined goals that represent a desirable, feasible reality, proved too inflexible for dealing with spatial planning issues and in many cases did not keep pace with societal developments. Blueprint planning expresses the desired policy in terms of fixed objectives. Too often, its goals proved impracticable due to changing societal needs, and increasing knowledge and insight. Now, however "the blueprinting approach of allocative planning has been abandoned. Its place is taken by a broader concept of societal guidance" (Friedmann 1973; xix).

It was mainly the Incrementalists (see § 4.6 and, for example, Lindblom 1959, Braybrooke and Lindblom 1963) who pointed out that reality is not linear by nature but consists of a multitude of sub-processes involving many actors. This concept of a complex interwoven reality contributed to the shift away from object-oriented planning towards process-oriented or procedural planning. Because goals must be continually modified as society develops, there is increasing focus on planning *processes*. The emphasis has shifted from the *end* to the *means*, i.e. from the *goal* itself to the constantly evolving planning processes that are *oriented* towards it. In the 1970s, 'iterative planning' and 'process planning' were generally accepted as methods that were compatible with the systematic correction of shifting goals, usually at regular intervals.

This continual change is clearly illustrated in Ozbekhan's planning process (see Fig. 4.4). For Ozbekhan, the starting point for government planning is a political question: '*What ought we to do*?'. Political goals are then translated into strategic goals: '*What can we do*?'. Strategic goals will eventually lead to operational goals: '*What will we do?*'. This final step will result in the expectation that a real change will be achieved whereby the current situation moves closer to the desired situation. According to Van Houten (1974), Ozbekhan is taking the wind out of the Incrementalists' sails by building political decision stages into the policy system. By making political objectives (referred to in § 4.3 as desired situations or concepts) 'leading', the planning process works back from a desired situation in the future to the present (see Fig. 4.4). In such cases, planning is "the exercise of human will, both in the conception of a preferred future and the implementation of the steps that are established to lead us to that future" (Ozbekhan 1969, in Van Houten 1974; 72).

A 'second-generation' systems theory subsequently evolved, partly as a result of the Incrementalist critique. The new theory retains the regulating and corrective functional mechanism but qualifies its basic assumptions. System delineation is less fixed, and there is greater emphasis on the time factor and on aspects such as the self-regulating effect of systems (Kreukels 1980; 26).

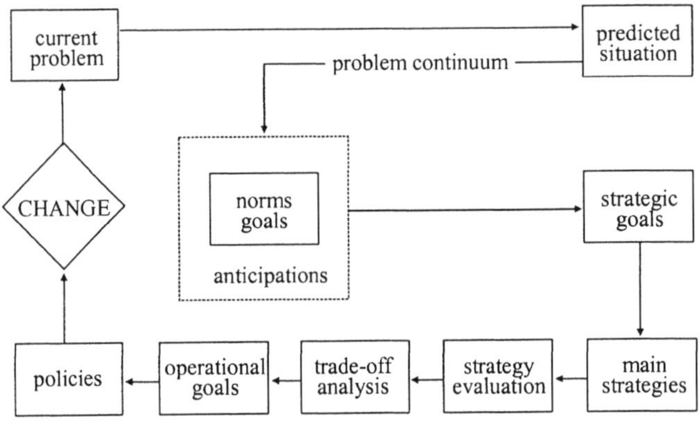

Figure 4.4 A planning process according to Ozbekhan
Source: Van Houten 1974; 72

Here, too, we can draw a comparison with planning-theory discourse. In the 1970s, the 'strategic choice approach' evolved in order to place greater emphasis on feedback mechanisms in the planning process, and on its dynamics (Friend and Jessop 1969). "This approach is based on the principle that decisions are not taken until it is strictly necessary. This expresses the explicit acknowledgement of uncertain factors and the importance of a flexible approach..." (Voogd 1995b; 69, see also Faludi 1986). The realisation also grew that decision-making depends to a great extent on the conduct of the actors involved, whose actions are based on their experience, perceptions, interests and power. In most cases, objects and events are assessed subjectively. The interplay between subject and object produces a perceived environment, to which policy is then adapted: "The human actor does not react to an environment, he enacts it" (Weick 1969, in Kramer and De Smit 1991; 48). This has also resulted in a shift in policy focus from the characteristics of the objective environment "towards the characteristics of the decision-making process, with which the policy system selects and uses information" (Kramer and De Smit 1991; 49).[24]

Effects of Planning

Dekker and Needham (1989) point out the importance of *goal effectuating* with regard to policy intentions and actions. The plan-based, systematic chain of goal-oriented action results in the realisation of intentions and concrete measures, and in the use of instruments. This, in turn, results in the full or partial achievement of objectives, in direct or indirect effects, intentional or unintentional secondary effects, short-term or long-term effects and non-occurring effects or contrary effects.

The desired effects are the foundation for linking the desired situation and reality as perceived at the time of decision-making. The effectiveness of the planning process can be measured by the extent to which an effect actually meets an expectation.

In addition to determining the effects of planning on the material object, it is also important to identify the underlying causes (predicted or otherwise) of an effect (desired or otherwise), or the lack of effect. The importance of this is twofold: on the one hand it makes it possible immediately to correct the actions being performed and, on the other, information is gathered that can be used to evaluate corrective intervention in the decision-making phase in the longer term. This last aspect is important because actions are by no means limited to the desired effect. Series of effects may also arise, for example the effects of actions that belong in the environmental cause-and-effect chain (see Chapter 2), that may become apparent after a period of time or only produce an effect on the material object in the longer term.

The fact that the desired effects of policy are only partially realised, or not realised at all, and the fact that modifying goals is equally unsatisfactory, may be due to the complexity of the issue that is the subject of intervention. The question then arises of whether, in such cases, policy should only be geared towards measures within a stable framework that have the desired effect, or whether another approach is desirable or necessary. In the case of the latter, the question is then: which approach?

Single and Multiple Objectives

Up to this point, this study has not specified whether we are dealing with a single problem or several, for which single or multiple objectives are required. If we assume a traditional planning structure for identifying and defining problems, for decision-making and recording, and for implementation and evaluation, this will usually relate to a single problem for which an objective must be formulated or a goal-oriented process defined. Even in today's decision-making culture, in which the emphasis is placed on integration and 'complex decision-making' (§ 4.7), a great deal of emphasis is still placed on single objectives deriving from a single problem by which a measurable result can be achieved.

The emphasis is shifting, however. A great deal of attention is being paid to participation and interaction by the various actors in the decision-making process (see also § 4.7), rather than to achieving a fixed goal or focussing on achieving a desired situation. In addition, much emphasis is placed on the idea that the decision-making process must also be seen as a learning process.

However, it is not sufficient to use 'learning process' as an argument for paying less attention to the goal-oriented approach if objectives cannot be realised. In planning, therefore, the goal-oriented approach must not be seen as an element of planning-based action that can be marginalized. The theory of Cohen, March and Olsen (1972) is particularly interesting for complex situations. They state that problems, solutions and goals can be successfully combined if they are brought together at the right moment. If the time is right, there is scope for integration or linking. In this type of decision-making, an issue is not considered in isolation but is placed in an ongoing process involving a series of problems and issues. Cohen, March and Olsen refer to this as a 'garbage can', into which decision-makers deposit their problems, solutions and goals. 'Joining in' by co-ordinating, integrating and linking these issues at the right moment can help to optimise the decision-making process.

According to De Bruijn and Ten Heuvelhof (1995), three conditions must be met before the linking of issues can serve a practical purpose:

- Problems and solutions should not differ too much. "If the distance between problem and solution is too great, no definitive decision-making will take place. In terms of management, this implies that the formulation and dimensioning of problem and solution must be very precise" (1995; 26).
- The stage at which, and the extent to which, actors are involved in a subject. "A connection will only be made if the actors involved simultaneously focus on the same issue within the same network" (1995; 26).
- The link must be satisfactory for all actors.

De Bruijn and Ten Heuvelhof's summing-up presumes a more or less horizontal network of actors (see § 4.7 on institution-directed action), within which each actor has insight into specific problems and incorporates that insight in the decision-making process. Yet the above conditions would hold true even if the combination and linking of issues were to be placed in a more vertical, hierarchical structure – for example that of central government which, instead of adopting a sectoral approach, adopts a more integral, region-specific, function-specific approach, which it then imposes on lower levels of government. Both situations nevertheless require a sound strategic foundation as a precondition for linking issues.

An integral strategy should emphasise the different issues, among other things, and reflect the frameworks and preconditions of policy. The actors involved should be identified, and – last but not least – common goals will have to be defined. This approach involving the linking of issues, the 'bundling' of solution strategies, and multiple goal realisation by combining different issues within a single solution strategy is referred to here as a *multi-objective approach*. In the goal-oriented dimension of planning-based action, this approach is seen as the opposite extreme to a single, fixed goal (see Figs. 4.8, 4.9 and 4.10). However, it is not an approach that can simply replace the single fixed goal in the planning process for every case.

Fixed goals are intended to reflect a maximum desired result. In such cases there are high expectations of success. This will not apply to a multi-objective approach which is the common denominator of a number of separate goals on which an integral process or common solution strategy is based. The emphasis is no longer on the elements of an issue, but on the context that various issues have in common, and which can link them. Multi-objective goal formulation and a multi-objective approach no longer focus on goal maximisation but on *making use of opportunities*. In this sense, multi-objective goal optimisation is a secondary aim. The approach to dealing with individual goals within this larger entity is another step that has not yet been discussed here. The approach will be developed in the following chapters and, in Chapter 7 in particular, assessed on the basis of practical experience.

The timing of decisions will become more and more significant as planning systems are increasingly seen to be complex, as context becomes more important, and as the results of planning are increasingly expressed in multi-objective terms. In complex situations, the timing of decisions is no longer determined solely by predefined steps or procedures. The need for flexibility, the presence of uncertainty, the role of actors, and the growing emphasis on complexity in goal-oriented action have resulted in a shift of emphasis from design mechanisms for planning to optimising the timing of decisions in the planning process. The moment at which choices are made is increasingly influenced by factors within and outside the planning process. Complex planning systems consist of "feedback networks, by which they respond in a self-regulating and self-modifying manner to stimuli outside the context"[25] (Mastop 1987; 156). This refers 'by definition' to the open nature of policy networks and systems, whereby the context of the networks and systems is the determining factor. This is logical, given the fact that a policy system does not exist purely for its own sake. Klijn (1996) points out that "for some time there has been a trend in policy science to give the context of policy a more central place" (1996; 24). This leads us to ask what 'context' is taken to mean, and how it relates to the policy system.

Context

Emery and Trist (1965) refer to 'the causal texture of the environment', particularly with regard to organisations. Kramer and De Smit (1991) refer to this 'causal texture' as the *contextual environment* of a system. This comprises not only the administrative environment, but also the material object of planning and the context of that object (see also Fig. 4.1). They distinguish between "four 'ideal types' of causal texture, approximations to which may be thought of as existing simultaneously in the 'real world' [...] though, of course, their weighting will vary enormously from case to case" (Emery and Trist 1965; 28). The four types of contextual environment vary in complexity and dynamics: (1) the *placid, randomized environment* is the least complex environment. It is stable and has a limited and random effect on the system; (2) the *placid clustered environment* denotes an environment in which clusters and accumulations of influences occur, from which a selection can be made that may or may not lead to an optimal starting position in the system; (3) the *disturbed reactive environment* is similar to (2), but comprises several equivalent systems that may compete with each other; (4) *turbulent fields* is the most complex contextual environment. It is similar to (3), but is strongly dynamic and apparently predictable in nature. An obvious example of this is the intangible context of the developments discussed in Chapter 3, which partly underpin the concept of the compact city but can also be directly at odds with the premises of that concept.

Emery and Trist point out that "turbulent fields demand some overall form of organization that is essentially different from the hierarchically structured forms to which we are accustomed [...]. To achieve a degree of stability [...] the strategic objective [...] could no longer be stated simply in terms of optimal location (as in type 2) or capabilities (as in type 3). It must now rather be formulated in terms of *institutionalization*. [...] Social values are here regarded as coping mechanisms that make it possible to deal with persisting areas of relevant uncertainty" (1965; 33-34).

The 'communalisation' of decision-making (see Chapter 5) makes the policy environment more complex. This does not make it easy to distinguish the policy system from the context to which it is geared. The complexity of a goal-oriented system depends not only on its context, but also on the position or 'behaviour' of the policy system towards that context, and how it incorporates the actors involved.

Kaiser et al. argue that "the world is going through major transformations. Turbulence has become a global condition with powerful local effects" (1995; 25). Kaiser et al. (1995) and Rosenau (1990) attribute this to a number of main factors: the transition from the industrial age to the post-industrial age, characterised by mass communication and micro-electronics, the growing importance of trans-national themes such as the greenhouse effect and the depletion of the ozone layer, the increasing influence of action groups on less-dominant governments and, in particular, the increasing ability of citizens to analyse and mobilise. In this perception, increasing complexity and stronger dynamics are linked to the growing importance of the role of actors (see also § 4.7). This relationship is increasingly seen as the determining factor in the interplay between an open system and the contextual environment to which it relates or is oriented. If we follow this argument to its logical conclusion, we are a small step away from the assumption that a complex environment, whether it be an administrative, social or spatial environment, will require a complex and dynamic policy system. However, this perception is not completely correct, given the reactions in current policy. Decreasing government intervention, increasing belief in self-regulation, and specialisation and prioritisation show that a complex dynamic environment needs a *different* policy approach rather than a complex policy approach.

4.6 Rational Theories for Decision-Led Action

According to Voogd, the approach referred to here as 'decision-led' is not based on "the concrete object, but on the choices that have to be made" (1995a; 24). Whereas goal-oriented action is based largely on continuous interaction between the planning system and the reality it seeks to change, decision-led action is separate from this interaction and is much more geared towards decisions themselves. Decision-led action thus appears to have much in common with the 'decision-centered view' of planning developed by Faludi (1987), which initially placed a great deal of emphasis on functional rationality. In the present study, communicative rationality – which at first sight appears to be directly at odds with functional rationality – is also seen as a theoretical foundation for decision-led action, in addition to functional rationality.

Rationality as the Basis for Planning

What we have described here as decision-led action shows a clear development path over time. Kreukels (1980; 31) concluded some time ago that the focus has shifted from problem definition and problem selection to the solution strategy. The increasing emphasis on *problem formulation* in the 1970s and 1980s followed on from a period of object-oriented planning, during which it became clear that there was insufficient scope for knowledge-building with a view to solving the problems, which were becoming increasingly complex. The focus shifted from object-oriented planning that emphasised the quality of information and knowledge, towards a planning system that took more account of the quality of decisions (Faludi 1987), i.e. a shift from the material object and objectives of planning towards the administrative object of planning and the planning process. In addition, discussing and reaching consensus on a commonly perceived problem became more of a goal in itself, due to the growing awareness of the influence of intersubjective behaviour. In the 1990s, this development was reinforced by the increasing emphasis on problem co-ordination and integration, 'bundled' solution strategies, communication, participation and interaction.

Because decision-making is seen by many as the most elementary part of the planning process, it is not surprising that most theories relate to it. The development of planning theory is largely characterised by the need for a *rational foundation* for decision-making and decision stages. Friedmann concludes: "if there is one theme that runs through all the discussions and debates on planning, it is that of rationality" (1987; 97). Here, then, rationality is the idea that there is an intellectual and consistent[26] theoretical framework that the subject uses to explain his perception of reality.[27] The most influential form of rationality is causality: the idea that one thing must necessarily follow on from another.

As early as 1955, Meyerson and Banfield introduced a *rational planning model* or *rational actor model*: "a planned course of action which is selected rationally is most likely to maximize the attainment of the relevant ends. [...] By a rational decision, we mean one made in the following manner: (1) the decision-maker considers all the alternatives (course of action) open to him; i.e., he considers what courses of action are possible within the conditions of the situation and in the light of the ends he seeks to attain; (2) he identifies and evaluates all of the consequences which would follow from the adoption of each alternative; i.e. he predicts how the total situation would be changed by each course of action he might adopt; and (3) he selects that alternative the probable consequences of which would be preferable in terms of his most valued ends" (1955; 314).[28] In short, this rational approach "demands the systematic consideration and evaluation of alternative means in the light of the preferred ends they are to achieve" (Alexander 1984; 63).

With the emphasis on 'means' and 'ends', rationality becomes a technical criterion and its significance lies in its function as a method and an approach, "to have criteria of success laid down in advance" (Sagoff 1988). This is a functional form of rationality: the 'best' course of action is chosen for a given goal. Dryzek (1990) refers to 'instrumental rationality', Healey (1983) refers to 'technical rationality', Faludi (1987) refers to 'procedural rationalism', while Friedmann (1987) and Verma (1996) use the term referred to by Mannheim (1940) and Webber (1963): 'functional rationality'. Although the authors named above use different terms to define their own individual perceptions of rationality, the underlying concept is the same (see also Berting 1996). For a long time, alternative theories about decision-making, decisions and choices have been put forward as reactions to this rational approach. Other theories arose out of the criticism to which this interpretation of rationality has been subject...

Limitations of the Functional-Rational Model

As early as the 1960s, the functional-rational line[29] as the main principle of planning came under discussion because it is a line of thinking that presumes that everything can be known and comprehended. Making 'rational' choices implies that a great deal is known about the problems concerned. Simon (1967), among others, disputes this. In most cases, it is simply not possible to possess complete information and, at the same time, make a well-informed and balanced comparison between all alternative solutions. Simon describes this as 'bounded rationality'. March and Simon (1958) point to a number of constraints with which planners are immediately confronted:

1. ambiguous and poorly defined problems;
2. incomplete information about alternatives;
3. incomplete information about the baseline, the background of the 'problem';
4. incomplete information about the range and content of values, preferences, and interests; and
5. limited time, limited skills, and limited resources
(March and Simon 1958, in Forester 1989; 50).

There is also increasing acknowledgement of the fact that decision-makers do not always have unequivocal preferences that, equally, do not precede action, and are mostly adaptive and easily subject to unintentional manipulation (Friedberg 1993; 50). Rittel and Webber (1973; 161) argue that: "The information needed to understand a problem depends upon one's idea for solving it". The subject's ability to consider and evaluate also plays a role: "Accordingly, people 'satisfice'; that is they discover and consider options one at a time, using as their evaluation criterion for adoption or rejection of a flexible aspiration level rather than rigid goals" (Alexander 1984; 63). Decisions cannot be based on information alone. Etzioni (1968) claims that planners – human as they are – intuitively employ 'variety-reducing techniques' to organise a complex reality by combining "a broad scanning of a particular problem field in low detail" with "a high-detail scanning of selected areas which at the broader level appeared to be of particular interest"[30] (Chadwick 1971; 337). In these terms, comprehensiveness in planning is a 'selective comprehensiveness'.

Lindblom (1959) and his followers – the Incrementalists referred to above – presume that most decisions only diverge from the *status quo* to a minimal extent, and thus establish a useful link with *institutional behaviour* in policy development[31] (see § 4.7). They argue that the planning process does not deal with fundamental questions, and that major errors are avoided by implementing changes in a piecemeal fashion. They point, above all, to the policy process as a process of complex interaction, in which the actors push their own strategic goals to the fore and attempt to protect them. This *strategic actor model* has overridden the *rational actor model* to a certain extent[32] (Klijn 1996).

The descriptive concepts of Lindblom and his followers focus mainly on the intersubjective role of actors in making decisions within the planning process. The concepts still largely ignore the role of the *contextual environment*. Berry (1974; 358-359) has shown that the functional-rational model, although 'technically sophisticated', requires a 'rather stable' environment. Friedmann also sees a paradox: "when planning is least needed – as under conditions of relative calm and stability – it can afford to be 'rational'; but when level-headed rationality is desperately wanted – as under pressure of an extreme crisis – planning is given the least scope for exercising its manifest function" (1973; 15). Bennett is also sceptical about the degree of certainty offered by functional-rational methodology: "rationalism can give no more than a false security that does not work in practice, and it is necessary to look more deeply into the situation and to recognize that uncertainty and hazard must always be taken into account" (1956; in Goudappel 1996; 66). Berry, Friedmann and Bennett correctly point to the relationship between the type of issue, the situation in which it arose, and the role of the functional-rational decision method. In the systems terminology used by Emery and Trist (§ 4.5), the functional-rational approach is primarily suited to a 'placid, randomized environment', i.e. a stable environment that is not complex.

The functional-rational, cause-and-effect, or hypothetical-deductive model presumes scientific progress based on a generally accepted knowledge structure, which looks at certain elements of an entity in isolation, while assuming that all other elements and relationships remain unchanged. This method will only lead to a full understanding of the element of reality in a very small number of cases. This vision, almost naturally holistic and fact-based, was only popular in spatial planning for a short period of time and to a limited extent.

From the moment the decision-making process begins to the moment of the final decision, many factors intervene, including social, political, technical and economic factors. There are constraining factors (space, time, money, manpower, information, etc.), factors change continually (change has become 'institutionalised' in our society), and the context in which the issue arose also plays a role. Because an issue cannot usually be considered out of context, that context will also affect the end result. When an issue has to be seen in the context of change, it is certainly more difficult to follow the principles of a functional-rational model for contextual reasons. In such cases, rational action means "accepting indeterminacy of social processes (uncertainty) and, consequently, the need to have only partial, sometimes temporary knowledge and insight" (Berting 1996; 17). The context-dependent relationship applies increasingly to complex issues, whereby certain factors are less reliable, calculable and weighable (see Galbraith 1973 and Kastelein 1996). Furthermore, irrational and emotional human behaviour is increasingly becoming a decisive factor.[33] This type of issue involves a great deal of uncertainty.[34]

Despite this critique, it is still useful to aim for decisions based on a functional-rational approach, with consistency, logic, evaluation, etc. (striving for certainty) as a counterpoint to (inter)subjectivity, intuition and emotion (accepting uncertainty). Planning is, then, still a systematic, goal-oriented action, guided partly by goal rationality[35] (see also Kreukels 1980; 58). Decision-led action as an elementary component of planning is an intellectual process, which is partially based on the functional-rational principle. The value of the principle depends to a large extent on the nature of the issue.

Objective Knowledge and Intersubjective Interaction

In the previous sections, functional rationality as a 'reliable guiding principle' of the intellectual process underlying planning-oriented action came under a certain amount of criticism. The main criticism relates to the limited ability of the subject to gather full and objective information on which to base choices. These choices are usually made in an intersubjective, social – and therefore constructed – reality. The influence of intersubjective reality in decision-making was also subject to criticism. Intersubjective co-ordination and communication are increasingly recognised as the driving forces behind social processes and developments. The idea is that many of these processes and developments can only be understood by reasoning based on communication and social interaction. This requires a different type of rational model.

In the words of Harper and Stein (1995; 237): "Rationality is a thin concept, similar to the concepts of justice, truth, and goodness [...]. To say that something is rational is just to say that we can give good reasons for it; that it has been arrived in an impartial manner; that it is consistent and coheres with other beliefs; that it is sensitive to relevant evidence, within a particular context, and relative to our interests". There is a continuous discussion on the clarity and scope of rationality and what its underlying perspective and principles should be. Nevertheless, rationality continues to be seen as the 'methodical programming' for the intellectual process, which identifies the facts and attempts to relate them to each other in order to understand them. However, the 'functional' interpretation is no longer the only interpretation on which the scientific discourse is based.

Mannheim (1940, 1949), the German sociologist "whose writings during the nineteen-thirties and -forties laid the basis for all subsequent thinking on the subject" (Friedmann 1973; xvii), identified two forms of rationality. He called the first form 'functional rationality' – the rationality that has been central to planning-theory discourse and relates to how stages in the planning process and the deployment of resources lead in a logical, causal way to the desired goal. Functional rationality is often related to positivist knowledge acquisition and the related deductive approach.[36] The second form of rationality that Mannheim identifies is *substantive rationality*.

Here, he calls for the reasoning underlying chosen goals, and the process that leads to them, to be taken into consideration. This form of rationality should explain why planning-oriented action is necessary and desirable, and why it is desirable and necessary to intervene in the social and physical reality (see also Berting 1996). "Mannheim argued that it takes a knowledge of the total situation to decide, for instance, whether to use a bomb to blast an entire city or, indeed, whether a bomb of such frightening power should be built at all"[37] (Friedmann 1973; 30). According to Faludi, this relates particularly to "the distinction between theory *in* planning (substantive theory) and the theory *of* planning (procedural planning theory)" (1986; 3).

While functional rationality is seen as the 'formal technique' for gathering as much objective knowledge as possible about an object of study, substantive rationality is increasingly seen as "the rationality of societal knowledge" (Verma 1996; 6). Weaver et al. (1985) refer to a rationality that is based on "the fundamental properties of social choice" (Weaver et al. 1985; 148). The terms 'social' (Van der Cammen 1979), 'collective' (Elster 1983), 'communicative' (Healey 1992) and 'collaborative' rationality (Healey 1997) are used in addition to substantive rationality. Healey refers to "making sense together while living differently" (1992; 143). This 'communicative' rationality relates to the 'logic' of behaviour, intersubjective choice, consensus and social action.

Communicative rationality is increasingly seen as significant in functional-methodological terms; it is the intersubjective that must be understood, not the objective. Because more and more emphasis is being placed on the 'theory *of* planning', it is appropriate no longer to see communicative rationality – as a theory for explaining intersubjective action – as substantive rationality. Communicative rationality, then, focuses on cause-and-effect relationships inherent in intersubjective conduct. It is logical that such relationships involve a great deal of uncertainty because they concern causal aspects of human behaviour. In this sense, communicative rationality is a differentiation of a form of rationality that is based on more indirect causal relationships.

According to Mannheim's distinctions, we are dealing here with a communicative rationality that is not so much a substantive as a largely functional-rational, albeit intersubjective, rationality. At the same time, this form of rationality involves a great deal of uncertainty. Here, communicative rationality differs from functional rationality in a positivist sense, whereby the functional rationality is largely based on direct causality. In the discourse on functional and communicative rationality, theories and lines of reasoning on objective and intersubjective knowledge acquisition are usually diametrically opposed.[38] This dichotomy is largely a reaction against the unilateral use and absolute belief in functional rationality in the period before the Second World War and the decades that followed it.

Communicative Rationality

According to Goudappel, we should learn to recognise that *complex and uncertain situations* require a different approach and methodology "than that which involves hard and/or exact 'facts'. [...] The distant, or 'objective' approach is not feasible in such cases. The subjective seems to be increasingly accepted as a legitimate factor in assessment, and it is even acknowledged as an indispensable element of policy preparation, decision-making and policy evaluation" (1996; 62). Friedmann (1973) states that "the problem is no longer how to make decisions more 'rational', but how to improve the *quality of the action*" (1973; 19). Kastelein agrees with this: "the nature and extent of 'certainty' or 'uncertainty' inherent in the task has (or should have) implications for the structure and management of the process. In cases that are 'certain', 'decision-oriented instruments, structures and strategies are appropriate, while more open-ended exercises are appropriate in 'uncertain' cases'\" (1996; 103).

In decision-making on complex social issues, then, there is an increasing shift from a 'closed' to a more 'open' form of decision-oriented action and from objective to more process-oriented or institution-oriented planning, with emphasis on intersubjective relationships. It is logical that, given the need to understand and explain decision-forming processes under these circumstances, there has been a shift from functional rationality to communicative rationality (see, for example, Dryzek 1990, Healey 1992 and Innes 1995). The focus is no longer on the problem itself but on its definition and the degree of consensus for that definition. Here, the starting point is "not so much the knowledge which logically results, but the knowledge about which a group of involved parties has reached agreement" (Woltjer 1997; 50). Communicative rationality therefore has little to do with logical-positivist theories from which functional rationality derives.

Decision-oriented action in planning involves not only dealing with the object or goals, but also determining who is to be involved in the decision-making process. The idea of rationality is adhered to, whereby insight and predictability is inherent in reasoning and, even in situations involving intersubjective interaction, it is possible to select and relate the relevant facts, and explain what has given rise to them. This also implies the expectation that certainty, albeit of a different nature, can be obtained through a communicative rational approach. Communicative rationality must therefore be seen as an essential part of the decision-oriented aspect of planning.

The shifts in planning theory and systems theory (see Flood and Jackson 1991) were heavily influenced by the work of the German philosopher Habermas and his theory of intersubjective orientation. According to Habermas (1987), "far from giving up on reason as an informing principle for contemporary societies, we should shift perspective from an individualised, subject-object conception of reason, to reasoning formed within intersubjective communication" (in Healey 1992; 150).[39] Rationality seen from a communicative-intersubjective perspective is no longer a matter of definitions, proposals, plans, and scenarios as the *starting points*, but rather as the *outcomes* of decision-making processes. Naturally, the role of planners will also shift accordingly, focus will shift from object-oriented goals to optimising interaction and participation, and the decision-oriented aspect intersects with the institution-oriented aspect of planning-oriented action (see § 4.7). However, Amdam correctly points out that "the instrumental and communicative rationales can be just as utopian, but that should not prevent us from making an effort to achieve them" (1994; 14).

We need rationality to provide an intellectual structure for coping with reality and simplifying it so that we can select empirical facts, identify relationships between facts and understand them. Functional rationality (the rationality of necessity) has definite constraints. Rationality of uncertainty and probability (Toffler 1990), and interrelationship have become counterparts. In decision-oriented planning action, the emphasis has shifted from a pronounced functional goal rationality within a hierarchic system based on control to the quality of action and interaction, and to learning processes for individual decision-makers in a social and therefore complex and dynamic context. We are beginning to realise that planning involves a great deal of uncertainty, that it has to deal with constant change and that, all too often, it has to take place within an 'ongoing process' of policy development.

The intellectual process that characterises decision-oriented planning action is therefore also geared towards the motivation ('from knowledge to action') of goal-oriented as well as institution-oriented planning action. Rational reasoning will not only help to explain action taken in relation to the question of *how* to evaluate the object of decision-making and planning, but also helps to explain *who* will be involved in the decision-making and *what* the results of the involvement will be. Discussions on rationality and planning deal with planning-oriented action not only as an intellectual, goal-led approach, but also as a societal and institution-oriented process.

4.7 Institution-Oriented Action: Interaction and Networking

Traditionally, interaction between actors is seen as a societal issue.[40] However, interaction also relates to the efficiency of decision-making and planning. Planning efficiency is explained above all by the extent to which the deployment of available resources contributes to achieving the defined goals. Planning efficiency depends on good organisation and communication, and co-ordination in striving to achieve common goals, in other words: *efficiency through intersubjective interaction.*

The Awareness of Intersubjective Conduct

Almost every human activity is an expression of institutional behaviour. Individuals (i.e. subjects) are only able to express themselves as actors in an institutionalised (i.e. social) environment, within which they participate according to predominant patterns of communication and interaction. In institution-oriented action, therefore, the emphasis is not so much on the individual as on the collective. Giddens (1984) assumes that individual behaviour is based on the individual's will, which can never be separated from its social context. He describes this process of continuous interplay between behaviour and structure as 'structuralisation'. In Buiks' words, "the interplay between the thinking and behaviour of a community of people [...] constitutes particular patterns of thought and action that have a prescriptive significance, the continuity of which no longer depends on individual members of that community" (1981; 262). In other words: institutionalisation. Wells (1970) describes institutions in terms of normative regulation of human behaviour. According to Buiks, this normative character expresses itself in "the types of sanction applied to non-institutional behaviour in certain situations, and in the form of a certain recalcitrance that results from attempts to change or undermine the institution" (1981; 262). This partly explains why decisions deviate only to a limited extent from prevailing thoughts and principles (§ 4.6). Institutionalisation is, then, one of the main principles underlying the development of a society's cultural and normative character that can explain the intersubjective behaviour of actors.

Institution-oriented behaviour is seen as an essential aspect of planning and is emphatically taken into consideration (see, for example, Friend, Power and Yewlett 1974). In recent years in particular, planning-theory discourse has focused increasingly on communicative planning (see § 4.6), and institution-oriented action plays a central role in this. Discussion of the issues between the parties involved is seen to be more relevant than achieving pre-defined goals and following agreed procedures and processes (Healey 1996, Sager 1994). The main factor influencing the decision-making process is how the actors reach consensus within the constraints imposed by institutional complexities. The use of concepts such as 'participation' and 'support and commitment' (Arnstein 1969), and the interest in consultation and consensus-building (see Innes 1996) indicate a shift towards communicative planning.

Communicative planning is based on the assumption that communication and interaction in the decision-making process do not need to be restricted to a single decision-maker (in this case the government). Friend, Power and Yewlett (1974) long ago pointed out the significance of interactive relationships within institutional structures. Kreukels (1980) also pointed out the importance of "establishing which developments arise in [... the ...] institutional complex, how the institutions within the complex relate to each other and, finally, what control and supervision processes exist" (1980; 97). Kreukels correctly observes that planning should be understood in the context of these complex institutional entities (1980; 93). For example, companies, action groups and individual citizens should be involved – and have a real sense of *being* involved – in decision-making processes. According to Gill and Lucchesi: "it is argued that when citizens have been actively involved in the decision-making process they are more aware of the possible problems and are more willing to live with the consequences than they are when decisions are imposed from outside. Thus, through active participation, citizens are educated to political realities, become more aware of problems, and tend less toward explosive solutions" (1979; 555). According to Woltjer (1997), it is important that "all 'powerful' interested parties work together, that problems are clearly defined, that interests are co-ordinated, that knowledge and information are shared, and that an initial period of time-consuming and intensive work will prevent delays in the longer term" (1997; 47). The interest in this broad, active involvement in decision-making processes developed during the 1990s.

This understanding of intersubjectivity, institutionality and institutional context is a recognisable thread in the evolution of systems theory, and is closely related to it. The increasing importance attributed to the roles of actors involved in policy development in an institutional context has substantially undermined the mechanical view of systems theory. Goal-oriented systems theory is less successful when it is applied to dynamic and unstructured problems in which human behaviour is the underlying factor. There is a growing realisation that "in describing the human activity, institutional or personal, the goal-seeking paradigm is inadequate" (Vickers 1968; 66). This is because the influence of subjective, limited and selective actors is too great. Once mechanical thinking had been replaced by 'corrective' thinking, there followed a shift towards 'interactive', actor-oriented and perception-oriented thinking. These theories emphasize the role of actors who operate within and outside institutional contexts in order to formulate, structure and implement (e.g. policy). This is known as 'relation maintaining', a concept "which uses decision making and policy making as its examples but is basically concerned with the nature of human understanding and human value judgement" (Vickers 1968; 66).

A consequence of focusing on interpersonal relationships in systems theory is that it has become more difficult to monitor and control system aspects. Interpersonal and institutional relationships are, after all, partly determined by unpredictable human behaviour and the conduct of organisations, whereby individual value judgements, attitudes and perceptions have a complex effect on policy relationships. The emphasis is shifting from the ability to devise solutions and make predictions to "*understanding* the social process which characterizes human affairs" (Checkland 1991; 67). This is not simply a question of "reducing uncertainty, but *managing uncertainties*" (Teisman 1992; 42). It is not the system *per se* that is central, but the system as it is perceived by those who set it up, develop it, and wish to use it. Individual characteristics of those who develop or wish to use the system also play a role, as well as the perceptions and individual interests of the actors and institutions that are part of the system or can be influenced by it. The key is not to think only in terms of 'what *is* it?', 'what *will* it be' or 'how *should* it be?', but also in terms of 'how *could* it be' and 'how will this be *perceived*?'. The interactive *network* is a system-theoretical concept that is used in the administrative context for answering these questions.

Actors and their Role in Organisational Relationships

Organisations and institutional relationships are increasingly seen in terms of information, communication and co-ordination networks. Planners work in organisational relationships within which the various actors are dependent on each other for knowledge, information and resources. Forester follows a "social problem-solving view in which 'social' is narrowly construed to mean 'organizational'" (1989; 30). He refers to "informal networks, steady contacts, and regular communication [which] keep planners informed" (1989; 30). Networks that follow formal organisational relationships determine not only the *informal* networks, but also, to a large extent, the effectiveness of the planning process. The question, however, is to what extent an institutional structure and its related interactive information and communication network should be geared towards, for example, informal relationships and relationships that exist outside the formal structure, given the nature of the problems with which decision-makers are confronted.

Insofar as there is communication and co-ordination via channels that have a very formal organisational structure, Vroom (1981) identifies six important and closely interwoven aspects: hierarchy, centralisation, formalisation, standardisation, specialisation and routine. Despite the fact that the terms 'hierarchy' and 'centralisation' are used synonymously in many studies, they are in fact two different things. It is possible, for example, for centralisation to take place in a context that has no hierarchy. *Hierarchy* implies a 'vertical' distinction in the relationships between actors; a distinction that does not necessarily have to be present when centralisation takes place. The extent of the hierarchy is primarily important for the decision-making and implementation structure of an institutional relationship, and therefore also for the degree of independence along the vertical axis of the decision-making structure within an institutional context. *Centralisation* can express 'an orientation towards the centre' in a vertical as well as a horizontal structure, without there necessarily being a hierarchy between actors. *Formalisation,* (i.e. the recording of decisions) guarantees a certain amount of continuity in the policy process. It also means an institutionalised form of control focusing on the discrepancy between desired and actual results. Far-reaching *standardisation* of norms, regulations and procedures, as can be seen in environmental policy, is usually an indication of a tightly controlled institutional context that is not conducive to flexibility. Vroom (1981; 302) points out that "We must [...] nevertheless conclude that standardisation is one of the central steering mechanisms of an organisation". *Specialisation* of actors within institutional contexts means narrowing the range of focus so that actors can increase their knowledge of specific problems and possible solutions. The concept of *routine* is closely related to specialisation (Rondinelli, Middleton and Verspoor 1989). Giddens (1984a) points out that a large proportion of activities are based on routine. Institutional structures lead to automatisms and routine-based activity, whereby institutions contribute to the stability of the order of which they are a part (Parsons 1951). It is the *extent* of routine-based activity that is important, in other words: "Is there sufficient knowledge [...] to [...] perform tasks efficiently, or is creativity needed on each occasion in order to find a solution?" (Vroom 1981; 303). In the light of the critique in section 1.2, it is also a matter of whether routine-based action based on generic principles is sufficient to deal satisfactorily with particular situations, or whether, given the nature of the problems, creative solutions are required in certain circumstances.

In addition, there is the concept of *performance* (§ 4.3), as mentioned above, which can be used to express the suitability (and, as an extension of that, the effectiveness) of planning and decision-making in more or less formal institutional contexts. Whereas in planning the underlying nature of goals was still a reaction to mechanical thought, the 'performance' of decision-making is a phenomenon in planning that clearly derives from growing recognition of the role and position of actors in various institutional contexts. In previous sections we have already referred to the subjective, autonomous, restricted and selective conduct of actors. It is also known that actors anticipate measures being implemented by not reacting as policy-makers expected or wished. During the decision process, decision-makers must take account of how others will react to their measures, and take into account "the possibility that others may act differently in anticipation of those measures" (Dekker and Needham 1989; 5). The performance and effectuation of high-level decisions are obstructed by the following: the partial or full rejection of a decision made at a higher administrative level, varying interpretations of abstract or strategic statements, the operational content of which is insufficiently sound, the contribution – which is difficult to assess – to general strategic statements made by individual practical actions, and/or the implementation – intentionally or otherwise – of opposing measures. These factors can ultimately limit the effectiveness of the entire planning system.

However, re-interpretation of decisions at a lower level may also lead to an improvement in or optimisation of the intervention required to solve a local problem or issue. At higher levels, where the decision has been taken, there is rarely sufficient information available to guarantee that the decision will produce the desired result. It is also conceivable that decisions are designed to be universal (e.g. standards in environmental policy) while, in practice, their diversity has not been taken into account. This in turn means that measures or decisions are not appropriate to the problem or issue (§ 1.2 and 1.3). It is also impossible accurately to predict every effect. The 'alarm' function in the implementation process is therefore essential, particularly at the stage where policy becomes 'real' in a physical or societal context. It is nevertheless the case that the ultimate success of an initial decision, however excellent, depends on a series of actions and decisions at other administrative or governmental levels.

The institutional context outlined here has an unmistakeable influence on planning-oriented action, on decision-making and on the issues that planning-oriented action and decision-making are intended to solve. Issues, their descriptions and their definitions are 'constructed'. According to Kooiman, this implies that "the formulation of problems by problem owners should be seen as an essential part of dealing with the problem" (1996; 34). In this light it is not surprising that, in planning-theory discourse, there is a shift in emphasis from solution strategies towards problem formulation, as we have seen in section 4.6.

Institutional Networks

The term 'performance' refers to the links between actors and institutions. The institutional links are the channels through which decision-making is adapted to the level at which the effects of policy should become tangible. This involves structuring links that bring together the various actors in a societal, organised context. These structures can be seen as various forms of *institutional networks*, which are social and interactive. Hufen and Ringeling (1990; 6) refer to such networks as social systems within which patterns of interaction develop between actors.

Institutional networks can be seen as system-functional models of institutional action, within which actors and institutions are bound together by a certain degree of interdependence. This *interdependence* exists when actors and institutions require each other's support in order to realise their goals (Scharpf 1978, Teisman 1992), and rests primarily on knowledge, resources and positions.[41] In our complex institutionalised society in which actors, organisations and institutions continuously set themselves goals that they cannot realise independently, interdependence is the rule rather than the exception. This applies in particular to complex issues that affect many people, involve many interested parties, and are strongly interwoven with the context in which they arose.

The realisation that interdependence is all too often an essential aspect of achieving specific goals is a clear indication of network-based institutional relationships. According to Teisman, “every new decision-making game will be [...] played by a unique constellation of actors, and will therefore follow a unique process” (1992; 50). This also assumes that problem-dependent patterns within networks will take on a largely unpredictable form. However, it remains to be seen whether this is a workable hypothesis for structuring decision-making processes in all cases.

The discussion above assumes that communication and information networks will be partly geared towards formal organisational relationships. There are vertical networks (of which the Dutch system of government is a good example) that include many fixed structures for decision-making and procedures. Put negatively, this is known as bureaucracy. Networks are not necessarily based on the best actor in the best place, allowing each actor to be seen from a horizontal and more or less equal perspective. Each institutional relationship has its own selection criteria and common denominators (Kooiman 1996; 35). Moreover, the composition of an institutional network is partly determined by historical and formal frameworks. Institutional decision-making techniques will also evolve out of existing structures and experience that are not always easy to break away from or refute and which could therefore stand in the way of a necessary change towards a more effective and efficient decision-making and executive organisation. Networks will also develop as a result of the routine behaviour of actors and institutions, whereby their participation and the role they fulfil within the network are largely based on limited, incomplete, inaccurate, or incorrectly interpreted information.

The extent to which a structure of institutional networks can be adapted to the nature of a problem depends on a number of factors. Some of the factors are not related to the problem, for example the formalisation of the institutional structure. The historical constellations within which institutions are used to functioning are also a determining factor, and experience and knowledge of various forms of planning also play a role. Problem-related factors include the realisation, knowledge and perception of a problem, the nature, extent and effects of a problem, and the interdependence of actors and institutions in defining a problem and dealing with it effectively. In an existing system, such as the system of government, problem-related factors are discussed as soon as a problem arises, with the aim of dealing with the problem as satisfactorily as possible. If it is not possible to respond adequately to the problem within the existing system, the institutional network will undoubtedly come under discussion, too.

Government is seeking to affect a shift in the way in which attempts are made to realise goals through decision-making and planning (see Kickert 1986). During the last decade in particular, the aim has been to shift the emphasis away from vertical, hierarchical decision-making to a more ‘horizontal’ form of decision-making based on shared governance for achieving goals defined by the government (§ 1.2 and 1.3). This means a change in institutional action on the part of the government. Many governmental actors will also be required to change their approach. The changes are complicated and do not happen automatically, as we shall see in Chapter 5 with regard to environmental policy. Decentralised decision-making and shared governance will alter the focus on decision-making. Non-governmental actors and institutions will be increasingly involved in the decision-making process in an active way (institution-oriented). In addition, a shift will take place in the way common societal goals are pursued (goal-orientation).

The increased involvement of non-governmental actors and institutions in decision-making and policy formulation is a decision in itself. The aim is not only to control in a peremptory way, but also to do so on the basis of shared responsibility and commitment. This development partly reflects the prevailing long-term government trends (see Chapter 1), and influences specific areas of government policy. It has a strong influence on environmental policy and the interface between environmental policy and spatial planning policy. However, this is not without consequences, as we shall see in Chapter 5.

Because non-governmental parties are involved in decision-making, there are relationships between as well as within institutional networks. *Intra*-institutional networks (e.g. between different governmental organisations, but more often between governmental and non-governmental parties) have noticeably less organised governance, and results are therefore less predictable than is usually the case within separate institutional contexts. By contrast, there is greater freedom with regard to defining problems, discussing responsibilities and strategy, and allocating resources (Vroom 1981).

Within that constellation, interdependence, interactive communication and information-sharing play an important role. An increasing number of actors are actively participating in policy development and decision-making to represent their own interests, and contribute their own specific expertise and possibilities. More than ever, policy development and decision-making has become a negotiating process (Kickert 1993; 26) that increasingly takes place between various governmental authorities, but also between governmental and non-governmental participants. This applies not only to environmental planning, but also to spatial planning (see also Chapter 3 and Korthals Altes 1995). This means that decision-making is becoming more complex from the point of view of an institutional and decision-oriented approach. It is precisely within these more horizontal decision-making structures that interactive networks occur. Such situations require other forms of decision-making and planning than in the conventional hierarchical approach (Kickert 1993; 26).

Three Theoretical Perspectives on Governance

Teisman (1991) presents three 'perspectives on governance', which he uses to define institutional roles that are characteristic of vertical governance, horizontal governance and the third form he identifies: pluricentric governance. Teisman summarises institutional role delegation in a vertical governance structure as 'unicentric'. The unicentric approach focuses on resolving societal issues from a single central point of governance. Teisman refers to institutional role delegation from a horizontal perspective as 'multicentric'. The multicentric approach does not focus so much on governance, but rather on the stimulation of societal progress through the exchange mechanism by actors whose function is primarily that of market player. The third approach to governance is known as 'pluricentric'. In this approach, central government and local actors/institutions are 'interwoven' and policy relating to societal development and the development of the separate entities is determined on the basis of interdependence. The relationship between what Teisman calls unicentrism and multicentrism is also referred to as the relationship between classical/traditional models versus market models (De Bruijn et al. 1991). The relationship between unicentrism and pluricentrism is referred to as 'top-down versus bottom-up' (Hanf and Sharpf 1978), and 'intervening versus calculating government' (Van Tatenhove 1993).

From the unicentric perspective, national government is the controlling body that dictates how society should develop and guides it in the required direction. The top level of the government can act in the general interest and specifies policy content and procedural regulations for implementing that content. Local government must follow the policy set at national level, and act accordingly. The emphasis is on a general policy that does not focus on specific local issues. Planning is an important tool for formulating policy in a step-by-step process aimed at clearly delineated and predefined goals. The 'unified decision maker' is, of course, hypothetical: "policy formation and policy implementation are inevitably the result of interactions among a plurality of separate actors with separate interests, goals and strategies" (Scharpf 1978; 346). Nevertheless, a society that is governed by a strong belief in its own makeability will be characterised by a dominant 'unicentric' top-down approach.

A society characterised by a multicentric approach to government will have a strong belief in the market mechanism. In such a society, self-interest – rather than the collective interest – is uppermost. National government is not familiar with the needs of individual local actors and institutions. However, multicentrism assumes an uncertain and uncontrollable context within which actors – grouped/organised or otherwise – respond to their constantly changing environment. The credo is not control but adaptation. There is nevertheless a strong belief in the self-governing ability of society, so there is a great deal of scope for actors and institutions to take autonomous decisions. Governance takes place through the market mechanism, and societal progress is based on competition that is the result of supply and demand from virtually autonomous players on a level playing field.

Teisman's pluricentric approach is a reaction against extreme unicentric and multicentric approaches. These are theoretical views based on positions that can be seen as opposite extremes in the continuum of perspectives on governance (see also Fig. 4.6). In Teisman's vision, policy systems consist of central and local entities, which are based neither on hierarchy nor local autonomy. The parties involved are interdependent and policy evolves through a process of interaction. This is a result of the high level of functional differentiation in western society, which in turn means that society is characterised by a multiplicity of specialisms and a high level of professionalism (De Bruijn and Ten Heuvelhof 1995). According to Teisman, in such circumstances, the 'real' decision-making structure is determined by true interaction between actors within and outside formal organisations: "The co-ordination and development of goals is not studied as an intellectual process at central level, or as a accumulation of local preferences, but as the outcome of a policy struggle between local and central entities. The entities must have a common interest in the policy. The communal interest arises from interaction and manifests itself after interaction has taken place" (Teisman 1992; 32). There will be increased interest in the organisation of policy processes – not with a view to controlling or adapting, but with a view to co-ordinating the actors through interaction.

Interactive institutional action increases significantly with the number of actors: "Not only the scope for action increases, but also the possible unintended consequences of action. Social interaction in larger groups is always characterised by a high level of uncertainty and dynamism" (Klijn 1996; 52). Attempts are made "to solve problems by focusing on the relationships between actors. In other words, the essence of the problem is localised in the relationships themselves and not within the entities involved" (Kooiman 1996; 41-42). The cause of a problem is hardly considered (Kooiman 1996), mainly because social reality is too dynamic and still developing. Neither will it help to collect information in advance: "In principle, complex processes of interaction [...] involve uncertainty and learning" (Klijn 1996; 52). Where interaction and participation are concerned, the goal is therefore "to generate new and increasing knowledge, and thereby initiate self-reinforcing learning processes in which are developed a mutual understanding of the present situation and shared visions of a better society, more realistic strategies as to how these visions can be achieved, and more comprehensive practical activities" (Amdam 1994; 5). According to Teisman, institutional network theory, which focuses on interactive relationships, is an appropriate model for pluricentric governance because the pluricentric approach presumes interdependence.

With regard to the 'centric' perspective, Teisman describes a number of theoretical approaches to governance and identifies a third approach between the two extremes which, in his opinion, involves interdependency. At one extreme we see a system based on central governance geared towards maximum control of society. This requires accepting the position of all those involved, and the – largely 'asymmetric' – interdependence derives from that position and related role. At the other extreme, every form of governance is excluded and society develops to a large extent according to a supply and demand mechanism. Actors are connected in a commercial sense but behave autonomously. By contrast, network approaches are based on the assumption that all actors are interdependent in reaching a common goal, and none of them is dominant. But it is much more likely that the various perspectives and related forms of dependence will occur together. The German social scientist Scharpf, the leading protagonist of interdependence, refers to this when he defines a network as "the ensemble of direct and indirect linkages defined by unilateral or mutual relationships of dependency" (Scharpf 1978; 362). A network is an 'ensemble' of dependencies, and these also occur in spatial planning: "Land use planning and decision-making resemble a high-stakes *competition* over an area's future land use pattern [...]. The players are locked together in a framework of interdependence in which they must gain agreement from other players in order to achieve their goals. Thus, the process's competition is tempered with the need for *cooperation* as well" (Kaiser et al. 1995; 6).

Pluricentrism is based not only on interdependency, but also on a *large degree of equality* in terms of knowledge, resources and role. The parties involved need each other, and regulation – let alone force – will not achieve the desired goal. This is a theoretical situation in which decision-making and policy development for the welfare of the individual *and* society are central. In such situations, there is not necessarily a pressing general need. Neither is the focus solely on the interests of individual actors and institutions. Pluricentrism emphasises common interests that are shared by a number of actors but seen from different perspectives.

In such cases, consensus-building must result in 'win-win situations', whereby each participant aims for an optimum rather than a maximum result (De Bruijn and Ten Heuvelhof 1995). Here, we are talking about "the complexity, dynamics and diversity of problems and their solutions as the *inter*active and, in problem-definition processes in particular, the importance of the *inter*subjective process that must be followed by processes and their owners. In this context, interchange is an essential concept" (Kooiman 1996; 47). This involves "a dynamic process that relates to the outcomes of interactions and the strategies of the actors involved. Given that each actor follows a different strategy to achieve a goal, the final result is the product of a complex interplay of strategies and actors. The result is difficult to predict, and the interaction process, which is reflected in changing goals, strategies and outcomes, is highly dynamic" (Klijn 1996; 17).

Decision-making and policy do not come into being through predictable predefined processes but evolve within network-like structures. This makes it difficult to make decisions and predict consequences. However, we can go further than this observation on the contribution of interactive behaviour and on the complexity of decision-making. The three theoretical perspectives on government defined by Teisman are each recognisable and usable but do not occur as such in practice. In practice, situations occur in which the elements and characteristics of the three theoretical perspectives arise in conjunction and in different combinations. For example, "developers are constrained by both land planning and market demand. To succeed, their projects must pass both a government test and a market test. [...] They operate in a market of buyers and sellers that is influenced by public plans and service programs but not driven by them" (Kaiser et al. 1995; 9). In that sense, Goudappel (1996) refers to different relationships between 'one' and 'the other': "processes and developments which display changing relationships [...] between the (f)actors involved" (1996; 79).

The institutional context within which decision-making takes place is relatively complex. However, we should note here that the complexity depends partly on the evaluation of the issue and the related goals that have been defined. It is important to note that interactive networks are appropriate for decision-making processes that are highly complex and uncertain, and bound by a number of rules. The pluricentric perspective on governance is based, for example, on a more or less symmetrical interdependence between the various actors. Partly as a result of this, the complexity of an institutional context not only leads to uncertainty, it is also a complexity that is at least knowable. The methods for dealing with this will be discussed in the following sections.

4.8 Complexity as a Criterion for Planning-Oriented Action

Being human, we tend to translate the reality in which we find ourselves into abstractions that are as simplified as possible. In the words of Berting: "Social reality is always extremely complex and we can observe, somewhat ironically, that complexity in itself is not the problem, but acting on the basis of simplifications of the social reality" (1996; 24). We should therefore ask how these abstractions can be used to organise our actions in order to make the society in which we live as comprehensible and manageable as possible.

The question is also how we can underpin these abstractions in a consistent way. The current academic method is to acquire knowledge in a structured way by organising a complex, dynamic reality by means of ordered, systematic and consistent theories, concepts, structures and models. After all, "no substantial part of the universe is so simple that it can be grasped and controlled without abstraction. Abstraction consists in replacing the part of the universe under consideration by a model of similar but simpler structure. Models, formal or intellectual on the one hand, or material on the other, are thus a central necessity of scientific procedure" (Rosenblueth and Wiener, in Kramer and De Smit 1991; 16). It is therefore a matter of what value the three approaches to planning action have as abstractions on which to base statements regarding decision-making and planning methods.

The discussions on the three behavioural aspects of planning emphasise 'complexity'. The question arises whether complexity is simply a metaphor for our inability to construct a workable and acceptable perception of reality. Or can much greater significance be attributed to the phenomenon 'complexity'? Can complexity function as a criterion for planning-oriented action? Table 4.1 is an example of how this could be done. The table summarizes the arguments on planning-oriented action from previous sections and links them on the basis of 'complexity'.

Complexity and the Discourse on Planning Theory

A number of arguments can be advanced in favour of complexity as a criterion for planning-oriented action. One argument has already been mentioned, namely that complexity is one of the mainsprings of planning-theory discourse. There is a second, more empirical argument, which will be discussed in Chapter 6. It is based on the observation that environmental/spatial issues vary in complexity and therefore require different planning strategies (Borst et al. 1995). Another argument for using complexity as a criterion is based on a relatively new scientific philosophy known as complexity theory.[42]

Before we discuss this theory and its related concepts in more detail, we will discuss the planning-theory argument that focuses on the relationships between the three perspectives on planning-oriented action. The idea of complexity as a criterion for making choices relating to planning-oriented action has not simply arisen out of nowhere. In each of the three perspectives on planning (see sections 4.5 to 4.7), there is a shift in emphasis from a vision that is simplistic, mechanical, object-oriented and strongly functional-rational towards a more differentiated and coherent view of reality with more nuances and increasing emphasis on human behaviour and interaction. The most notable element in the discussion on the goal-oriented perspective on planning action is the transition from fixed goals, via an iterative and goal-oriented approach, towards the formulation and realisation of multiple-objective goals. The structure of planning is also considered from this goal-oriented perspective based on systems theory. This reflection reveals a number of shifts in planning-oriented action. There is a shift from fixed linear processes to non-linear processes that are part of an ongoing process of policy developments and actions whereby, in addition to linear and cyclical actions, network-based multiplicity is also necessary to indicate which interaction will lead to real interventions in the physical and societal environments. As a consequence, there is a movement away from generalisation (i.e. general goals, standards and principles) towards specialisation, with each issue being assessed in its own individual context. There is also a shift in institution-oriented perspective on planning-oriented action, namely from a traditional, hierarchic governance structure, via shared governance, towards market forces. This is a shift from a vertical top-down structure towards a more horizontal and interactive network of actors who are interdependent to more or less the same extent.

This shift is also away from a vision of object-oriented governance towards a vision in which intersubjectivity is a decisive factor and actor participation is important. The decision-led approach to planning-oriented action reveals a shift of emphasis from direct causal relationships towards interactive relationships. This shift can be seen most clearly in the critique of functional rationality and in the growing interest in communicative rationality as a theoretical foundation for planning-oriented actions.

The shifts in the theoretical visions of planning are partly a response to the increasing complexity of the real world and to increased knowledge of and insight into reality and the world around us. The developments are probably also due to the shift in post-war society away from a functional orientation towards a greater appreciation of quality, which occurred from the 1960s onwards. Above all, the shifts appear to reflect the continued modification and piecemeal rejection of the logical or neo-positivist tradition, which for many years had been the foundation of the discourse on planning theory and systems theory, and strongly influenced scientific views of planning. In the post-war years, these views began to crumble as they became increasingly unworkable in complex planning situations. The development of planning-theory discourse is therefore a response to the increasing complexity of the issues that are the subject of that discourse *and* of planning-oriented action.

It would be incorrect to infer that traditional theories, which were popular in the early days, no longer have any value. "It is a widely misunderstood myth about scientific revolutions that a major 'paradigm shift' [...] means that all scientific knowledge that went before is immediately null-and-void as a result of the new paradigm shift" (Eve 1997; 275). If we assume that issues should be assessed in terms of their complexity, the traditional concepts will be valid so long as they are applied to the appropriate issues, i.e. the more straightforward issues. In this sense, visions that relate to more complex issues should not automatically be seen as replacing traditional approaches, but as providing a wider perspective on the issue.

The argument for seeing complexity as a key aspect for evaluating issues and making choices is developed further in Table 4.1. Developments in planning theory are grouped as goal-oriented, decision-oriented and institution-oriented planning actions. Table 4.1 shows that interaction is increasingly important in complex issues. However, it also shows that goal-oriented action is still very relevant in complex situations too. The insights presented here are also linked to the complexity of a planning issue, in a relationship between the aspects involved in a planning issue, its theoretical foundation and the strategy for dealing with it. The degree of complexity determines the links between the different features of the three perspectives on planning action (see also Figs. 4.8 and 4.9). Table 4.1 presents complexity as a criterion for developing effective and efficient planning strategies, and in particular as a criterion for *decision-oriented planning* or, preferably, a 'meta'-criterion that will affect goal-oriented and institution-oriented planning action (see § 4.9). Table 4.1 can therefore be summarised as a 'decision-oriented' instrument for establishing relationships between conflicts, complexity and decision-making.

Objective and Intersubjective Evaluation of Complexity

The complexity of an issue or object can be measured according to the number of elements and features it comprises, as well as their dimensions (heterogeneity), their relationships and coherence, the extent to which they are subject – individually and together – to change (stability), and the limiting conditions of the object of study. According to Berting (1996; 27), it is difficult to gain comprehensive insight into these aspects in most complex issues. Complex issues are usually characterised by the large amount of information required. That information is often incomplete, inaccurate, or inconstant, is difficult to assess realistically and is only given limited consideration "which results in complex processes and considerations being reduced to simple yes/no decisions or left/right resolutions" (Holland and Holdert 1997; 9).

Complexity is more than an object-related concept. It is also a subjective, relative and normative concept. The extent to which something is perceived as complex depends on the actors involved. Complexity, then, depends on how the subject (observer) assesses the object of study: "the observer-dependent nature of the simple-complex criterion" (Jackson and Keys 1991; 142). The fact that complexity is a normative concept means that, whenever it is used in a societal context, it acquires an intersubjective significance. It is no easy matter to find a commonly agreed definition for the nature, scope and objectives of a problem and for more complex issues this is all too often a problem in itself. If the elements and characteristics of the object are also seen in a societal context, and are therefore partly assessed in an intersubjective way, there is very little scope for object-related considerations. Fuenmayor describes this as 'interpretive complexity' (1991; 234).

Table 4.1 A typology of planning-oriented action (§ 4.5, 4.6 and 4.7) based on three categories of complexity

Degree of complexity of planning issue ➔	Orientation towards object Effectiveness of planning	Orientation towards rationality Choices relating to efficiency and effectiveness of planning	Orientation towards intersubjectivity Efficiency in planning
	A. What has to be achieved?	B. How can it be achieved?	C. Who is involved?
	Scope of goal and action structure Emphasis on effects and decision stages	Justification of decisions Emphasis on choices	Actors and institutional links Emphasis on interaction
	Goal-oriented action	← **Decision-oriented action** →	**Institution-oriented action**
Relatively straightforward	- Emphasis on constituent parts of the whole (closed system) - Fixed goals (blueprint planning) - Linear mechanical regulation process - Fixed decision stages - Decision-making process has clear beginning and end	- Full or extensive knowledge - Few or no uncertainties - All-embracing - Control of the whole - Functional rationality - Direct causal (*causa proxima*) relationships predominate - Reductionism - Strongly delineated issues - Main aim is predictions and solution strategy	- Central governance - Vertical network - High degree of formalisation, standardisation and routine - Policy-maker is decision-maker - Hierarchical interdependence - For a collective that is not actively involved - Tightly controlled institutional links with clearly defined tasks and responsibilities

Relatively complex	- Emphasis on whole *and* constituent parts in an open system - Shifting goals (iteratve planning) - Linear phased cyclic planning process with feedback, correction and self-regulation - Decision stages are process-dependent - Beginning and end of decision-making process varies	- Knowledge insufficient; limited and selective availability - Uncertainty due to continuous assessment and discontinued feedback - Selective scope - Co-ordination in terms of the whole - Bounded rationality - Behavioural interpretation - Holism - Diffuse delineation of issues - Strong emphasis on problem definition and problem selection	- Decentralised shared governance - Local network - Mix of formalisation, standardisation and specialisation - Role of policy-maker is part of collective decision-making - Symmetrical interdependence within context framework - Collective, local and individual interests are given equal consideration - No hierarchical local autonomy, but shared responsibility and commitment
Relatively very complex	- Emphasis on whole, on constituent parts *and* contextual environment - Linked or integrated problems, solutions and goals (multiple-objective approach) - Information cycles - Decision stages as a dynamic, interactive part of ongoing process - Nature of decision-making process is continuous	- Knowledge acquisition in a dynamic and interactive ongoing process - Uncertainty is a constant, together with autonomous variable factors - Context-dependent - Adapt to context - Communicative rationality - Interpretative analysis (*causa remota*) is predominant - Expansionism - Issue is part of a larger whole - Problem co-ordination / integration and bundling of strategies	- Interactive governance - Horizontal network - High degree of specialisation and flexibility - Role of policy-maker is 'socialised' - Symmetrical interdependence, varying interests - Local and individual interests are basis for development - Highly variable and problem-based institutional links with responsibilities that are difficult to identify

A number of context-related factors can be added to this summary of factors (which is not intended to be exhaustive) that determine the complexity of an issue. What is important here is the extent to which the object's elements and characteristics are open to, interact with and are influenced by the contextual environment. Emery and Trist (1965) have defined various contextual environments, which range from a stable, more or less unchanging environment to 'disturbed reactive environments' within which 'turbulent fields' can be observed (see § 4.5). In the constituent parts that form the whole, and also in the object's contextual environment, there will be more or less autonomous indefinable and chaotic elements that could increase the level of unpredictability. The continuous interaction between the object and its context, and the presence of autonomously developing elements, result in a process of evolution of the object or system over a given time span: "many of the problems we 'solve' do not stay solved because the problems themselves change" (Ackoff 1981). Policy geared towards complex issues and complex decision-making is therefore usually little more than a set of well-considered measures for dealing with change in a rapidly evolving environment (Nijkamp 1996, Albers et al. 1994).

A Systems-Theory Perception of Complexity

Despite the fact that the complexity of an object or issue depends largely on the observer, it is nevertheless possible to make concrete statements about complexity, for example by viewing the object or issue from a systems-theory perspective, as we have done above. In this chapter, two methods from a systems-functional perspective are brought together to produce a simplified and systematic insight into reality. One of the methods is based on the constituent parts that make up the whole and contribute to the knowledge of the whole. The other method relates to the contextual environment and its influence on the object of study.

It has also been argued that an approach that focuses on the constituent parts results in knowledge but does not guarantee adequate insight into the whole. It is not only the parts in themselves that are important, but also the relationships between them. Furthermore, in a number of cases, the coherence of the individual elements that form the object of study are perceived as complex and subject to change (*internal complexity*).

The need for detailed knowledge of constituent parts will diminish as a result of continuous change that is not always unequivocal due to unstable contextual factors. Thus, the unchanging environment in which an issue arises is only a theoretical starting point, and a view based on that starting point is no more than a first step towards comprehension, insight and knowledge. The reductionist idea that knowledge of reality and reality itself come together at a point that can be reached by increasingly detailed study of the object therefore has limited validity and depends to a large extent on the nature (i.e. complexity) of the issue (see also Nijkamp 1996; 134-135).

It is, however, standard practice to delineate an issue or object of study. In doing so, the choice is made, intentionally or not, to separate the issue from – or even to compare it to – other problems and underlying societal developments (Kooiman 1996; 45). If the problem is linked to the contextual environment in which it arose, it is logical that influences from that environment will affect the problem and, to a certain extent, the assessment of the problem (*external complexity*). In such cases, the decision-making and planning processes may be influenced by corrective feedback from the contextual environment.

The level of detail of the study (or solution strategy) will decrease as the coherence between the individual parts of the whole becomes more extensive and diverse, if the object is affected by its contextual environment, and if the whole is subject to change. By contrast, there will be an increasing need for information on the changing relationships between the main elements and for knowledge of external influences (see Fig. 4.5).

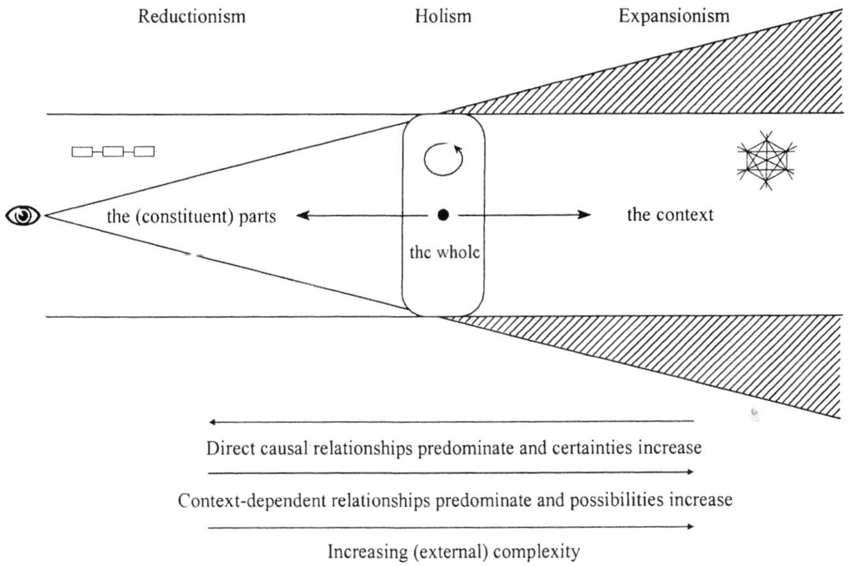

Figure 4.5 An overview of the parts, the whole and the context
Note: See endnote 12 in this chapter

Also important in this connection is a view on universal and self-organising systems deriving from stable and instable principles, respectively. According to Prigogine (1996) stable systems are those within which "small changes in the initial conditions have minor consequences" (1996; 27). In an unstable system, small changes in the initial conditions "will inevitably diverge exponentially over time" (1996; 27). It is also unlikely that a method applied to an unstable system will produce the same results twice. We must think in terms of probability rather than certainty.

We can deduce from these arguments that the effects of a possible method are difficult to predict without actually implementing it (Vemuri 1978). Interventions in reality will not result in fully controlled effects. Instead, interventions through various steps could have unpredictable results and lead to unwanted effects.

"Any attempt to use a quantitative approach to aid the solution of a problem can [...] only give information about the likely effects rather that the exact effects of a proposed solution" (Jackson and Keys 1991; 143). Complex issues are probabilistic and consequently non-recurring to an increasing extent (see also Goudappel 1996; 71).[43]

According to this system-functional description, the degree of complexity of an object or issue depends on whether the subject chooses to focus on certain aspects of an issue when developing a solution strategy and whether the subject chooses to allow for influences from the contextual environment of the issue when devising a strategy. If the strategy focuses only on the individual parts of the issue, the level of uncertainty surrounding the issue will depend on its internal complexity. If contextual influences are considered, the number of factors influencing the issue will increase substantially, as will its complexity. External complexity, which measures context-dependence, will then be a determining factor (see Fig. 4.5). The greater the direct causal relationship between the constituent parts themselves, and between the parts and the context, the greater the extent to which the strategy becomes a 'given' rather than a choice, and the more predictable the end result will be. If the issue is mainly characterised by more unequivocal relationships, the level of uncertainty will rise and the predictability of the possible end result will decrease. Such cases are referred to in this study as having a greater degree of complexity.

Complexity Theory as an Explanation for Development and 'Progress'

The complexity of issues has not remained an entirely implicit element of planning-theory discourse. In the 1990s, a trend towards management-science thinking evolved and quickly gained ground in the Netherlands and a number of other countries. This trend is referred to as 'complex decision-making' (De Bruijn and Ten Heuvelhof 1995, Klijn 1996, Teisman 1992), a term which denotes a decision-making strategy based on the interaction between actors within network-based constellations. Teisman (1992) sees this (§ 4.7) as a pluricentric approach, whereby actors are interdependent in terms of knowledge, resources or power. Here, the term 'complex decision-making' refers above all to specific network-based governance strategies. The phenomenon 'complexity' is developed further in this context, and variations in the level of complexity are not considered.

Reference is made above to the relevance of the *level of* complexity. We assumed that complexity could be seen as a variable with a given range. This is more or less the central assumption in what is known as 'complexity theory' and relates to formal discussion on a different, more abstract level than the level on which planning-theory discourse takes place. At the same time, complexity theory in its current form not only transcends but can also connect the different scientific disciplines.

The theoretical discussion of complexity that attempts to grasp reality evolved partly out of the Chaos theory of the 1980s (see Gleick 1987). Chaos theory offers a mathematical perspective on system properties to which Euclidian geometry could not easily be applied. Chaos theory seeks to explain, among other things, apparently unstable systems that can have a drastic effect on relationships. Mandelbrot's (1982) fractal theory and Feigenbaum's convergence in non-linear transformations (1978) were significant impulses in the move towards perceiving 'order in chaos'. The theoretical discussion of complexity evolved from developments in physics, chemistry and biology on the one hand, and model-building and economics on the other. It reflects an increasing acceptance of the fact that equilibrium is the exception rather than the rule in human society and in the world of physics and biology (Casti 1995, Cohen and Stewart 1994, Coveney and Highfield 1995, Kauffman 1995, Lewin 1997, Mainzer 1996, Prigogine and Stengers 1990, Waldrop 1992).

Complexity theory assumes that development and 'progress' cannot be expected in a world in which the Newtonian dynamic of a never-ending cycle of repetition prevails. Newton's world is a world in perfect equilibrium. Development is only possible in a situation of disequilibrium when circumstances are complex and certainty and predictability are replaced by uncertainty and probability. Given the initial conditions, there is no longer an unequivocal development that leads to an unequivocal outcome. By contrast, we are faced with a series of possible routes that will result in different outcomes. A process that is repeated on the basis of the same initial conditions will not produce the same outcome every time and, if it does, this is merely coincidental. There is diversity instead of uniformity. This diversity does not necessarily lead to the degeneration of the process – in physics reference is made to increasing entropy – but may lead to increased complexity, which may even take on a 'chaotic' form.

In his essay, Waldrop (1992) discusses the result of computer-simulation research, from which it can be inferred that there is a continuum of simplicity, complexity and chaos. And Waldrop's message tells us more.[44] Our dynamic reality, including life itself, occurs 'at the edge of order and chaos', "where the components of a system never quite lock into place, and yet never quite dissolve into turbulence either" (1992; 12), in which a complex reality can be found that reflects "the right balance of stability and fluidity" (1992; 308). Pirsig (1991), in a metaphysical argument, points out that 'the right balance between stability and fluidity' is the result of constant tension between static and dynamic quality,[45] which he sees as the driving force of development.[46] The message from both these authors is that the world is never in balance,[47] but there is nevertheless a knowable reality. This is possible because, in addition to the inevitable and continuous presence of chaos, a certain amount of order results from developments in the direction of stability and equilibrium. The chaos can also be constantly elevated to a new order, ultimately resulting in development and progress.[48] Simplicity, complexity and chaos are apparent opposites "that, at second sight, should be considered more in terms of their complementarity", as Goudappel (1996, 76) concludes. The question that remains is how we should deal with this view of reality.

From this metaphysical standpoint, complexity is *a comprehensible reality and an essential given* for development. A situation of increasing complexity, that appears to be moving towards chaos, will not necessarily be subjugated by chaos but can evolve into clear structures and relationships at a higher level, where other structure-defining values and norms exist.[49] Here, an antithesis can lead to a similar argument. A whole divided into its constituent parts can appear to be straightforward, but at a lower level each part may have its own different order of complexity.

Thus far, the discussion on complexity has taken place primarily within the 'hard' scientific disciplines. External influences, which break into 'stable' systems, contribute to the dynamics and render processes increasingly uncertain, are studied in physics, meteorology, chemistry, biology, ecology and economics (Kauffman 1995, Prigogine and Stengers 1990, Waldrop 1992). Now, social scientists who use qualitative constructions are also studying complex external influences, dynamics and uncertainty, increasingly being seen as factors that cannot be ignored (Eve, Horsfall and Lee 1997). It is a matter of seeing "the facts of social complexity. Too many social processes, such as the fluctuations of the global market, seem to operate behind our backs and out of our control" (Bohman 1996; 152). Complexity hardly features at all as a theme in the discourses on sociological theory and planning theory, yet a number of aspects of complexity theory could be relevant to sociological and planning theory.

In the first place, it is relevant because the phenomenon of complexity leads to a greater understanding of the world around us: "we can at least now understand what a chaotic system is doing, and how it is doing it, when we see it" (Eve 1997; 278). Furthermore, there is a relationship between simplicity, complexity and chaos, not only in the worlds of physics and biology, but also in social reality. Table 4.1 is an example of how a vision based on complexity has been developed for planning-based action. In the third place, we should remember that chaos is not necessarily unpredictable and therefore problematic. In the fourth place, we can point to the fact that sociological issues have an unbounded stochastic character that is inherent in their complexity. This need not be problematic because the loss of certainty is compensated by the increased possibilities for dealing with the issues. "Here, probability and necessity are not two irreconcilable enemies, but each acts as the other's companion..." (Toffler 1990; 21). The important thing is to make optimum use of the opportunities that issues present. It is also a matter of formulating a technique or framework for identifying possibilities, developing them and assessing their feasibility.

Even in a complex reality, straightforward issues can arise for which a complexity-based approach is neither useful nor necessary. However, complex issues should be dealt with according to their complexity. To a certain extent, the internal complexity, external complexity and delineation of complex issues are inherently unclear. This is especially the case when it is difficult to determine the complexity of an issue or object of study, and the complexity or object of study is subject to change. As complexity increases, the aspects 'complexity' and 'change' become a given rather than uncertain factors.

In this chapter and the previous chapters, we have taken the first steps towards formulating a model for dealing with complexity in planning. Chapters 2 and 3 discussed the fact that the interface between the natural and spatial environments can be viewed from different standpoints. On the one hand, this interface presents itself very simply, as in the cause-and-effect relationship of environmentally harmful activities. On the other hand, its complexity is difficult to encompass in appropriate policy concepts. The compact-city concept is a good example of this. This chapter deals primarily with planning-theory arguments that have been categorised according to goal-oriented, decision-oriented and institution-oriented approaches to action. The practical result of this categorisation can be seen in Table 4.1, in which a distinction is made between 'relatively simple', 'relatively complex' and 'relatively very complex'[50] planning issues. This vision of reality will be used below as the foundation for a coherent theoretical framework for planning-oriented action.

4.9 Towards a Coherent Theoretical Framework for Planning-Oriented Action

In section 4.3 we saw that, in order to form a well-founded vision of the world around us, it is not necessary to consider a single perspective. Indeed, several parallel perspectives should be considered. This suggests a pluriform vision based on a complex and dynamic physical reality, the sociological process, the interaction between these, and the policies for intervening in that reality. In the previous sections, we discussed three perspectives on planning action and linked them in terms of complexity. We argued that the method for decision-making and the content of solution strategies depend on the degree of complexity attributed to an issue. In this section, this idea is developed further into a coherent vision of planning. In concrete terms, this means that the choice of planning strategies for issues arising at the interface between the natural and spatial environments can be explained by the individual complexity of those issues, which are considered from three perspectives: decision orientation, goal orientation and institutional orientation (see Chapter 5). These three perspectives are the foundation for the model of planning-oriented action that is the aim of this study.

In planning-theory discourse there are several pluriform approaches to planning. In contingency theory, pluriformity is even seen as an elementary part of planning-oriented action. But in a sense it is also an approach that has been developed in a conventional way. The approach places strong emphasis on the goal orientation of decision-led action, which will be discussed below. We will then discuss the 'pluriform' role played by actors in decision-making and governance processes, and in particular the institutional orientation of decision-led action. In order to construct a functional model for planning-oriented action based on a pluriform combination of planning visions, we will also have to consider the relationships between those visions. Friedmann's (1973) 'typology of allocative planning styles' will be used as the example and basis for this. Together with the contingency approach and the institutional orientation, Friedmann's typology provides a number of abstractions for structuring the final theoretical framework for planning-oriented action.

The Contingency Approach: a Goal-Led Approach to Decision-Led Action

Pluriform approaches to planning are by no means unknown in planning-theory discourse. Bryson and Delbecq (1979), for example, believe that there are several approaches and strategies, each of which makes its own contribution to the planning process and the realisation of goals. Although they point out that the "validity of the existing schools of planning thought, or 'one best way' approaches" (1979; 177) cannot be denied, the relationships that link the different approaches are much more interesting, particularly when planning issues are characterised by a high level of uncertainty and arise in an unstable context. 'One best way' approaches "do not exist in isolation. They fit into a contingent planning framework" (Bryson and Delbecq 1979; 177).

The *contingency approach* (Sharpf 1978) evolved out of discussions on 'planning contingencies' that began in the 1960s on the possible relationships between various planning concepts and theories (Rondinelli, Middleton and Verspoor 1989). This approach can be seen as an empirical reaction to the normative 'one best way of organizing' visions that prevailed in management and planning (Kickert 1993; 26). Contingency theory is based on the idea that, however diverse issues may be, links can be established between sets of characteristics, elements and criteria (contingencies). In other words: "this approach posits that the appropriate range of choices regarding organizational structures and process is *contingent on* any number of relevant factors" (Bryson and Delbecq 1979; 167). It is also assumed that "choices regarding planning phases and tactics are seen as dependent on planning goals and contextual variables" (Bryson and Delbecq 1979; 167). The goals of the planning process are regarded as more or less fixed and are implicitly simple (see also Table 4.2). Such assumptions are described above as functional-rational, a perspective that was discussed in section 4.6.

The context of a planning issue is important in the contingency approach and is assumed to be a decisive factor for the planning process (see also Bryson, Bromiley and Jung 1990). There is no 'catch-all' planning method that guarantees success because that success partly depends on the context (see also Klijn 1996; 28). It is the specific situation rather than the general theory that determines the planning strategy and the course of the planning process. There is therefore a "situation-specific interplay of structure and process" (Rondinelli, Middleton and Verspoor 1989; 45) that leads to a strategy or a change in strategy. These ideas are fully in line with a pluriform approach, dealing with context and change, and with complexity. The question as to what contingencies can be based on is left unanswered. Nelissen asks, rightly: "Who or what will consider contingency?" (1992; 40).

If we are to search for an answer to this question, with a view to bringing together the three defined perspectives on planning action into a single contingency vision, then the contingency approach by itself is not enough. The contingency approach evolved during a period in which planners were seen as specialists and the main architects of the planning process. Intersubjective thinking, the institutional perspective on planning action, and communicative interaction in planning are therefore not basic elements of the contingency approach. The planner organises the actors, but is not one of them and does not interact with them. This last point is criticised by Teisman (1992; 113-114), for whom it is a reason to introduce the alternative of 'interactionism'. With regard to the pluriform approach discussed here, what Teisman calls interactionism is not so much an alternative but complementary, 'contingency-plus', as it were.

Interaction: an Institutional Orientation towards Decision-Based Action

The role of actors in the decision-making process is largely determined by the relationship between the issue and the prevailing constellation of governance strategies.

The movement between planning strategies is denoted by arrow A in Fig. 4.6. A downward shift along line A represents a shift from central governance towards a market-centric approach, involving an increase in the number and diversity of actors participating in the decision-making process (line B in Fig. 4.6). The downward movement also represents a shift away from the collective interest towards the individual interest. The movement along line A is not necessarily a transition from one theoretical perspective to the other, but above all a *shift in emphasis* within the pluriform whole of planning perspectives. The characteristics and elements belonging to one perspective will not necessarily become redundant but will become less important in relation to characteristics and perspectives of the other theoretical perspective.

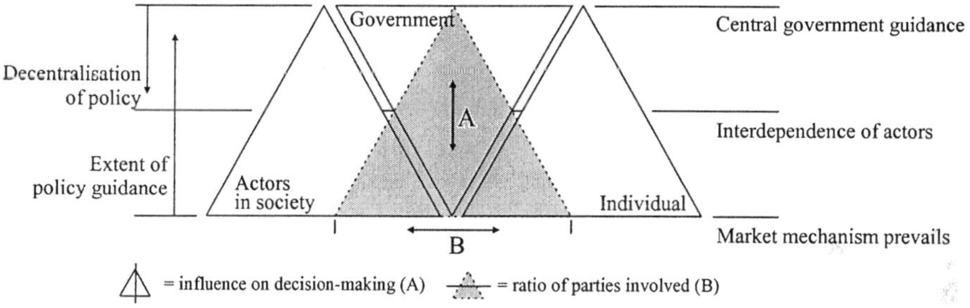

Figure 4.6 A schematic representation of the nature and effect of relationships between governmental and non-governmental actors/institutions on societal processes

Note: For A and B, see text

Among other things, a shift in emphasis between planning perspectives means that one perspective becomes less influential than another. Now that the boundaries between the public and the private sectors are becoming increasingly blurred, and planning strategies (according to De Bruijn and Ten Heuvelhof 1995) are evolving into an 'ongoing process', have become 'changeable and fluid' (1995; 17), and are characterised by irregularity and the lack of a clear beginning and end, these shifts will manifest themselves in various forms.

Nelissen (1992) identifies four types of shift from hierarchical to horizontal network structures. He refers to a 'market shift pattern' when private companies predominate in networks, to a 'corporatist shift pattern' when intermediary, non-profit organisations start to establish a profile for themselves, to a 'citizen shift pattern' when individual citizens become more assertive, and to a 'statist shift pattern' when government becomes increasingly dominant within a network. In environmental planning, the distinctions between the various non-governmental actors are still fluid and, depending on the type of issue, they are encouraged by the government to participate in the decision-making process (see also § 5.4 and 5.5).

The idea that shifts in emphasis between planning perspectives have important consequences for the relationship between the defined perspectives on planning action does not exclude the study of individual theoretical and ideal situations, or aspects thereof. But here, too, a pluriform approach to planning will mean that changes to certain elements of planning strategy must be seen in the context of the relationship between the strategy in question and other planning strategies. A consequence of contingency is that everything must be seen in relation to everything else.

For Teisman (1992), the main question is which organisational proposals could improve the functioning of what he calls a unicentric, multicentric and pluricentric organisation (§ 4.7). If the assumption is that the solution to decision-making issues does not lie in theoretical planning perspectives as such, but should be sought in the relationship within which characteristics and elements of the planning perspectives manifest themselves in a planning strategy adapted to a specific situation, it is no longer a matter of matching an issue to an appropriate theoretical perspective on planning. This involves bringing together the different characteristics and elements of the three theoretical planning perspectives to form an effective and efficient pluriform strategy. This also has consequences for the composition and participation of the parties involved, and the possible interactions between actors during the decision-making and planning processes.

Friedmann's Typology of Allocative Planning Styles

This viewpoint on strategies and their relationships is in line with the 'typology of allocative planning styles' drawn up by Friedmann (1973). He identifies four elementary styles of planning: command planning, policies planning, corporate planning and participant planning, which should not be left out of the discussion here (compare Rondinelli, Middleton and Verspoor (1989) for management strategies). Command planning "is associated with strongly centralized systems of governmental power" (1973; 71). This typology corresponds to what is referred to in Table 4.2 as rational, functional-rational or centralised planning. Strict goals are imposed, and compliance is essential. "The command system comes closest to the formal decision model of allocative planning. Its information requirements are extraordinarily demanding. Complete, accurate, and timely information must be obtained about all aspects of the system to be planned, and the controls must cover all the variables relevant to achieving specified performance levels" (1973; 72). 'Policies planning' assumes a more flexible organisation within which obtaining full information and control is no longer a realistic option. "Associated with weakly centralized systems of government, its method is to induce appropriate actions through statements of general guidelines and criteria for choice, the provision of material incentives, and the dissemination of information for decentralized planning" (1973; 72-73). Goal formulation remains important. A comparison with 'comprehensive planning' as referred to in Table 4.2 appears justified. There is indirect control. "Policies, for instance, are meant to make some allocative choices impossible while increasing the probability of other, more desirable ones" (1973; 73). Corporate planning emphasizes the planning process rather than the realisation of targets. "More specifically, the results of negotiations through which corporate planning is sustained are not determined in advance; they crucially depend on the distribution of effective power among all the participants in the bargaining process and on their comparative skill in using this power" (1973; 74). The fourth planning style is known as participant planning, which "occurs under conditions where power to implement decisions resides in community forms of social organization and, consequently, is dispersed" (1973; 76). The process is central to this typology too, but, in contrast to the corporate planning model, participation is greater and more diverse. Participation is more important than results.

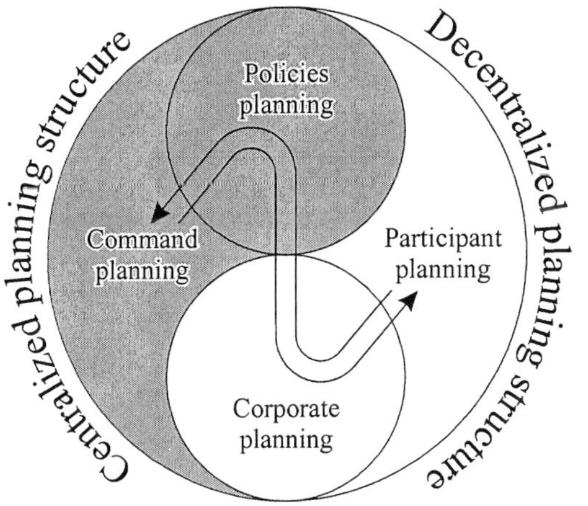

Figure 4.7 The planning styles distinguished by Friedmann, and their possible relationships

Source: Friedmann 1973; 80

Friedmann remarks, interestingly, that "each of the four styles of allocative planning described may be thought of as one component that must be joined to others to make a guidance system work" (1973; 79). In short, a typology is theoretical and cannot be seen entirely separately from 'the other' in practice. In Fig. 4.7, Friedmann's diagram shows the relationships between the styles he has identified. The most obvious relationships between planning styles and the curved arrow that passes through the 'yin-yang' structure denote the shift in emphasis from centralised to decentralised planning. Here, it is important that the language Friedmann uses is appropriate for the shifts identified in planning-theory discussions. Using Friedmann's view, these shifts can be translated into relationships between the different visions and styles of planning. The discussion below will focus on the relationship and links between the different styles of planning. The relationship is based on the three approaches to planning action discussed above.

A Functional Framework for Planning-Oriented Action

Models "must be inclusive and comprehensive and also allow a maximum of generality. And yet, they should be simple and structured so that they can be readily applicable for practical purposes" (Lim 1986; 76). Friedmann (Friedmann and Weaver 1979) uses the terms 'comprehensive', 'consistent' and 'coherent' as criteria for ideas that are combined into composites (Van den Berg 1981). Midgley (1995) refers to 'methodological pluralism', which in effect is "a meta-theory to identify the strengths and weaknesses of different methodologies, which are thereby viewed as complementary" (1995; 62). This means, among other things, that when the cohesion between different visions of planning can be increased, the resulting pluriform perception of reality is more consistent than was previously the case. There should also be consistency in its basis, comprehensiveness and contiguousness. Such a pluriform perception of reality will not necessarily lead to one standpoint or two – extreme or non-extreme – standpoints, possibilities or criteria that are decisive for planning-oriented action. It is much more likely that there will be many (theoretically the number is infinite) viewpoints, possibilities and criteria. It is therefore logical that there will be several strategies that could be appropriate for dealing with an issue. In such cases, we speak of 'scenarios'. From a practical point of view, the pluriform approach has certain constraints with regard to manageability (number of strategies) and recognisability (diversity of strategies). There is a level beyond which a pluriform picture of reality or a pluriform method loses its cohesion: "Pluriformity then becomes atomisation and then a chaos that consists of incompatible components"[51] (De Bruijn and Ten Heuvelhof 1991; 170).

The functional model for planning-oriented action should incorporate goal-oriented (i.e. effectiveness-related) and institution-oriented (i.e. efficiency-related) characteristics that, depending on the decision-oriented choice between central and decentralised governance, should allow the positioning of issues and the appropriate solution strategies. Figure 4.8 shows a framework within which the axes are characteristics of planning-oriented action. One axis represents the *scope of goals* (relating to goal-oriented action) and the other represents the *scope of relationships* (in terms of institution-oriented planning). Figure 4.8 shows the framework in its most basic form.

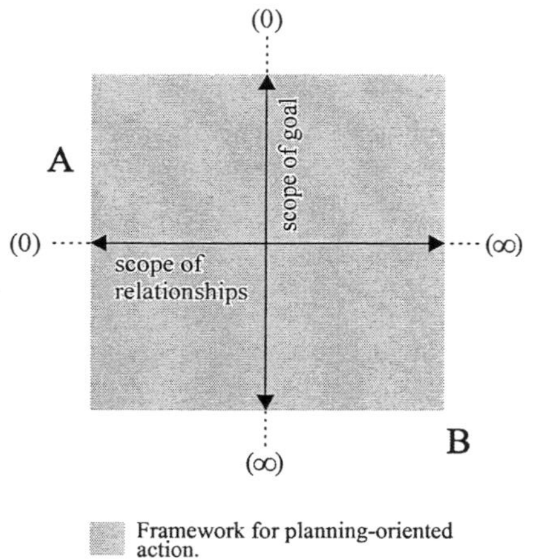

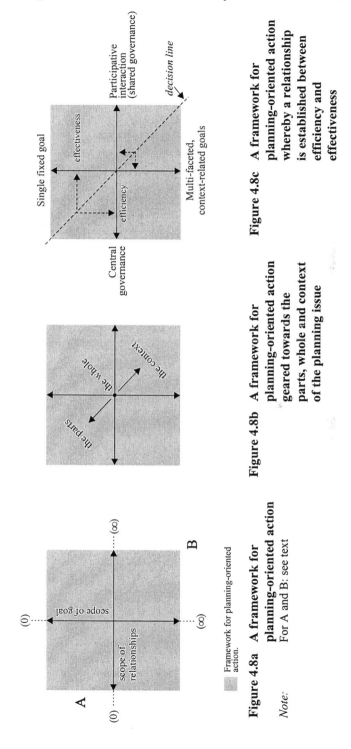

Figure 4.8a A framework for planning-oriented action

Note: For A and B: see text

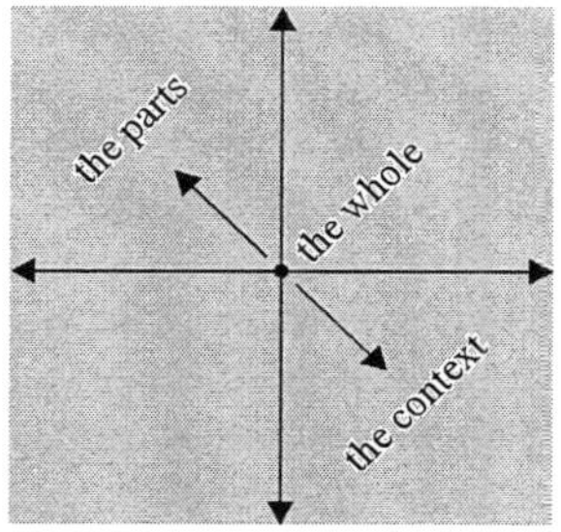

Figure 4.8b A framework for planning-oriented action geared towards the parts, whole and context of the planning issue

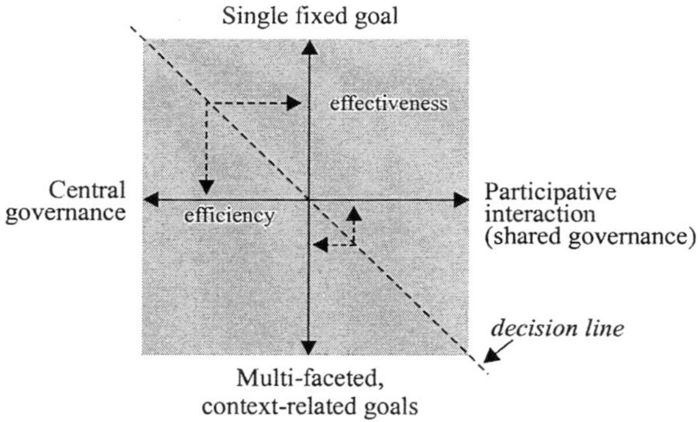

Figure 4.8c A framework for planning-oriented action whereby a relationship is established between efficiency and effectiveness

Table 4.2 A typology of planning-oriented action

Decision-oriented ↓	**Actors / Scope of goals**						
	Single			Multiple			(∞)
Simple	Object	small no. of parts	(functional-) rational and central planning	Object	several constituent parts	e.g. comprehensive and scenario planning	
	Rationality	direct causal		Rationality	feedback		
	Inter-subjective	hierarchy		Inter-subjective	compromise		
Complex	Object	parts and context	e.g. contingency planning	Object	parts and context	integral interactive and participative planning	
	Rationality	feedback		Rationality	interactive		
	Inter-subjective	co-ordination		Inter-subjective	consensus		
(Chaos)							

Note: Uses structure of Table 4.1

The scope of relationships (i.e. the possible number of interactive actors participating in the planning process) and the scope of goals of an issue are plotted as opposites here. The axes of both variables cover a theoretical continuum from zero to an infinite number of actors and goals. In figure 4.8a, A indicates the area of full control.[52] B is the area within which full control is lacking with the market mechanism functioning in its purest form. The grey shaded area represents institutionalisation (see also Teisman 1992; 51). This is the area within which planning-based action can occur. Nomothetic theories, typologies or planning strategies that have proved their worth under specific conditions can be placed in this framework. These theories may have a neo-positivist (emphasis on constituent parts) and interactive (emphasis on the context) basis, depending on their position in the framework (Fig. 4.8b). The idea is that the theories are not mutually exclusive, but overlap or complement each other in accordance with pluriform thinking.

Table 4.2 shows a typology[53] in which complexity (i.e. the decision-oriented perspective on planning, is set against the number of actors or institutions and against the scope of goals associated with an issue. This is a structure in which goal-oriented, decision-oriented and institution-oriented action are determining factors in the characterisation of planning issues and identifying appropriate planning strategies. This scheme of alternatives for planning-oriented action is not exhaustive. That would not be possible because the structure is too simple. The scheme's significance lies in the fact that it can be used to determine how characteristics and features of the three perspectives on planning action can be used to position theories, conceptions and visions on planning and to develop planning strategies.

This scheme emphasises the fact that rational planning, functional-rational planning and central governance are not a thing of the past but, depending on the nature of the issue, may still be appropriate, supplemented by other governance strategies as necessary. It is a nuance of the visible shift in current planning discussions, which again places strong emphasis on the planning process on the assumption that participation and the support base are of more value than the mechanism that can lead to goal realisation. In this context, Woltjer (1997) points to normative considerations as if it is no longer appropriate to develop plans and keep them confidential until they are finalised. Centralised governance is 'out' and participation in planning is 'in'. In the words of Snellen, it is possible to recognise a product of this principle. He argues that we must "move away from the idea that the government is the only central institution that directs society. The relationship between government and society is not one of governed object to governed system, as governing subject to governed object [...]. Instead of a subject-object relationship, a subject-subject relationship should be established between government and society" (1987; 18). However, in the light of the discussion above, this one-sided approach is doubtful. This study therefore defends the principle whereby combinations of different governance strategies are applied to the decision-making process, depending on the nature and scale of the issue. After all, when the construction of a new road is being discussed, the colour of the traffic lights is never under discussion.

Figure 4.9 is an abstract representation of the relationship between goal-oriented, institution-oriented and decision-oriented planning action (see also De Roo

1995, De Roo 1996, Miller and De Roo 1997) that incorporates complexity as the criterion for decision-oriented action, and therefore as an element linking the various perspectives on decision-led action. Depending on the degree of interaction and the goals towards which it aims, planning issues can be characterised as simple, complex or very complex. Determining the degree of complexity – should parts of the issue be dealt with or should its context also be considered? – is a decision-oriented choice in the planning process. The decision-oriented choices therefore constitute the diagonal that extends from the upper-left quadrant to the lower-right quadrant (see also Fig. 4.8b).

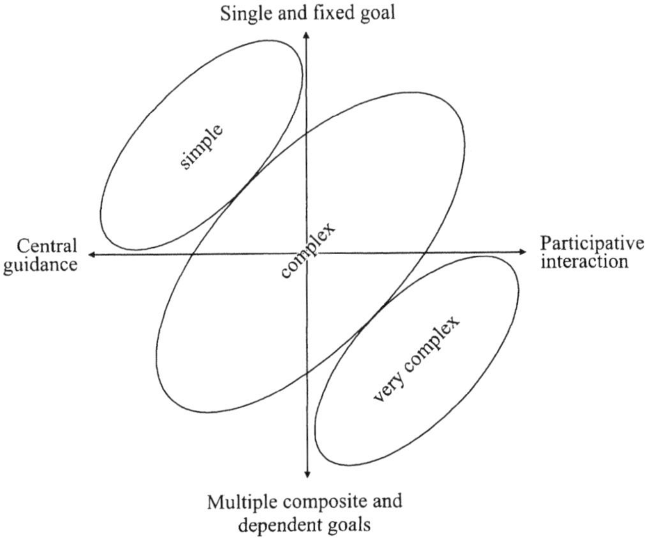

Figure 4.9 Framework for planning-oriented action, in which the relationship between planning goals and interaction is based on complexity

The imaginary diagonal axis represents the degree of complexity, and will therefore also determine the relationship between interaction and the scope of the goal(s). This framework answers Nelissen's question 'who or what will deal with contingency?' (1992; 40). It is also a response to the comment made by Bryson and Delbecq that developing strategies is not a question of the different approaches to planning in themselves, but of the relationships between them.

When the framework for planning-oriented action is applied, emphasis will be placed on (1) unravelling the nature and scope of the issue and determining its complexity (see Fig. 4.9) in order to find appropriate planning-oriented measures (for a detailed representation, see Table 4.1), (2) the changes to which the issue may be subject (see Fig. 4.10), (3) selecting the appropriate planning strategies, using the scheme of alternatives (Table 4.2) as necessary, and (4) selecting the appropriate instruments.

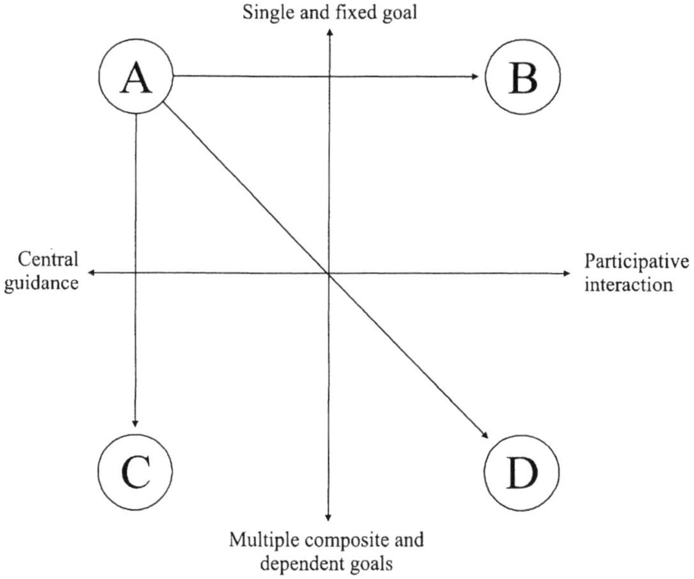

Figure 4.10 Relationship between instrument A and other instruments in the framework for planning-oriented action

Note: For A, B, C and D, see text

This framework based on theoretical conclusions should form an 'umbrella' for the various strategies and instruments used in environmental policy. The framework is also intended to provide insight into the structural change processes that occurred in environmental policy during the 1990s. In section 6.1, these changes are described as being manifestly visible in the development and acceptance of the instrument known as integral environmental zoning (IMZ) and the ROM designated-areas policy. Integral environmental zoning (A in Fig. 4.10) was designed as a hierarchical instrument whereby framework-setting standards were enforced to guarantee an acceptable level of environmental quality around complex industrial locations. During the few years in which the instrument was tested in practice, however, the method itself and its principles were increasingly criticised (§ 5.4). Other instruments were proposed, based on a different management philosophy. After a period of initial scepticism, the ROM designated-areas policy (between B and D in Fig. 4.10) was better received than integral environmental zoning instruments (§ 5.4). The ROM policy is designed for implementation at local and regional levels on the basis of participative decision-making. It also involves locally defined goals that relate not only to the natural environment, but also to other aspects of our physical surroundings. Another interesting instrument is the Amsterdam 'Bubble Concept' ('*stolpmethodiek*', C in Fig. 4.10). This approach has a number of inherent preconditions, which means that it is diametrically opposed to environmental zoning within the spectrum of goal-oriented action. The essence of the Bubble Concept is that sectoral norms are no longer seen as a method for separating environmentally harmful activities from environmentally sensitive areas, but as a method for achieving an integral level of environmental quality within a given area (§ 5.5). A broadly defined overall goal thus replaces sectoral goals. In section 4.5 this is referred to as the multi-objective approach.

In Chapter 5, these instruments will be placed in the framework presented above for planning-oriented action. That chapter will focus on the development process of environmental policy, which will be described in the context of the framework for planning-oriented action and the related approaches to planning action.

4.10 Conclusion

The issue of 'sustainable' separation and interweaving of environmentally harmful and environmentally sensitive functions and activities is a type of issue that anyone can comprehend. Industry and traffic bring risk and nuisance, which in turn lead to environmental impact in local residential areas. This is a good reason for maintaining a certain distance from such activities. However, developing a satisfactory solution for such issues is quite another matter. Environmental/spatial conflict, which can be so easily explained in terms of the causal relationship between harmful activities and sensitive functions that results from contaminating emissions, cannot always be represented as a direct cause-and-effect relationship. There are many social and economic interests involved in the process that leads to the release of harmful emissions and in the consequences of such emissions. The spatial-functional context of environmental/spatial conflicts (e.g. the compact city) partly results in the fact that maintaining sufficient distance between harmful and sensitive functions is not necessarily the logical consequence of environmentally harmful emissions. In practice it is not always easy to implement such an apparently simple solution because many other factors influence and determine the nature and scale of an environmental/spatial conflict – not least the re-evaluation of principles, policy changes and political standpoints, and shifts therein.

For a long time, environmental/spatial conflicts have been considered as relatively simple issues that could be resolved through environmental standards, an instrument based on a functional-rational policy vision. We have already seen that, in practice, the inflated evaluation of environmental standards works in favour of a participative approach, i.e. a communicative-rational approach. This development could lead to a situation in which interaction between actors becomes more important than the conventional objectives designed to protect the environment.

In order to be able to assess this development in terms of environmental planning and analyse its implications for decision-making methods, three approaches to planning action have been identified and developed:

- a goal-oriented approach;
- a decision-oriented approach;
- an institution-oriented approach.

The implications of these approaches illustrate an interesting shift in planning theory from nuances, modifications and constraints on the logical-positivist tradition in planning towards planning theory based on participation and communication. This development – at least in principle – can be compared to the critique of the policy-based approach to environmental/spatial conflict. Above we have seen how this parallel is a stimulus to using planning-theory discourse as an aid to formulating arguments when following and evaluating new policy developments. This relates to:

- the shift from fixed targets to ongoing, integral goals;
- the shift from functional to communicative rationality;
- the shift from hierarchic structures to more horizontal, interactive networks.

The shifts in planning-theory discourse have resulted in greater emphasis on problem definition, the planning process and the involvement of actors in that process. Interaction in planning has increasingly become a goal in itself, while there is decreasing emphasis on the eventual result of planning on the physical and social environment. By contrast, there is an increased awareness that the results of interactive planning strategies can be predicted only to a limited extent, that the degree of certainty of an issue is limited and that, all too often, the ultimate result of planning will depend on the number of parties and interests involved. In short, more consideration is given to the 'complexity' of an issue or conflict.

Here, however, there is a move away from the idea that the principles underlying technical, centralised and functional-rational planning are 'out' and should be replaced by other principles, which can be used to explain complex issues and be developed into a successful planning strategy. We must guard against 'over-commitment' to the intersubjective and interactive elements of planning. As decision-making processes become more participative, goal-oriented action remains appropriate, albeit in a different form. Goal-oriented and institution-oriented planning measures are neither mutually exclusive nor diametrically opposed to each other. Rather, they should be seen as complementary.

The study has used the method of classifying the features of goal-oriented, decision-oriented and institution-oriented approaches to planning according to their 'complex' character. Table 4.1 is the result of this classification, and can be used as a guideline when determining the complexity of issues and selecting appropriate planning-oriented measures. The classification is reduced to its essential elements in Fig. 4.8. In the terminology of Etzioni (1986; 8), a framework has been established of 'higher order, fundamental' abstractions, with which 'lower order, incremental' abstractions can be combined that arise from and/or lead to abstractions at a higher level. It is a framework for planning-based action, whereby features of the goal-oriented and institution-oriented approaches to planning are linked by complexity as a decision-oriented criterion for planning-based action.

Complexity as a criterion for planning-based action is discussed primarily in terms of systems theory. Depending on the complexity of an issue, the emphasis is placed either on its constituent parts, on the whole issue in relation to its constituent parts, or on the relation of the whole issue to its context. As the emphasis shifts towards direct causal relationships between constituent parts, a functional-rational approach – command-and-control planning – is readily available and, given the initial conditions of the issue, the ultimate result of planning will be reasonably predictable. Where there are less clear relationships between the various relevant aspects of an issue, and contextual factors of the issues are also seen as determining factors, the complexity of the issue will increase, as will the uncertainty with regard to the end result. However, the possibilities for dealing with the issue will also increase and this is an important argument in favour of formulating a solution strategy that is not only more area-specific, but also devised within a broader administrative framework – situation-specific, area-oriented planning that evolves towards shared governance. Such solutions would then be based on a more communicative-rational approach. Taking this argument further, we can say that, with complexity as a criterion for planning-based action, the somewhat dualistic relationship between functional rationality and communicative rationality can be bridged.

The cohesive framework for planning-based action allows a connection to be established between two apparently diametrically opposed instruments in environmental planning: integral environmental zoning and the ROM designated-areas policy. The question remains as to how the planning-theoretical conclusions can be applied in practice, in this case the developments that are taking place in environmental policy in the Netherlands. This question is central to the following chapters, which will discuss the empirical evaluation of the planning-theory developments identified in this chapter, the related arguments, and the abstractions deriving from them. The discussion in the next section will focus on developments in area-specific environmental policy and the consequences of those developments.

Notes

1 Unless stated otherwise, the term 'actors' is used in this study as a collective term for individual 'actors', groups of 'actors', institutions and institutional groupings.

2 Voogd (1986, 1995a) makes somewhat similar distinctions: spatial orientation, decision-making orientation and action-orientation.

3 According to Van Vught, the distinction between policy and planning lies in "the observation that planning is a means of underpinning policy. Planning gives policy a systematic character" (Kreukels 1980; 77). Van Vught describes policy as the search for goal-oriented answers to perceived problems, while planning can be seen as "the systematic reinforcement and co-ordination of decisions that are intended, [...] to deal with problems. [...] Both 'policy' and 'planning' are geared towards solving problems" (in Kreukels 1980; 77).

4 This emphasis on choice in planning does not mean that the definition, implementation and evaluation of policy are not equally important elements in the cyclical planning process (see, for example Faludi 1987 and Kreukels 1980; 56).

5 Pluralism is also a term used in sociology and political science to refer to the American empirical democratic model (conventional pluralism) that stood for 'corporate capitalism' in the 1950s and 1960s. This meaning does not apply here.

6 Post-modernism is a reaction to the modernist movement that heavily influenced

Western thought following the Renaissance in the 17th century. 'Modernism' is the faith in development and progress by means of technology and science, and the belief in absolute truth. According to Harper and Stein (1995; 233), "Overemphasis on this aspect led to scientism (the claim that science and its method is the only source of knowledge), positivism (the claim that only empirical knowledge is valid), foundationalism (the belief that absolute foundations for universal truth can be established), and absolutism". Milroy identifies four dichotomies that characterise post-modernism: "It is *deconstructive* in the sense of questioning and establishing a sceptical distance from conventional beliefs and, more actively, trying both to ascertain who derives value from upholding their authority and to displace them; *antifoundationalist* in the sense of dispensing with universals as bases of truth; *nondualistic* in the sense of refusing the separation between subjectivity and objectivity along with the array of dualisms it engenders including the splits between truth and opinion, fact and value; and *encouraging of plurality and differences*" (1991, in Healey 1992; 146). The present study will concentrate on the latter two aspects, without establishing any further link to post-modernist thought.

7 See Chapter 3, endnote 1.

8 In addition to planning *concepts*, there are also planning *doctrines* (Alexander and Faludi 1990, Korthals Altes 1995). With concepts there is a general awareness of assumptions, which are usually open to discussion. A concept could be said to have become a doctrine when those assumptions are accepted to such an extent that they are considered as implicit and given. Doctrines are therefore assumptions that have become prejudices and automatisms, and have become part of the basic principles that are broadly supported by society. Faludi and Van der Valk take the Dutch planning tradition as an example: "the Dutch [...] love order and neatness and, if the need should be, they call upon the authorities to maintain them [...]. Order is the rule, and rulers are called to maintain order. This is the key to Dutch planning [...]. The point is, rule and order is not imposed from above, it pervades the Dutch way of doing things" (1994; 7-8). Dror speaks of 'megapolicy assumptions' (in Alexander and Faludi 1990; 7): "policy belief systems of an order higher than the discourse about operational policies, plans, programs and/or projects". These beliefs are not usually articulated, but they influence the decision-making and planning processes. The term 'doctrine' refers to implicit beliefs that are more or less fixed in terms of the planning process. Chapter 1 points to the fact that *change* is also often seen as implicit, for example in terms of the cyclical movements that influence the abstract principles applied by central government.

9 Lim (1986) applies a classification that is more or less comparable to the 'dimensions' applied by Zonneveld. Lim proposes that planning be seen as a "threefold activity requiring technical, intersubjective, and critico-ethical dimensions of theory building and practice" (1986; 75). He restricts the subjective dimension to the *ethical* aspect, while Zonneveld keeps to the *aesthetic* aspect. In this study we discuss the *rational* aspect of planning.

10 In addition to this basic theory, various authors distinguish between different levels as a foundation for reality, issues and visions (see, for example, Boulding 1956 and Goudappel 1996). Pirsig (1991) also refers to levels, however, he does not refer to the levels of systems but to levels of quality: material, biological, social and intellectual quality. Braudel (1992; 21-22) distinguishes between timescales: 'geographical time',

the much shorter 'social' timescale, and the shortest timescale: 'individual time'. The psychologist Maslow identified a hierarchy comprising five levels of needs that motivate individual behaviour. Basic needs are met first, followed by 'higher' needs. Flood and Jackson (1991) apply Habermas's 'theory of knowledge-constitutive interest' (1972) as a basic principle of systems theory. Forester (1989), for example, uses Habermas's theories of intersubjectivity and communication as the foundation for pragmatism, a philosophical theory developed by Blanco (1994) for planning-theory discourse. Others adopt the ideas of, for example, the French philosophers Derrida and Foucault. Foucault's work focuses on the relationship between knowledge and power (see also Harper and Stein 1995, Innes 1995). The ideas of Habermas and Foucault in turn derive from "Aristotle's notions, especially about politics and practical reason (praxis and phronesis)" (Dryzek 1990; 9). Those 'notions' still apply, and are central to 'critical systems thinking' (Flood and Jackson 1991).

11 In his 'integral' vision of planning, Gillingwater applies a threefold approach that he sees as "an outline, or more appropriately, a sketch of what might constitute a more adequate theory for the nature and practice of planning, especially public planning. These subject areas taken together are, firstly, the concept of the process of planning ('internal'), secondly, the concept of the planning domain ('internal-external'), and finally, the concept of the social change ('external')" (1975; 60).

12 'Comprehensiveness' denotes the 'holistic' viewpoint that is common among town planners. It is related to holistic philosophy. Holism maintains that the constituent parts of an entity or system can be better understood by referring to the entity or system as a whole, rather than to its separate parts (Lui 1996). 'Comprehensiveness' denotes not only holism, but also reductionism and expansionism. Reductionism is the principle of analysing complex entities into simple constituents. Expansionism is the principle of explaining something by referring to a greater whole, i.e. the supra-system or milieu in which it exists (Kramer and De Smit 1991, see also Fig. 4.5).

13 See, for example, Friedmann (1973) and Van der Cammen (1979) for a discussion of this point.

14 Partidário and Voogd (1997) point out that "especially in environmental planning there is a growing concern that environmental problems cannot be treated in isolation, but that interrelationships between environmental problems as well as with social and economic problems should be included" (1997; 2, see also Miller and De Roo 1997, Voogd 1994). They refer to "advocates of holistic approaches such as Senge (1990) and Marietta (1995)" who argue that "only when we have a better grasp of the systems as a whole we can begin to understand how we can organize possible solutions for environmental problems" (1997; 2).

15 The planning process does not always follow a fixed sequence of formulation, recording, implementation and feedback. Korsten (1985), for example, refers to the distinction between forward-oriented policy development and backward-oriented policy development. This involves the regulation of actions by means of 'feedforward' or feedback (see also Noordzij 1977; 42). In the case of future-oriented action, first the policy intention or a policy goal will be proposed or defined, then the possibilities for implementation will be examined. A 'backward' approach begins by ascertaining "the potential for implementation and possible difficulties" (Dekker and Needham 1989; 4), followed by policy formulation. When planning relates to 'ongoing' processes, it is often not even possible to define a specific planning structure.

16 The Nijmegen planning community in particular places a great deal of emphasis on policy implementation (see, for example, Wissink 1986, 1987). The question of *what is required* to implement planning decisions is considered an important aspect of implementation. Emphasis is placed on the relationship between objectives, instruments and effects (Dekker and Needham 1989; 10, and Ringeling 1987).

17 Here, Mastop refers to the coordinative/communicative perspective.

18 From a completely different perspective that focuses on a government that learns as opposed to a government that takes action, Van der Knaap (1997) arrives at an interesting and comparable 'threefold theoretical approach: (A) learning through system-cybernetics, (B) cognitive learning, and (C) socio-collective learning. This categorisation also reflects the basic concept referred to in section 4.3 (see Fig. 4.2), according to which reality can be viewed. System-cybernetic learning primarily involves the feedback of information on discrepancies between policy objectives and results. Cognitive learning is primarily a question of improving understanding and acquiring knowledge by assimilating external stimuli into existing knowledge and accommodating (i.e. adapting) knowledge to external stimuli (Piaget 1980; 103). "Cognitive development is, as it were, driven by the repeated friction between external stimuli and internal cognitive schemes" (Van der Knaap 1997; 59). In short, it involves the constant inclusion and modification of theoretical frameworks that are subject-based, consistent and rational. The social dimension of learning involves learning *from* others and learning *with* others through interaction and communication. Van der Knaap's theory of a 'learning' government is based on the theoretical concept of social constructivism, which itself rests on the belief that there is no such thing as an objective reality; reality is a generally accepted subjective perception.

19 Despite the fact that this is becoming increasingly negotiable, in the first place this is usually the government, or parts of it, where necessary in relation to other participants in the social dialogue.

20 According to Kramer and De Smit, "systems theory leads to a world view that is governed by 'purposeful' conduct" (1991; 5), whereby probability, choice and the influence of a particular context are the key to understanding phenomena. From the point of view of methodology, systems theory is above all an approach: the systems approach. According to Von Bertalanffy, the systems approach is "attempting scientific interpretation and theory where previously there was none, and higher generality than that in the special sciences" (1968; 14). According to Noordzij, systems theory "identifies the similarities, and therefore also the differences, between various systems, and this has implications for the type of model or strategy used to analyse and resolve problems at different levels" (1977; 20-21). Von Bertalanffy has attempted to formulate a 'general systems theory' that functions as a form of 'umbrella' discipline that focuses on entities.

21 Cybernetics is the science of control and 'steering' (cybernetics derives from the Greek word for 'steersman': *kybernetes*) by means of control mechanisms and communication, with the emphasis on the transfer and interpretation of information and related feedback, in short: the 'behaviour' of systems independent of their environment (see Ashby 1956, Chadwick 1971, Kramer and De Smit 1991, Wiener 1948). Information and control are central to cybernetics. However, "by the 1960s cybernetics had lost its luster" (Horgan 1996; 207), because the strong belief in complete control through information was not borne out by actual results. Today, only the disciplines of

information technology and logistics apply cybernetics theory with any degree of success.

22 The systems approach considers an object not only in terms of its components, but also as an entity. It focuses on the arrangement of and relations between the components that connect to form the entity. The approach also seeks to explain open systems in terms of how the components and the entity interact with their environment. Systems theory is a departure from the positivist idea that an entity can be understood by reducing it to the properties of its parts or elements.

23 'Hard system thinking', as the opposite of 'soft system thinking', can be described as a system with clearly structured components and attributes that can be delineated, defined and quantified. This system can be used to formulate solutions and control methods, and generate forecasts. Emphasis is placed on optimising this system (see also Waring 1996).

24 One of the consequences of this development is that the material object of planning is too often overlooked. However, the fact that this study emphasises goal-directed action in addition to the other forms of action will help to ensure that the material object is considered in every case of planning-based action.

25 The concept of policy networks originated in the 1960s and 1970s in interorganisational sociology and political science, and was 'discovered' by management science in the 1970s (see Klijn 1996).

26 Elster argues that "consistency, in fact, is what rationality in the thin sense is all about: consistency within the belief system; consistency within the system of desires; and consistency between beliefs and desires on the one hand and the action for which they are reasons on the other hand" (1983; 1).

27 The empiricist Locke stated that "[...] there appear not to be any ideas in the mind before the senses have conveyed any [...]" (in Störig 1985; 53). This does not mean that knowledge is the observation of a fact 'in itself'. The intellect, or the mind, and – provided it relates to consistent theories and tenets – rationalises or 'welds together' facts into experiences, insight, understanding, and ultimately knowledge. Kant pointed to the intellectual processes that connect facts and phenomena; logic has expressed those processes in a basic form since the time of Aristotle. According to Kant, all experiences are established through thought forms by means of which raw materials provided by the senses (including causality) are processed; it is therefore clear that we will encounter these forms in all types of experience (Störig 1985; 66). Rationality is therefore the functional-consistent theory that bridges the gap between facts and conceivable, real concepts, relationships and – if necessary – solutions.

28 For a critique, see for example: Etzioni 1968, Faludi 1987, and Popper 1961.

29 This rationality relies heavily on the idea of the need for a cause-and-effect relationship. This idea found a strong foundation in Newtonian dynamics, a view based on physics that strongly influenced the discourse on the social sciences (see De Vries 1985) despite the fact that, during the twentieth century, the belief in direct causality as the all-determining principle in physics eroded as a result of new developments and insights. The second law of thermodynamics was not fully compatible with Newtonian dynamics. At the microscopic level, quantum mechanics is replacing the laws of Newtonian dynamics and, at the level of the universe, relativistic physics is pushing out classical dynamics. Heisenberg's Uncertainty Principle makes it clear that probability exists in addition to necessity, while the second law of thermodynamics shows that classical

dynamics is limited to equilibrium situations. These conditions are the exception rather than the rule, in a world that is complex, often unstable and rarely in equilibrium. In addition to a 'mechanical' approach, physics also appears to require a 'statistical' approach. In most cases, we are only able to predict probable situations (statistic causality) (see also Prigogine and Stengers 1990). This method has been used for a long time in the social sciences, not so much on rational-theoretical grounds, but on practical grounds, 'in order to be able to make statements.'

30 Etzioni has developed this concept into a step-by-step strategy that is now known as the 'mixed-scanning approach' (Etzioni 1968). Mixed scanning is "a hierarchical mode of decision making that combines higher order, fundamental decision making with lower order, incremental decisions that work out and/or prepare for the higher order ones" (Etzioni 1986; 8, see also Goldberg 1975; 934).

31 In Charles Lindblom's work 'The Science of Muddling Through' the author discusses the incremental nature of policy: policy decisions are continuously reviewed, and the policy formulation and implementation is systematic only to a limited extent. Cohen, March and Ohlsen (1972) refer to this method as 'garbage-can decision making' (see also § 4.5): decisions are usually made by chance and in retrospect it is difficult to identify who intended the outcome.

32 Hofstee (1996) points out that citizens are taking over the lead from policy-makers when it comes to influence, largely because the intelligence of the general public is undergoing a 'spectacular' increase, and labour extensivisation is creating more leisure time. The influence of the individual is decreasing, however, as the assertiveness of the population as a whole increases as a result of general opinion-forming and collective decision-making (1996; 50). Personal and psychological factors also play a role, although, according to Hofstee 'personal idiosyncrasies' remain a hidden dimension in decision-making processes. "Some people are more intelligent, more aggressive, better informed, more fortunate, and more assertive – in short, have more influence – than others" (1996; 50).

33 Harper and Stein point out that rationality and irrationality should not be seen as absolute dualisms. They suggest that "this wide conception [of rationality and irrationality] can be viewed as a continuum [...]. Moreover, the various instances of rationality (and irrationality) are not fixed by an essence and will form [...] an overlapping family of resemblances" (1995; 237). Goudappel (1996) also assumes that the irrational and the uncertain are 'natural' phenomena with their own characteristics and laws. This is why we should not be satisfied with "the claim that the logic of decision-making can fail" (1996; 62).

34 Uncertainty diminishes as the decision-making process progresses from concept, through conclusion, to implementation (Kastelein 1996).

35 Goal rationality is the idea that choices can/should be based on logic and make a positive contribution to achieving a logically defined goal.

36 In this form of rationality, the planner is seen as an expert and technician who, according to Friedmann (1973), sets out, manages and controls the elements of the planning process in an elitist manner. Underlying this process is a functional step-by-step, cause-and-effect mechanism.

37 This requires not only an eye for detail, but also an eye for the whole and the context in which it arises. With this in mind, Mannheim calls for a "holistic view through planning [which] would lead to greater sanity and order" (Friedmann 1973; 30).

38 For a discussion of this point, see Van der Cammen (1979), Faludi (1987), Friedmann (1987) and Verma (1996).

39 An important criterion for communicative rationality is the acceptance of a shared problem definition. According to Kooiman (1996; 41), this has the disadvantage that 'potential alternatives' are ignored or defined away. De Neufville and Barton (1987) share this view. They argue that, with regard to decision-making and problem definition, there is a level of 'granted knowledge', which originated from myths and ideas about 'public issues' and "shared moral evaluations, events and possible solutions to problems" (1987; 181). Every subject, every group of subjects and every organisation has its 'own' rationality (see also Berting 1996) and irrationality, which are based on historical, aesthetic and ethical (i.e. cultural) preconceptions. This individual rationality is not always in line with society's perception of rationality (Barry and Hardin 1982) or with outcomes that are subject to political rationality (Friedmann 1971). Faludi (1987; 52) also refers to the political influence on decision-making, and the political role of planners in the process. In the words of Forester (1989): "if planners ignore those in power, they assure their own powerlessness" (1989; 27) and "to be rational, be political" (1989; 25). These are very pronounced opinions on the political/administrative factor in planning. But the opposite is also true: "Planners often find themselves acting as educators for elected officials who are unaware of the potential impacts of their planning decisions" (Kaiser et al. 1995; 16, see also Lucy 1988).

40 Sociology has adopted the phenomena 'institution' and 'institutionalisation' as central elements in the discipline. The French sociologist Durkheim describes sociology as a science of institutions, of their evolution and functioning. Sociology focuses on all theories and patterns of behaviour that are 'constituted by a collectivity' (Durkheim 1927). According to Smith (1963), institutions are "patterns of interaction and social relationships that characterise given social groups and social systems" (in Buiks 1981; 266, see also Hufen and Ringeling 1990).

41 Here, 'position' can be explained in terms of role (functional task) and influence (functional performance).

42 This philosophy transcends the scientific disciplines but also connects them. It is a universal philosophy of development and progress, with Nobel Prize winners among its strongest adherents (Gell-Mann 1994, Prigogine 1996, Arrow, in Waldrop 1992).

43 Complex systems are not necessarily less stable than relatively simple systems. Complex systems can also be stable. The theory is based on a 'global' stability consisting of irregular patterns, with a degree of local stability (Gleick 1987; 50-51).

44 In his essay 'Complexity; the emerging science at the edge of order and chaos' (1992), M. Mitchell Waldrop describes the creation, visions and progress of a multidisciplinary team of researchers comprising economists, informatics specialists, biologists and physicists. Focusing on the 'science of complexity', the team works under the name of the Santa Fe Institute for Studies in the Science of Complexity (http://www.santafe.edu/sfi/aboutSFI.html). Waldrop's essay brings together ideas and principles on complexity. However, the book lacks a sociological perspective. The research described in the essay reveals a universe in which order and chaos occur simultaneously, interchanging and interacting. Development and progress are to be found precisely at the 'edge of chaos, i.e. the point between chaos and stasis: "it turns out that the maximum fitness is occurring right at the phase transition" (1992; 313).

45 In his book *Lila*, Pirsig, who is more of a philosopher and essayist than a proponent of complexity theory, discusses the interaction between static and dynamic quality as a condition for development and progress. He argues that static quality can be found in fixed lifestyle patterns, and represents stability and experience (Giddens and Vroom refer to 'routine', see § 4.7). However, where a dynamic quality is lacking, static quality will ultimately lead to rigidity. Dynamic quality – the 'undefined best adaptation', to use a Darwinian phrase, is the phenomenon that presents itself to us as the source of everything that is new and *ad hoc* – represents creative ability, change and evolution. Dynamism and static quality are mutually dependent because "a step forward [...] is meaningless unless a static pattern can be found to prevent the progress from reverting back to the conditions that prevailed before the step was taken" (1991; 160).

46 According to Waldrop, the complex reality of static and dynamic quality (Pirsig 1991) is "the result of simple rules unfolding from the bottom" (Waldrop 1992; 329). This reality can be understood by searching for a number of simple rules that distinguish it from chaos.

47 Philosophers have pondered equilibrium and change for centuries. Heraclitus wrote that the unity in our world is born of diversity; a harmony resulting from a tension of opposites that leads to a world in which we can expect a never-ending cycle of change (in Russell 1995; 71). Plato, on the other hand, claimed that our perception of the world is based on a perception of an unchanging reality: the idea, or *form* (Aufenanger 1995; 31-32).

48 The economist Boulding (1956) expresses these higher levels of development in a system hierarchy, which he uses to argue that mechanical and linear theories do not explain the existence of cells, human beings and societies. He identifies nine levels of increasing complexity. The lowest level of the hierarchy is the static framework. On the second level, dynamism is added through motion. The remaining levels are control systems, self-maintaining and organic systems. The levels finally converge into social systems as we know them. This is a systems-theoretical view that will only find limited support in the discourse on complexity, which argues that each level has its own order, dynamic and complexity (see also Bahlmann 1996, Chadwick 1971, Jackson and Keys 1991, Kramer and De Smit 1992).

49 On the basis of such views, Bahlmann (1996) argues that unstable – and therefore complex – situations do not necessarily have a disastrous effect on decision-making and organisations but can also lead to development. "A high level of disequilibrium is characterised by contrasts and conflict, not by harmony and consistency. Disequilibrium can therefore generate new perspectives and forms of behaviour" (Bahlmann 1996; 92, see also Stacey 1993) and can be a source of innovation that is lacking in static systems.

50 In colloquial speech, the term 'chaos' usually has a broader meaning than in the context of complexity theory. The term 'very complex' will therefore be used instead.

51 Here, De Bruijn and Ten Heuvelhof are assuming a pluriform society, whereas this quote is used to support the vision of the pluriform simplification (and structure) of a complex society.

52 This is a theoretical extreme that is apparently not completely utopian. The most inspiring example is in the United States, where cities appear to be in the grip of the car and its users. But there is also Disney World, Orlando, 'the capital of unreality' and probably the most centrally planned and governed city in the world (Kunstler 1993).

53 Clear parallels can be drawn between the planning styles advocated by Friedmann and

the planning strategies referred to in the scheme of alternatives for planning-oriented action. The parallels exist in spite of the different purposes underlying the categorisation by style and strategy. Friedmann's structure was devised "using the distribution of power in society as the principal criterion" (1973; 83). The discussion in this study of the three perspectives on planning action focuses on another theme: the degree of complexity. Within the constellation of individual perspectives on planning action, 'power' is a theme that can be categorised as institution-oriented and it influences decision-oriented and goal-oriented aspects.

PART C
INTERACTION AND CHANGING GOALS IN AREA-SPECIFIC ENVIRONMENTAL POLICY

Chapter 5

The Standardisation and Institutionalisation of Environmental Policy

From a Technically Sound Policy to Policy Based on Shared Governance

> The challenge, therefore, is to obtain reasonable assessments of the complexity and stability of the environment, and to design project management processes and organizational structures to fit more closely with environmental conditions. Although judgmental, the assessments can indicate the degrees of uncertainty, stability and complexity... (Rondinelli, Middleton and Verspoor 1989; 49).

5.1 Introduction

Over three successive decades, structured environmental policy in the Netherlands has developed into a full policy area with its own characteristics and interrelations. As we have seen in Chapter 1, these interrelations give environmental policy a goal-oriented character. Environmental policy therefore has a dominant influence, particularly on spatial planning. It is not surprising, then, that this meets with resistance from spatial policymakers, for whom striking a balance between the various interests involved has almost become a goal in itself. Partly for this reason, the goal-oriented approach to environmental policy has come under pressure – "standards systems have always been *the* instrument for shaping environmental policy" (Kuijpers, in Bakker 1997; 5). The charm of the simplicity that characterised environmental standards policy had to 'hold its own' against the undeniable complexity of environmental/spatial conflict.

The interaction between environmental regulation and spatial planning is open to discussion – and rightly so, given the objectives of environmental and spatial policies. Both, after all, aimed to ensure that the quality of the physical environment remained as high as possible. It was therefore logical to 'bundle' the capacities of these policies. Government and authorities saw the co-ordination and integration of spatial planning and environmental policy as a logical and useful development, within which the principle of subsidiarity became a *leitmotiv*. In addition, there was increasing recognition of the need for greater involvement of interested parties (participatory policy implementation) and local circumstances would carry more weight (area-specific policy implementation). These principles were not unknown in spatial planning policy. This is not true of environmental policy, however, which would have to undergo the greatest modification in terms of co-ordination and integration.

The environmental planning approach of the 1990s – so the government and authorities had jointly decided – should be communalised (§ 1.2). At the same time, steps were being taken to decentralise and deregulate policy (§ 1.3). Environmental policy was also influenced by experiences of actual environmental/spatial conflicts at local and regional levels (see Chapters 2, 3 and 6). As a result, environmental policy in the Netherlands appeared to move from one extreme to the other in a relatively short space of time, namely from 'a technically sound policy to a policy built on social consensus ...'.

The process of co-ordination and integration, and of communalisation and decentralisation, was implemented in the expectation that policy would be more effective. McDonald has pointed out, however, that integrating environmental policy and spatial planning would benefit the strongest most: "Mainstream planning – with its emphasis on economic development, infrastructure, land use planning, and development control – has an uneasy relationship with [...] emerging fields. [...] Environmental planners are responsible for piecemeal application of environmental impact assessment, pollution control, and environmental quality standards. Integration has suffered and the chance for integrating economic, social and environmental factors has reduced". (McDonald 1996; 232, see also Slocombe 1993; 290). This lack of enthusiasm for integration in the United States, where spatial planning and administration is quite unlike that in the Netherlands, should have been taken as a warning. Environmental and spatial planning (i.e. protection and development) cannot simply be bundled together in the same category.

Above we have seen how, in the Netherlands, the straightforward measure of maintaining a given physical distance between environmentally harmful and environmentally sensitive areas was hampered by the demand for space for a diversity of activities. In such a situation it is far from easy to improve co-operation and co-ordination between environmental and spatial planners. In some cases, restricting environmental impact to a permitted level in an urban area was simply not a realistic goal. It is not surprising, then, that the discussion on combining the capacities of environmental and spatial policy arose partly out of the rapid developments in spatial planning, which will continue far into the twenty-first century (see Chapter 3). There was a need for flexibility and nuances in environmental goals in order to safeguard spatial and economic development. Striking the right balance between these interests is not only a question of what co-ordination and integration would achieve, but also who would benefit.

This chapter focuses on development and decision-making in environmental policy – a policy that is in a process of transition. The change process was not – and still is not – a smooth one. It was and is therefore difficult to predict. However, a certain pattern can be seen in the changes. This chapter will also consider these changes in order to ascertain what they could contribute to solution strategies for environmental/spatial conflicts. The approach to environmental/spatial conflicts depends to a large extent on the changing way in which the policy fields of environmental and spatial planning are organised and co-ordinated.

Section 5.2 will discuss the first steps towards structuring environmental policy, taken at the beginning of the 1970s. Ever since a separate environmental policy was first implemented in the Netherlands, it has focused on environmental protection and the 'grey environment' (Chapter 2). This chapter deals not only with the general principles of policy, but also with the role of environmental standards. Standards are the keystone of Dutch environmental policy. Ambient standards largely determine the goal-oriented nature of environmental policy and set the framework for other forms of policy. Within a relatively short space of time at the beginning of the 1980s, it became clear that the first steps towards structuring policy required modification (§ 5.3). At this point, policy integration became an important theme. This development had virtually no effect on ambient standards. The spatial implications of standards systems and zoning did not become apparent until ambient standards, translated into integral zones, eventually became part of the policy integration process (§ 5.4). The introduction of the ROM designated-area policy placed greater emphasis on participation – including that of non-governmental agencies – and an integrated area-specific approach. ROM was considered a possible alternative to ambient standards, which were perceived as inflexible and restrictive. The standardisation and institutionalisation of environmental policy came under discussion in the light of these developments, which were concentrated in an 'area-specific' strand of environmental policy (§ 5.4). In the 1990s, this policy was prominent in discussions on the relationship between environmental policy and spatial planning. Spatial constraints imposed by inflexible and hierarchic environmental regulation have since led to new thinking that represents a shift in environmental policy. The new thinking is mainly reflected in changes in the regulation of noise nuisance and soil-remediation policy (§ 5.5). The question that comes to mind when we consider these developments is this: environmental policy is adapting to a changing world, but at what cost?

This analysis of environmental policy in the Netherlands is basically chronological. The material results of the analysis are structured according to the three approaches to planning action described in Chapter 4: decision-oriented, goal-oriented and institution-oriented approaches. The framework for planning-based action presented in that chapter is used to analyse the relationship between developments and new thinking in area-specific environmental policy. In line with the vision developed in Chapter 4, this study is based on the premise that changes and adaptations in environmental policy are due to the fact that the awareness of the complexity of issues (i.e. environmental/spatial conflicts) is greater than ever before.

5.2 The 1970s and the 'Limits to Growth'

In its controversial report, *The Limits to Growth* (Meadows et al. 1972), the executive committee of the Club of Rome stated that "It is only now that, having begun to understand something of the interactions between demographic growth and economic growth, and having reached unprecedented levels in both, man is forced to take account of the limited dimensions of his planet and the ceilings to his presence and activity on it. For the first time, it has become vital to inquire into the cost of unrestricted material growth and to consider alternatives to its continuation". The rapidly growing Dutch population had worked hard to erase the impact of the Second World War. Now that, after immense effort, war economies had been transformed into advanced social systems, economic and social progress had become the most important goal, to the greater glory of human ability.

More than ever, society's main aim was progress. However, as it was concluded in 1972, that progress had almost reached its limits. The rapidly increasing population was becoming more and more demanding, while the physical space in which society exists existed was increasing only marginally or not at all. Moreover, the natural resources used to produce material needs are finite. Consumption and production grow and change all the time, and produce mountains of waste and environmentally harmful emissions that have not only a local, but also a global impact.[1] The Club of Rome called on us to rethink and control our social systems, which are so heavily oriented towards development and progress and that they fail to consider the negative consequences: "We affirm finally that any deliberate attempt to reach a rational and enduring state of equilibrium by planned measures, rather than by chance or catastrophe, must ultimately be founded on a basic change of values and goals at individual, national and world levels" (Meadows et al. 1972; 187).[2] With the publication of the Priority Policy Document on Pollution Control, the Dutch government appeared to agree with this.

The Priority Policy Document on Pollution Control: Standards as a Foundation of Policy

In a letter dated 4 July 1972, the Minister for Public Health and the Environment, Stuyt,[3] presented the Priority Document on Pollution Control (VM 1972) to the Speaker of the Second Chamber of the Dutch parliament (*Tweede Kamer der Staten-Generaal*).[4] The publication of the Policy Document was preceded by the introduction of two sectoral environment acts: the Pollution of Surface Waters Act (WVO, 1 December 1970) (Smit 1989; 119) and the Air Pollution Act (*Wet Luvo,* 29 December 1970 - 18 September 1972, Michiels 1989; 157). These were the first of a whole series of sectoral Acts designed to protect the environment against the excessive negative consequences of human activity.

The Priority Policy Document set a framework for policy that had hitherto focussed only in part on environmental protection, and in an *ad-hoc* way. Environmental problems were not dealt with until they became fully apparent, and were primarily considered in the context of public health.[5] If the scale of a problem warranted measures, it was dealt with, but little consideration was given to 'cause and effect'. Policy geared towards such problems was described as 'caring for the land, air and water' (Grondsma 1984). As wealth increased sharply during the 1960s, production and consumption grew accordingly. Spatial planning, which at that time was oriented towards development, could not regulate the pressure of environmental impact on the physical environment.[6] The need for a structured environmental policy has increased as we have discovered more about environmental spillover and as the number of environmental problems continues to increase (see also Chapter 2). Despite the warning signals at the beginning of the 1960s with regard to excessive environmental load, De Koning points to the years between 1968 and 1972 as the period in which there was a real breakthrough in thinking on environmental protection and the consequences of pollution for man. This breakthrough led to the publication of the Priority Policy Document, which heralds the beginning of environmental planning.[7]

The Priority Policy Document identified a number of phases in the development of environmental policy. It stated that the *corrective* phase of policy had already progressed to a transitional phase characterised by a shift towards a more active environmental policy focusing on remediation policy. This policy would have to comply as far as possible with the 'Polluter-Pays Principle'.[8] At the same time, the policy was designed to prevent environmental problems occurring in the future, thereby anticipating the 'standstill principle' that later became a principle of environmental policy.

Despite the fact that the Priority Policy Document was based on the healthcare principle, the ecological context was already becoming clear. This was an important step in relation to the situation in the past. The approach to the ecosystem was compartmentalised into air, water, soil and the organisms they contain. This structure was chosen partly out of a sense of urgency, which was a further reason for adopting a more integrated approach (Mol 1989; 28). This compartmentalised approach was evident not only in the Priority Policy Document, but is also reflected in the legal framework of environmental policy. The WVO and the *Wet Luvo* had already been implemented and the Noise Abatement Act and the Soil Pollution Act had been announced.

The Priority Policy Document established the structure for the near future on the basis of a compartmentalised and – in line with public-health policy and in anticipation of the first EC environmental action programme[9] – standard-setting approach. "Concern for health as a basic principle, supported by balancing socio-economic costs and benefits on the one hand, and scientific advice and the priorities of society on the other can, globally speaking, be defined as the foundations of environmental standards. The further functioning of this standards system will be an important task in the years to come" (VM 1972; 2). In 1971, in a letter to the Second Chamber, Prime Minister Biesheuvel had already expressed the view that the environment should not be the exclusive responsibility of the Minister for the Environment and Public Health, although the coordination of the standards-setting procedure for acceptable levels of pollution and nuisance should be part of the policy area of his ministry (Tan and Waller 1989; 21). The Department of the Environment undertook to formulate general environmental standards with a view to creating a basic level of quality throughout the Netherlands. Here, environmental policy differed from policy relating to water management, spatial planning and nature, where area-specific policies were the order of the day (see also Biezeveld 1992). This approach had implications for the working relationship between the Department of the Environment as representative of the government on the one hand, and regional and local authorities on the other. The latter would play an executive role in realising general environmental standards formulated by central government. Nevertheless, the Priority Policy Document shifted the emphasis of environmental policy from *ad-hoc* intervention towards a compartmentalised policy based on general standards.

The 1976 Policy Document on Ambient Environmental Standards (PDAES)

The structuring of environmental policy in the Netherlands resulted in a growing need for environmental standards at lower levels of government, in industry and in environmental organisations. The standards policy, particularly when based on figures, was expected to ensure legal certainty and equity (VM 1976). This role was elaborated in the 1976 Policy Document on Ambient Environmental Standards (PDAES; *Nota milieuhygiënische normen*, VM 1976). This is an interesting document, particularly in the light of the heated discussions on standards during the 1990s, from which we can identify the objections to standards-based policy that were voiced back in 1976.

Opting for ambient quality standards, particularly when expressed in figures, automatically involves striking the right balance between many different and diverging interests when formulating policy: "The specified value, and therefore the degree of risk, is determined by evaluating the possible harmful effects of the pollution on the one hand, and the sacrifices society must make to reduce that risk on the other. This is a political choice," the Policy Document (VM 1976) points out. Furthermore, because this choice is political, the focus and location of standards (i.e. the limits to what is considered acceptable) will shift over time. The speed at which targets are tightened up and have to be realised can also be the product of a process of balancing various interests. Standards are defined as "general rules that are binding to a certain extent, expressed in quantitative terms or otherwise" (VM 1976). National *numerical standards* can therefore be seen as *general regulations expressed in quantitative terms*. The PDAES sets out the following policy principles in addition to – or perhaps in preference to – numerical standards:

- as far as possible, pollution must be prevented and/or dealt with at source;
- in general, a Best Available Technology approach must be adopted;
- the *standstill principle* must be applied.

On the basis of these principles, environmental-protection policy aimed to reduce environmental pollution resulting from human activity by "the co-ordinated use of a number of instruments, including physical regulations, which include ambient environmental standards" (VM 1976; 6).

The PDAES defines five types of standard designed to combat various aspects of pollution occurring between the source (i.e. the emitting installation) and the recipient (i.e. the object of protection):

- procedure and production standards: standards for fixed installations/products;
- discharge or emission standards: standards relating to emissions at source;
- immission standards: standards for the immission of pollutants in a recipient area, function or object;
- quality standards: environmental standards relating to the condition of an area, function or object;

- exposure standards: standards relating to the level of pollution to which individuals or populations are exposed (based on VM 1976; 6).

Procedure and production standards derive from the product-oriented and source-oriented thread of environmental policy (see § 5.3). Emission standards are also source-oriented in terms of policy. This group of environmental standards are not specifically intended to have consequences for spatial planning. However, this does not mean that the impact of emissions on the spatial environment is regarded as unimportant. Models are increasingly used to convert emission figures into immission figures, with a spatial and area-specific interpretation.

It was mainly the standards controlling immissions and quality that had consequences for spatial planning. Immission standards were designed to control the quantity of pollutants that encroach on an area, function or object in a given time period. Immission standards *à la lettre* are pollutant-specific, whereas quality standards reflect a desired situation for an area, function, object, individual or population. Immission and quality standards support and specify targets for environmental policy (see § 5.3).

The role of exposure standards in environmental policy has many aspects. Because these standards do not relate to ambient environmental quality, it is inappropriate to focus on emissions, immissions and quality. Exposure relates mainly to humans, e.g. exposure to pollutants in drinking water and food. Exposure to radiation is also important in this context.

Immission and quality standards – and, indirectly, emission standards – are used to monitor the levels of pollution to which areas, functions, objects, populations and individuals are exposed. This form of monitoring will, where necessary, result in source-directed and/or area-specific measures. Consequences for spatial planning can be identified on the basis of immission and quality standards – and possibly on the basis of emission standards – in order to define, aim for and attain the desired level of quality in the ambient environment.

Selecting standards as an environmental policy instrument and a criterion for human activity was a logical consequence of the growing awareness that many environmentally harmful activities should be restricted. Solutions were required for problems of general concern, but for which there was no policy infrastructure in place. In such cases, measures in the form of standards can be easily and quickly understood and implemented, and can provide short-term results for a specific problem. It is no surprise, therefore, that the Priority Policy Document and the 1976 PDAES focus on the goal-oriented aspect of policy. The statement in the PDAES that "Quality and immission standards are not only instruments of environmental policy, but can also be seen as goals in themselves" (VM 1976; 10) requires no further explanation. The most concrete and detailed application of this standpoint is probably to be found in the Noise Abatement Act (*Wet geluidhinder*).

The Noise Abatement Act: a Centralised Policy Framework for Defining Standards

After the first wave of environmentalism, which reached a high point in 1972, the tide ebbed somewhat due to the Oil Crisis in 1973 and the subsequent economic recession. Nevertheless, the legal framework for environmental policy continued to evolve rapidly. Before the 1970s, this framework had largely been determined by the Nuisance Act.[10] A Nuclear Energy Act had also been implemented, and the echoes of the fanfare announcing the implementation of the Pollution of Surface Waters Act (WVO) and the Air Pollution Act (*Wet Luvo*) could still be heard. These were followed by the Chemical Waste Act in 1976, the Waste Substances Act in 1977[11] and the Noise Abatement Act in 1979. The flood of legislation continued into the 1980s. It was centralised and compartmentalised into soil, water and air, and sometimes also geared towards specific threats to the environment (waste, radiation, and noise). The Soil Protection Act was not implemented until well in the 1980s (1986), despite the early reference made to it in the Priority Policy Document. The environmental legislation referred to here was framework legislation that was designed to be developed into specific regulations and standards, usually by means of a governmental decree (AMVB). The Noise Abatement Act was an exception to this.

It is somewhat strange to point out that "noise nuisance was first acknowledged as a problem in the residential areas close to Schiphol" (Tan and Waller 1989; 13). Even stranger is the fact that the Noise Abatement Act contains no provisions relating to aircraft noise, and that the National Environmental Policy Plans (NMPs) of the 1990s make no reference at all to the problem of aircraft noise. The political debate was characterised by an apparently perpetual love-hate relationship with the consequences of aviation-related activities for environmental planning, and environmental policy sidesteps the issue of what came to be known as 'Mainport Schiphol' in the 1990s (see § 5.4). On 28 December 1961, the Minister for Transport, Public Works and Water Management requested advice on aircraft noise. 1967 saw the publication by the Advisory Committee on Noise Nuisance from Aircraft (aka the Kosten Committee) on the consequences for public health of aviation-related noise nuisance. The report prompted Kruisinga, then Secretary of State for Social Affairs and Public Health, to request the advice of the Health Council of the Netherlands on how to combat all forms of noise nuisance (Grondsma 1984). Noise nuisance was increasingly recognised as a national policy issue.[12]

The instrument of national environmental zoning based on environmental standards[13] was introduced not in response to the Noise Abatement Act, but following amendments to the Aviation Act in 1978[14] (Van Kasteren 1985). This formalised – at least on paper – the direct relationship between environmental standards and spatial planning. The amendment of November 1978 – to Articles 25, 26 and 27 in particular – specified that airports should be zoned. In 1972, De Kroon still expected that "the short-term definition of noise contours will offer sufficient protection for local residents, so that noise nuisance can be kept within acceptable limits" (in Willems 1988; 102, see also Willems 1987). That the amendment could be accepted in all reasonableness is evident from a remark made in that same year by Kruisinga (State Secretary for Transport, Public Works and Water Management) to the effect that, at the insistence of the Second Chamber, he wanted to establish a zoning system in the short term (in Tan and Waller 1989; 22). The sense of urgency meant that Kruisinga did not wait for the Noise Abatement Act to be implemented, but instead submitted amendments to the Aviation Act. However, until well into the 1990s, no progress was made beyond indicative zones, presented as policy proposals by the Ministry of Transport, Public Works and Water Management. This meant that the KE standards could be used as a guideline where no zone existed, but the provisions relating to limit values within a zone did not apply (Otten 1993; 283). Consequently, the zoning policy for aviation-related noise remained extremely flexible for many years (see also § 5.4).

On 13 February 1979, the First Chamber approved the Noise Abatement Act.[15] This is a unique Act, for a number of reasons. It contains definitive standards, but also specifies how the standards are to be translated into spatial/physical zones, and which instruments can be used for this purpose. The Noise Abatement Act was a piece of environmental zoning legislation and was therefore unique in comparison to other environmental legislation, with the exception of the Aviation Act. The Act had a strongly programme-based approach, which was reflected in the indicative long-term planning and remediation programmes of the 1980s.

Environmental zoning in general, and noise zoning in particular, lead to conflict between environmentally harmful activities and environmentally sensitive functions. In theory there are two possible scenarios. First, the area identified as subject to pollution does not overlap with existing and/or potentially environmentally sensitive areas. In the second scenario these areas *do* overlap. The second scenario constitutes an environmental/spatial conflict and environmental protection measures are required in order to avoid spatial consequences (Borst et al. 1995).

The Noise Abatement Act was perceived by many as a 'monster of an Act' that set out the permitted levels of noise nuisance from traffic and industry (Braak 1984, *Commissie Evaluatie Wet Geluidhinder* 1985, Van Dongen 1983). The Act was heavily criticised.[16] Although it had only recently come into effect, the decision was taken to review the Act with the exception of one or two sections (Otten 1980; 234). The Noise Abatement Act strongly influenced the development of effect-oriented environmental policy in the 1980s. In 1989, Minister Nijpels (1987-1990) stated that the Act "anticipates other environmental aspects of the relationship with spatial planning and housing. The regulations relating to standards, zoning and planning constraints are very important for those policy areas and call for close co-operation and co-ordination" (1989; 3). There are good reasons why the Noise Abatement Act set the course for standard-based environmental policy in the 1980s:

- its standards were quantitative;
- its national standards were formulated by central government (i.e. formalised in a political or administrative context);
- its standards were adapted to different situations;
- its standards reflected the interests of a single aspect of the environmental policy sector;
- the standards were coercive and provided a framework for other forms of policy.

The Noise Abatement Act set out the consequences of non-compliance with standards and the requirements relating to the noise zone. Environmental requirements and the spatial consequences of noise-pollution emission were therefore brought together in a single Act (Brussaard et al. 1993, Neuerburg and Verfaille 1991). Although it was not the main purpose of the legislation, this aspect of the Noise Abatement Act acted as a stimulus to the subsequent aim of bringing spatial and environmental policy closer together through co-ordination and integration. The Act should therefore be seen as having contributed to subsequent developments in the internal and external integration of environmental policy (Nijpels 1989).

The departmental integration of environmental protection and spatial planning did not proceed smoothly. In the words of Kroese, former Director-General of the National Spatial Planning Agency (RPD): "It was a point of principle. Conventional thinking in spatial planning was based on the theory that balancing all the interests involved would produce the best result. At the Department of the Environment, however, policymakers believed that the maximum permissible limit value for noise was non-negotiable, which meant that it would have to be a boundary condition for the land-use plan" (in Tan and Waller 1989; 36). In accordance with the course set by the Policy Document on Ambient Environmental Standards (PDAES), the consideration that precedes the central definition of general standards became the subject of a political choice.

Because, by definition, mandatory standards are not adapted to specific situations, while the relationship between the cause and effects of noise *is* determined by local circumstances, the Noise Abatement Act specified a range of admissible values between the target limit value and the maximum permitted limit value. This allowed standards to be adapted to local situations, and the provincial government has the authority to set the 'higher' values within this range for individual cases. This procedure is known as the 'higher-value procedure' (see Chapter 7) and allowed local authorities a degree of policy flexibility. The Noise Abatement Act contains regulations for a large number of specific situations. The question remains, however, as to how specific general regulation can or must be in order to enforce or manage solutions and goals at a local level. This question relates to the efficiency and effectiveness of decision-making, and arises continually during the further development of the standards system.

Soil, Odour and Risk Standards

Despite the announcement, in the 1972 Priority Policy Document, of soil-protection legislation, sectoral environmental legislation dealt mainly with air, water, waste and noise.[17] It was accepted that air and water are 'sensitive' to pollution. The Rhine had been so badly polluted for so long that hardly any fish species could survive in it, and smog had been a frequent occurrence since the beginning of the Industrial Revolution. By contrast, the susceptibility of the policy compartment 'soil' to pollution was underestimated for a long time. This is evident, for example, in a ruling by the Netherlands Supreme Court mentioning the date 1 January 1975 as the date from which cases of soil pollution could be declared imputable (see also Bierbooms 1997). In the spring of 1980, the discovery at a new housing development in Lekkerkerk of soil contamination caused by PCBs and substances such as xylene and toluene caused an unprecedented scandal. The Lekkerkerk incident and its aftermath sent a shockwave through communities everywhere. Minister for the Environment Ginjaar, who held the view that the area was no longer habitable, and the subsequent publicity campaign launched in Lekkerkerk under Mayor Ouwerkerk, helped to bring the matter to the attention of parliament and the public in no uncertain terms. In this sense, the Lekkerkerk incident could be said to have instigated the second wave of environmentalism in the Netherlands. After Lekkerkerk (cost: EUR 80 million), many black moments followed (Bouwer, Klaver and De Soet 1983), with contamination at Volgermeer, Griftpark, Diemerzeedijk, Zellingwijk (Gouderak), Stadskanaal and De Kempen. In 1981 there were 4,300 reported cases of environmental contamination (Brink et al. 1985). An inventory carried out in 1987 by the Confederation of Netherlands Industries and Employers (VNO-NCW) identified some 100,000 functioning industrial locations as 'suspect' (Verschuren 1990), and estimated the cost at EUR 27 billion. In a ten-year soil-decontamination programme (Steering Group for the 1989 Ten-year Soil Decontamination programme) this sum was increased to EUR 45 billion. This figure did not even include the decontamination of river and lake beds – with which the Netherlands is richly supplied thanks to a fluvial network and estuary. An evaluation study commissioned in 1996 by VROM mentions 350,000 cases of contaminated river and lake beds, with the majority marked as 'urgent' (§ 5.5) (VROM 1996c). The report concluded that, as a result of the autonomous spread of mobile pollution, the cost of decontamination was increasing every year by approximately EUR 172 million. This is more than the state makes available every year under the WBB[18] for research and remediation (Roeters 1997). In any case, the cost is such that soil decontamination programmes will have to continue well into the 21st century.

The exponential growth in 'known' cases of soil contamination, including those in nature reserves and housing developments, led in 1983 to the Soil Clean-up (Interim) Act, which was a form of emergency legislation designed to cover the period until the Soil Decontamination Act came into effect. The interim Act specified what action should be taken at contaminated sites. The public outcry over soil pollution in the Netherlands resulted in legislation that, to a certain extent, can be compared to the Noise Abatement Act. Strict general standards were drawn up – although they were not included in the Act itself – as translations of goal-oriented soil-protection policy. The standards were based on the Soil Clean-up Guidelines (*Leidraad Bodemsanering*) issued by VROM (1983b). The guidelines defined three values: A, B and C, which were the foundation of soil-decontamination policy until the 1990s. The 'A' value was the reference value, i.e. the indicative level above which contamination becomes demonstrable. The 'B' value was the test value for further investigation. If a specific examination showed that the concentration of a pollutant exceeded the 'B' value, the guidelines indicated a possible exposure risk for man or the environment, and further examination was therefore necessary. If the 'C' value (control value for decontamination) was exceeded, preparations would be made for clean-up measures.[19] Investigation was therefore an important step in the process in order to determine the extent and concentration of the pollution. Standards specified the decision-making stages in the procedures to be followed, including the spatial-planning processes relating to the future use of potentially contaminated locations.

The relationship between environmental policy and spatial planning in terms of environmental standards extends beyond noise and soil standards. Policy also had to deal with the issue of odour nuisance, which was put on the political agenda at the end of the 1970s by the Environmental Hygiene Inspectorate (Bakker 1987). The Nuisance Act was the first appropriate instrument for dealing with the issue, and local authorities were ordered to deal with 'odour zones' around areas of intensive farming, according to national guidelines under the Act (Comissie Rey 1976). It is interesting that the proposal for measuring odour levels was set out in the 1984-85 Indicative Multi-Year Programme for Air (*IMP Lucht* 1984-1985, TK 1984), whereby odour concentration was related to the perception of odour by local communities. In contrast to other forms of environmental nuisance, the measurement of odour concentration is inextricably linked to the perception of the odour in question. Despite the fact that not everyone was convinced that odour was a problem (Bakker 1987), interim limit values were proposed for odour concentrations, which – it was assumed – would come into effect as a governmental decree based on the Air Pollution Act. This related to one odour unit per m^3 as a 99.5 percentile[20] and one odour unit per m^3 as a 98 percentile, which would apply, respectively, to new and existing polluting installations, and could be implemented by provincial authorities as licensing guidelines. Panels of 'odour assessors' were appointed to objectify perceptions of odour. The results could then be used to translate odour norms into spatial zones. This objectification did not mean, however, that a clear link could be established between the perceived noxious odour and its source. The more pollution-emitting sources there are at a location, the more difficult this becomes. An example is the Rijnmond area (the urban and industrial zones surrounding the port of Rotterdam) where 80 per cant of the 10,000 complaints received each year relate to odour (Bakker 1987). The Indicative Multi-Year Plan (IMP) for 1986-1990 states that formulating odour standards is by no means easy, which is why "it has not yet been possible to set an odour standard that guarantees a satisfactory climate for the majority of the population" (TK 1985; 72). Further proposals for regulating odour nuisance were not made until 1992, this time in the Policy Document on Noxious Odours (*Nota Stankbeleid*, TK 1992). By this time, however, the opposition to inflexible standards had increased substantially (see § 5.5).

Risk assessment became important with the advent of nuclear power stations in the 1970s (Van Kasteren 1987). There was also a dioxin accident in Seveso, Italy (see Chapter 2), which led to a European directive on safety risks in the chemical industry (the Post-Seveso directive). This directive led to the Prevention of Major Accidents Decree (BRZO) in the Netherlands. There was a growing need for greater insight into the safety aspects of high-risk functions and activities for their surroundings. 'Risk' became a policy principle, expressed as national zoning guidelines designed to limit the risk surrounding a number of specific sources such as LPG filling stations and natural gas pipelines. Policy dealt not only with actual pollution but also with the *probability* that it would occur. The policy document *Dealing with Risks* (*Omgaan met risico's,* TK 1985; Appendix 2, TK 1989) introduced the risk-assessment approach[21] as a general principle for drawing up standards: "The risk-assessment approach [...] is the foundation of effect-oriented policy, and is the instrument for measuring and predicting negative environmental impact by balancing the estimated quantitative risk against risk limits" (TK 1989; 1). With regard to deaths resulting from major accidents and exposure to dangerous substances and radiation, the maximum acceptable level (limit value) for the total risk was set at a mortality risk of 10^{-5}/year. This relates to the 'individual risk'. The negligible risk (target value) is, 'where theoretically possible', set at 1 per cent of the maximum acceptable level. In order to 'prevent social dislocation' (TK 1989; 13), a group risk was determined: the probability of an accident resulting in 10 deaths per activity is set at an average of once every 100,000 years, whereas an accident causing 100 deaths must occur no more than once every 10 million years etc. This form of group-based standard was a direct result of the frequent environmental disasters in and around urban areas (see Chapter 2). Individual and group risk can be translated into risk contours around locations such as industrial estates, marshalling yards, flight routes, roads and waterways.[22]

In addition to the development of odour and risk standards, there were several other developments. The ALARA principle[23] was introduced, in the first instance as a part of policy on radiation (TK 1985; 16). An ecological guideline was drawn up to limit the local impact of ammonia emissions from livestock farms in 'acidification-sensitive areas'.[24] Following the introduction of statutory noise-level maps, local authorities are being urged to indicate, in a spatial form, the total noise quality of an area (Westerhof 1989).

Emphasis is placed not only on a single source in relation to its immediate surroundings, but also on the total picture of environmental quality. The 'little green book' entitled *Bedrijven en milieuzonering* (Industry and Environmental Zoning), published by the Association of Netherlands Municipalities (VNG, 1986), introduced a method for "systematically assessing the distances that should be maintained between industry and environmentally sensitive areas" (VNG 1986; 17). This indicative method of assessment categorised industry according to a wide range of pollution categories and calculated the required distance in relation to environmentally sensitive residential areas. The relationship between an installation or industrial location and its environment was no longer considered in terms of individual environmental aspects, but in terms of almost all types of relevant pollution (see also Hutten Mansfeld and Zijderveld 1982). This was the first practical application of integrated environmental zoning[25] (see § 5.4).

The VNG method for integrated environmental zoning is now a common and generally accepted method for determining 'safe' distances between industry and homes in terms of environmental quality. The method is easy to implement, and therefore cheap. Because the specified distances were based on experience and were averages for the industrial category in question, the values were only indicative and not based on actual on-site pollution measurements. As industrial activities become more complex and increased in scale, the VNG method quickly becomes less practicable.

In the area surrounding DSM-Geleen, a large chemicals complex located close to large residential areas in the municipalities Geleen, Stein and Beek, local residents experienced a decline in liveability due to noise, noxious odours and safety risks (Van den Nieuwenhof and Bakker 1989; 8). At-source measures proved ineffective, so in 1984 it was decided to chart the actual environmental impact surrounding the DSM site, and take any necessary planning measures. This initiative of the Limburg provincial authority was designed to result in an integral zoning project for DSM (IZ-DSM), and was not entirely unexpected. This initiative went hand-in-hand with the introduction of statutory noise zones in 1986. The provincial authority, VROM, the Environmental Protection Inspectorate and DSM itself were involved in the IZ-DSM project, which resulted in risk, noise and odour contours based on statutory norms. It was proposed that the contours be translated into an integral contour by means of an integral environmental zoning method based on cumulative nuisance levels. Planning implications could then be attached to the contours. However, a number of residential neighbourhoods in Geleen were located in the zone between the 10^{-6} and 10^{-8} risk contour (Van Kasteren 1987; 15). The noxious-odour zone also covered a large area. The brakes were applied to the project because of reservations relating to the far-reaching integral proposals and the outcry raised by local communities and politicians. In the Minister's statement of 21 November 1987, only the noise contour was made official and formalised in a 'Royal Decree'. The safety contour was declared to be indicative, and the noxious-odour contour was set aside for the time being.

In 1988, in addition to the IZ-DSM project in Limburg, the Maastricht local authority also completed the Environmental Policy Integration Project (PIM). The purpose of this project was to obtain a full picture of negative environmental impact in the municipality of Maastricht, based on a large number of environmental aspects (Colstee-Wieringa 1988). This became necessary after residents in the Boschpoort neighbourhood unsuccessfully petitioned for the neighbourhood to be extended (VROM 1989d). The wide range of data collected for the project was then translated into coloured maps showing the spread of various forms of pollution within the Maastricht municipality. The separate maps were combined into a single integral map, with the colour red indicating an 'environment unsuitable for habitation'. The colours orange and yellow indicated 'areas requiring attention'. The aims were to gain insight into environmental quality in the Maastricht area, and to use this information to justify the choice of locations for spatial development. The IZ-DSM and PIM projects attracted a great deal of public attention in the Netherlands, and caused several parties to become interested in an integral approach. VROM then took up the concept of integral standards and zoning. In March 1988, VROM invited the provincial authorities to submit pilot projects for integrated environmental zoning under the Multi-Year Implementation Programme for Noise Abatement (VROM 1988b). The programme thus linked developments in environmental standards with the shift towards integration of the 1980s (§ 5.4).

Tan, then leader of the project on the Cumulation of Sources and Integrated Environmental Zoning (see § 5.4), was not the only person who believed integrated zoning to be a logical step that was in line with the development of integrated environmental policy (in Van den Nieuwenhof and Bakker 1989; 11). Integration was the most important theme in environmental policy during the 1980s (§ 5.3), but it had little influence on the development of sectoral standards during that period. Ever since structural environmental policy came into existence, environmental standards have been presented as a tool, a framework and also as a goal. Standards became a generally binding environmental goal and a restrictive framework for other areas of policy relating to the physical environment. In more general terms, environmental zones are the spatial translation of the quality of the 'grey' environment (i.e. environmental health and hygiene) and as such are a somewhat one-sided expression of the relationship between environmental policy and spatial planning. Because decontamination is a permanent issue in environmental policy, because policy has to deal with many other aspects apart from noise abatement, and because the relationship with spatial planning was gaining in importance, the system of environmental standards would continue to be extended as a goal-directed outcome of planning-oriented action.

The coercive character of environmental standards in respect of other forms of policy was in line with the concept of remediation expressed in environmental policy of the time. The Netherlands was polluted and had to be cleaned up. At the same time, new environmental issues arose. There was public concern not only about soil contamination, but also about acid rain, phosphates in detergents, the impact of major chemical accidents in or close to urban areas (see Chapter 2), the greenhouse effect and the depletion of the ozone layer. A logical consequence of these public concerns would be that the government take on the role of orchestrator, designer and conductor. Environmental policy still had a long way to go, and still had to earn its place in the pantheon of delineated policy fields, despite the fact that a thorough 'spring clean' was required throughout the Netherlands to reverse the negative impact of existing pollution and – above all – to prevent future problems. Against such a background, the coercive and hierarchical character of environmental regulation seems reasonable.

5.3 Integration: the Buzzword of the 1980s

As we concluded in Chapter 2, increased knowledge of the mechanisms that can lead to environmental problems provided insight into the complexity of environmental problems and the impact they have on many different segments of society. It became increasingly clear that the regulations of the 1970s did not answer the complexity of such issues and the extent to which they interrelate. Because the effect of sectoral legislation was largely restricted to the environmental 'compartment' at which it was aimed, it resulted in a sharp increase in deflection of environmental spillover[26] between the different compartments (De Jongh 1989), instead of the decrease it was intended to achieve. Examples include sludge, waste-disposal problems resulting from water treatment, and air pollution resulting from waste incineration (De Koning and Elgersma 1990). Here, 'deflection' is a collective term for the social factors and physical causes (mentioned in Chapter 2) that contribute to a greater or lesser extent to the ineffectiveness of environmental policy. Deflection is therefore not the only problem. A source of pollution can adversely affect several compartments simultaneously (Teunisse 1995; 10). Furthermore, it proved difficult accurately to assess the environmental impact in each compartment, although this did not become obvious until the effects of soil contamination began to manifest themselves.

At the beginning of the 1980s it was thought that environmental policy should be rendered more effective and transparent, and that the constraints imposed by its sectoral and above all compartmental structure should be removed. Until this time, compartmentalised policy had been effect-oriented and had focused on remediation. Existing problems were given priority but nevertheless appeared to worsen, and not enough preventive action was taken. Preventive policy had not yet been developed and proved ineffective and inefficient in the context of an effect-oriented policy.

The conviction grew that environmental issues should be considered together and in relation to their context. This would involve formulating a technical-functional policy and making administrative changes. While the interdependence with other policy areas became increasingly apparent (Ringeling 1990), in practice policymakers and policy implementers were still very much confined to their own compartment or sector. Policy cohesion and the holistic ecological approach were thus overlooked (Van Ast and Geerlings 1993; 164). Other reasons for restructuring environmental policy were inadequate enforcement and control and other ministries' lack of commitment to its development. Policy also had insufficient influence on other layers of government. Schoof (1989) attributes the problem of deflection to the fact that provincial authorities failed to enforce environmental policy. Critics who claim that steps must be taken to make environmental policy more effective should look closer to home.

A u-turn was required to preserve environmental policy. The implementation of the WABM – the Environmental Protection (General Provisions) Act – in June 1979 was an indication that the legislator recognised the need to restrict the development of sectoral environmental legislation and policy.[27] Environmental policy would remain under central control and changes would be made. In 1982, environmental policy came under the direct control of a minister for the first time, and under a new ministry: Housing, Spatial Planning and the Environment (VROM). Meanwhile, government departments had also adopted a new approach. *Deregulation* was the new panacea for over-regulation in all layers of government (Kickers 1986) and *integration* became a buzzword in environmental policy.

Within the space of two years VROM had drawn up plans for reshaping environmental policy. 1983 saw the publication of the Environmental Policy Integration Plan (VROM 1983) and the policy document on Deregulation in Spatial Planning and Environmental Management (VROM and EZ 1983). These documents outlined the proposals for restructuring environmental policy. In 1984 the policy document *More than the Sum of its Parts* (*Meer dan de Som der Delen*, VROM 1984a) and the IMP for Environmental Management 1985-1989 (IMP-M, VROM 1984b) were published. These documents argued above all for a renewal of the content of environmental policy. They represent a decision in favour of a policy process based on internal integration (i.e. the integration of the elements of environmental policy), and on external integration (i.e. integration with other government policies).

The process of *internal integration* was largely successful. This success was reflected in the publication of the first National Environmental Policy Plan (NMP-1) in 1989 and the Environmental Management Act in 1993. In the 1980s a number of forms of *external integration* were developed that focused on the relationship with spatial planning. These forms of integration involved (1) sectoral environmental zoning, particularly for noise, (2) sectoral area definitions for groundwater protection, soil protection and Quiet Zones, and (3) environmental impact reporting. However, these forms of integration were somewhat one-sided and were based on environmental policy. A more balanced and ambitious external integration – integration in the true sense of the word – did not come about until integrated area-specific environmental policy was introduced at the end of the 1980s (§ 5.4).

Environmental Policy Integration Plan

In June 1983, the Minister for VROM submitted the Plan for the Integration of Environmental Policy (PIM) to the Second Chamber. The purpose of the PIM was "to realise greater completeness and cohesion in the preparation, recording, implementation and evaluation of environmental policy" (VROM 1983). In the PIM, environmental policy was no longer seen as purely sectoral and intended only to protect and improve the environment. It became a *facet policy*, the aim of which was to develop a cohesive and consistent policy for all government activity relating to the environment. A direct relationship was established between environmental policy and spatial planning. The two policy fields were not only integrated within the same ministry, but were both oriented towards the same material object: the physical environment. The policy fields were described as "mutually supporting and complementary in aiming to ensure that the quality of the physical environment is as high as possible" (VROM 1983; 5). In the PIM, environmental policy and spatial planning were seen as equal, complementary and partially overlapping areas of policy. Integration was consequently seen as "a precondition for an effective and efficient environmental policy that takes account of the effectiveness and efficiency of the whole range of government policy" (VROM 1983; 6). In the plan, the system of environmental standards was seen as an important instrument for shaping the integration process.[28] Standards were seen as an elementary part of the *regulatory chain* on which environmental-policy planning was based at the time.[29] The regulatory chain consisted of five links: (1) legislation and regulation, (2) standards, (3) licensing, (4) implementation and (5) enforcement. There should be "a smooth transition between one link in the chain and the next, and the links should form a closed circle" (Winsemius 1986; 79). According to Winsemius (Minister of VROM at the time), the weak links in the chain were implementation and enforcement. He proposed developing the environmental-standards system for all areas of environmental policy so that the standards were generally binding, despite the realisation that this could result in "a degree of inflexibility" (VROM 1983; 25).

The PIM identified two alternative principles for the further development of the standards system. The first principle was to implement strict but realistic standards with genuine opportunities for flexibility or tightening up and the second principle was to implement more flexible standards that could be tightened up. The latter did not find a footing until the second half of the 1990s, by which time 'strict, but realistic standards' had proved too inflexible for the local and specific contexts in which environmental/spatial conflicts arose (see § 5.5). Initially, the decision was taken to develop the standards system along the lines of the first alternative, based on the principles of the 1979 Noise Abatement Act.

Despite the fact that policymakers chose the 'strict but realistic' option, there was already a discussion on how the standards system could be made more flexible. The PIM referred to *differentiated standards*, and the possible negative side-effect of unfair distinctions being made. The 'bandwidth' method (i.e. the introduction of upper and lower limits) was also discussed. This system was later developed according to the *progressive standards* principle.[30] Readers may also conclude from the PIM that permitting *justified exceptions* to the standards – a theme that was not seriously considered until the 1990s – was not considered worth looking into. The reference to the Bubble Concept[31] ('*stolpconcept*'[32]) as a means of increasing 'self-regulation' is also interesting.

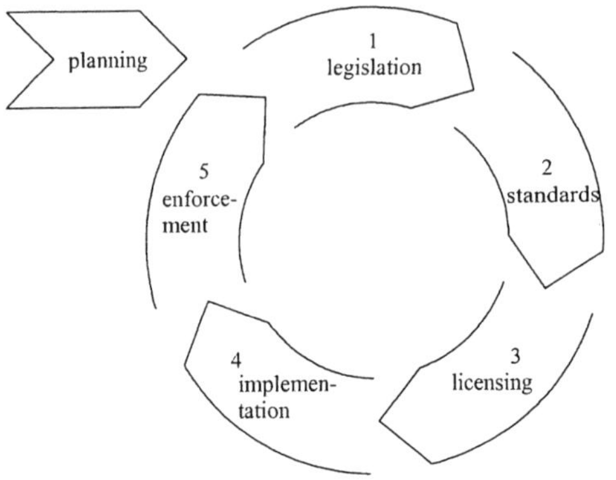

Figure 5.1 Regulatory chain
Source: VROM 1983

The PIM emphasised the need for a shift of focus from co-ordination to integration, but did not specify how this should be achieved. The action programme for the Deregulation of Spatial Planning and Environmental Management (DROM) (VROM and EZ 1983) initiated an integration process, partly with a view to streamlining, simplifying and reducing regulation and legislation. The programme was based on the licensing system and on legislation that was being drawn up, and retained the principles and objectives of spatial planning and environmental management. The PIM and the DROM reflected the need to integrate the various sectors of environmental policy. Co-ordination became integration, integration partly involved streamlining the regulatory process and was oriented internally and externally towards other areas of policy. This established the basic foundation for an environmental policy in progress.

More than the Sum of its Parts

The PIM assumed that, at government level, there was "a strong need for [...] a twofold approach based on compartments" as well as "groups of activities" (VROM 1983; 21). This is the 'target-group policy' that was developed in the National Environmental Policy Plan (NMP) (VROM 1989) and supplemented with strategies and policy measures. The PIM had already discussed the possibility of gearing policy towards changing the behaviour of various groups, categorised according to their specific environmentally harmful activities. The first policy document on environmental policy planning, entitled '*More than the Sum of its Parts*' ('*Meer dan de Som der Delen*', VROM 1984a), proposed the further development of a policy aimed directly at industry, agriculture, consumers, and the traffic and transport sectors, etc. The purpose of this was to tackle environmental problems at source and, where possible, prevent them by reducing harmful emissions. The document favoured 'shutting the stable door *before* the horse bolts', and rightly so. In this sense it was right to expect that a target-group approach would be more successful than an effect-oriented approach. This reasoning is in line with the shift of emphasis away from remediation towards prevention. So high were the expectations of the target-group approach that it overshadowed effect-oriented policy during the second half of the 1980s and the first half of the 1990s. The compartmental approach to the environment had definitely been dispensed with.

The '*Sum of its Parts*' policy document acknowledged that a division into components (air, soil, water and noise) was not possible without reference to the environment as a whole. Integration was therefore a main theme in the document. Two developments in environmental policy were also mentioned (VROM 1984a; 14) that relate to making environmental policy more effective and efficient. There is a clear shift away from reactive remediation-oriented policy towards anticipatory policy aimed at prevention. There was also a reorientation of the relationship between central government, provincial government and local government, with the emphasis on the greater involvement of lower levels of government in the environmental-policy planning process.

The developments described in the '*Sum of its Parts*' policy document were intended as a first step towards integral planning in environmental policy. A general planning system, it was claimed, would provide clarity for third parties but must not be hierarchical. A hierarchical structure "would not be compatible with the way in which tasks and responsibilities are delegated in environmental legislation. Neither would it be compatible with the planning approach on which this document is based, [...] i.e. oriented mainly towards the planner's «own» decisions" (VROM 1984a; 15-18). In this context we also speak of complementary administration. As an extension of this, the policy document discussed in detail the relevance of standards in environmental policy planning. The role of standards with the 'imperative character of general regulations', which were largely quantitative, was reaffirmed. Such standards should be binding for third parties and be implemented above all via legislative instruments (governmental decrees, bye-laws, etc.). The '*Sum of its Parts*' document represents a shift towards more qualitative standards as a basis for policy and as objectives. Environmental quality objectives were considered important in terms of the influence environmental policy exerts on spatial planning. It was proposed that, at the provincial level, environmental policy plans include area classifications based on the environmental quality of ecological criteria. When a balance had been struck between the interests involved, areas would be designated in the subsequent regional plan.[33] It was also recognised that rapid developments in water-management policy and environmental policy required a pragmatic approach to co-ordinating policy plans and aims. "The ongoing developments in the abovementioned subjects [...] could result in a process whereby the two forms of planning alternately take a step forward, so that water-management plans and environmental policy plans become involved in a game of leapfrog" (VROM 1984a; 45).

Other important proposals advocated strategic structures and processes for formulating environmental policy in the future. Third parties should be engaged in the planning process at the earliest opportunity. An open process was therefore preferred with a view to building public consensus. In addition, all the elements of environmental policy should be brought together and linked within a cohesive framework, particularly at central-government level. According to the '*Sum of its Parts*' document, the complexity of environmental issues should be rendered manageable by means of structures within which it was possible to establish links between the various elements of environmental policy.

These content requirements determined how a fragmental, sectoral environmental policy should evolve into an integral policy. If the requirements were met, policy would no longer be based on a compartmental approach. However, it would not be possible to dispense with the *compartmental approach* completely because remediation-based, effect-oriented policy had proved too important for reducing pollution, noise, waste, resource depletion and contamination of the soil and air. Furthermore, the volume of existing sectoral regulations was considerable and could not simply be dispensed with. The compartmental approach would therefore have to be one of several approaches for structuring and implementing environmental policy (see also Winsemius 1986). The *target-group* approach was becoming increasingly popular. In addition, a *substance-oriented approach* was adopted for a time. It was designed to deal with potentially hazardous substances. *Product and volume-oriented* approaches were geared towards reducing emissions. Integral chain management was the most influential theme of the product-oriented approach, and was intended above all to combat environmental impact resulting from the life cycle of products. The volume-oriented approach was designed to reduce emissions to the environment. These approaches dealt with recognisable aspects of environmental policy. They were valuable approaches in themselves, but must also be seen in relation to other approaches.

Despite the growing preference for a target-group approach and the reservations regarding effect-oriented policy, the '*Sum of its Parts*' policy document did not fully recognise the potential importance of the *area-specific* approach. This approach focused on "the environmental functions of types of area in a general sense (residential areas, nature conservation areas, lakes, etc.) and/or areas defined by spatial planners as requiring special measures (national parks, industrial complexes, etc.). The approach also related to administrative bodies (provincial/regional authorities, etc.)" (VROM 1984a; 27). It was not an entirely new approach. Initial steps had already been taken in the Noise Abatement Act (quiet zones and sanctuaries) and in water-management policy (water extraction areas). The policy document, however, underestimated the importance of the area-specific approach in co-ordinating policy with the spatial planning and water-management sectors. It was not until the 1990s that area-specific policy shook off its sectoral past that had been shaped by an effect-oriented approach and the prevailing target-group policy. From then on, it played an important role in the process of integration and co-ordination with spatial planning (§ 5.4).

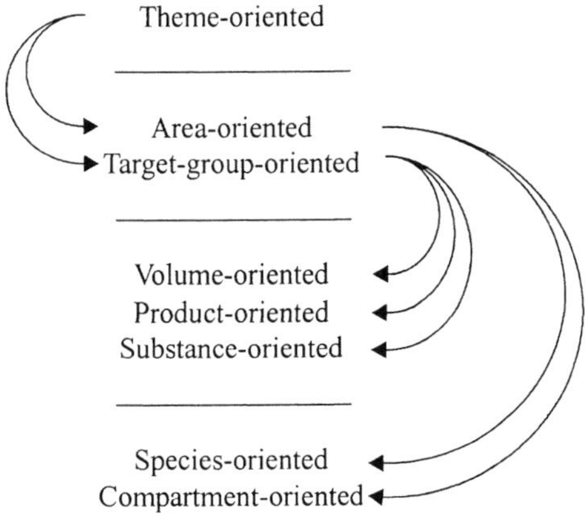

Figure 5.2 Approaches to environmental policy and their interrelationships

The '*Sum of its Parts*' policy document expected that the approach described above would render "the majority of relationships between sub-areas visible and manageable" (VROM 1984a; 27). It also pointed out that it was not possible to devise a 'holistic' approach that embraced all aspects of complex environmental issues. However, in certain circumstances a *thematic approach* could be effective when combined with other approaches.

Formal Steps Taken in the Indicative Long-Term Programme for the Environment, 1985-1989

In the same year that the '*Sum of its Parts*' policy document described an integrated environmental policy for the future, definite steps in this direction were being taken through the Indicative Long-Term Programme for the Environment 1985-1989 (IMP-M) (VROM 1984b), which was based on the approaches of the '*Sum of Parts*' document. However, in the IMP-M, the thematic approach was central rather than supplementary to the strategic formulation of environmental policy[34] (see Fig. 5.2).

The two-track approach of environmental policy (i.e. source-directed and effect-oriented) was also essential (see Fig. 5.3). This was a genuine breakthrough in the planning system for environmental policy (Mol 1989; 29). Although environmental strategy was based on a number of central environmental themes, subsequent detailed measures for dealing with environmental issues had to be formulated using the dual approach, which was both source-directed and effect-oriented.

The *source-directed approach* had always been the mainstay of environmental policy, as the IMP-M emphasised. But this did not mean that past policy had always focused on the effects of environmental pollution. The increasing need for consistent and transparent environmental policy that was also more effective highlighted the explicit role of source-directed policy (i.e. target-group policy) in the integration of environmental policy. It was stated that "such an «integrated» method of policy formulation" (VROM 1984b; 76) should be largely directed at "groups of sources that are homogeneous in policy terms" (VROM 1984b; 76). From an environmental point of view, this means 'homogeneous' groups that could cause pollution (VROM 1984b; 77). Branches or sectors of industry were given as examples. The national government saw these categories of target groups as an important foundation for source-directed policy. However, because central government "is not usually directly involved in the licensing of individual sources" (VROM 1984b; 76), it should draw up guidelines for each target group for the benefit of local authorities and water boards. The guidelines, so the reader is given to understand, should be drawn up in close consultation with local and regional authorities and "lower level of governments should as a rule be allowed a certain amount of flexibility in implementing the policy" (VROM 1984b; 76). An advantage of the target-group approach – it was stated (VROM 1984b; 78) – was that it was also used in other policy areas such as economic policy, which would facilitate cohesion with other forms of policy. The purpose of the approach was also to encourage the belief in 'internalisation' (Winsemius 1986) by influencing the behaviour of individuals, groups, organisations and businesses through education and information. Nelissen (1988; 207) argued that following this path, with central government acting as a 'promoter', had a high potential for success. In addition to existing regulatory instruments, incentive measures would have to be developed. The government also wanted to appeal to target groups to take responsibility: "all groups are expected to accept responsibility for the environment and act accordingly" (TK 1989; 13). Increased involvement and shared governance became important themes in the 1990s, and formal agreements between target groups and the government became increasingly common.

The *effect-oriented approach* was intended to "prevent and/or restrict negative impact on the physical environment by defining levels of pollution below which, based on existing knowledge, it can reasonably be expected that the impact in question will not occur" (VROM 1984b; 56). Here, the importance of quantitative environmental standards speaks for itself. A disadvantage of the compartmentalised approach had been the deflection of problems from one compartment to another. Under this approach it was difficult to take a holistic view of the environment in an ecological and spatial sense. It was therefore decided that policy should be based on "the expected and negative effects of exposure to substances, physical phenomena and physical encroachment"[35] (VROM 1984b; 56). Central to this approach was the *probability* that a negative effect would occur (see Chapter 2 and § 5.2), and risk therefore became a parameter. Policy would therefore be geared towards "specific physical or chemical agents (substances, noise, radiation) that affect compartments and geographical areas" (VROM 1984b; 57). IMP-M 1985-1989 still placed strong emphasis on substance-oriented policy. Compartment-based policy would have to be developed in the sectoral IMPs. However, the IMP-M stated that there was "insufficient information on which to base an area-specific policy, insofar as the government can be assumed to be responsible for this" (VROM 1984b; 57). The document explained 'area-specific' policy as follows: "an effect-oriented policy for an area focuses on the cohesion of measures with a view to protecting specific environmental qualities and/or the specific significance of that area" (VROM 1984b; 57).

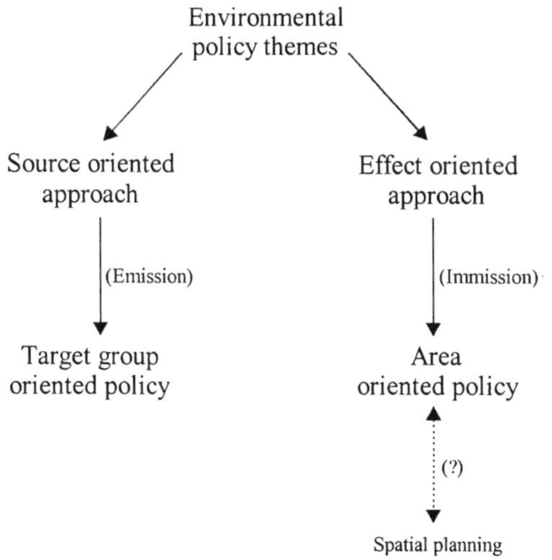

Figure 5.3 The structure of integrated environmental policy, showing the clear relationship to spatial planning

Effect-oriented and area-specific policy rapidly became more structured following the IMP-M 1985-1989. In the IMP-M 1986-1990, the significance of the 'local environment' became clear for the first time: "the physical environment [is] understood to be a system of four very closely related compartments: air, soil, water and organism, whereas, from the point of view of nature and landscape protection, the physical environment is the result of interaction between nature and culture. The insights and experience gained in recent years both here and in other countries have made it clear that, in order to protect our physical world, we should perceive it as a cohesive structure of ecosystems on different scales, from local/regional to global" (TK 1985; 10). The specific nature of each individual scale, including the local scale, was therefore recognised.

Environmental-protection areas and environmental quality requirements were named as two specific instruments of effect-oriented environmental policy (TK 1985; 116). The environmental quality requirements were twofold, namely general and specific. *General quality* referred to a basic level of quality introduced in the Priority Policy Document (1972) that was to apply throughout the Netherlands. An example of this was the preferred limit value in the Noise Abatement Act. The general quality of the environment in the Netherlands should be expressed in environmental quality requirements, as set out in the 1976 Policy Document on Environmental Quality Standards. Area-specific policy was considered necessary to protect *unique environmental qualities* in the many areas of the Netherlands where the 'local' level of environmental quality is appreciably higher than the general level (TK 1985). The clearest examples are quiet zones, sanctuaries, and groundwater and soil-protection areas. This specific form of effect-oriented and area-specific environmental policy was designed to clarify "the objectives relating to the quality of the environment in the Netherlands as a whole or in given areas, and the responsibilities deriving from those objectives for target groups" (TK 1985; 12). Above all, a dialogue should be established between decision-making, plan development and implementation and other forms of policy relating to the physical environment, in order to ensure that the desired environmental quality of a given area was in line with the desired functions and uses identified by spatial planners and nature conservancy organisations, and in water-management policy and traffic/transport policy (see also Biezeveld 1992; 3). This area-specific orientation in environmental policy provided the foundation for an integrated area-specific policy: the ROM 'designated areas' policy (§ 5.4).

According to the IMP 1985-1989, the integration process should be initiated by VROM. Other ministries would then follow the course and structure it set for environmental policy. The time was not yet ripe for vertical policy co-ordination and national plans based on a long-term vision (Mol 1989; 29). This new development would have to be formalised in a National Environmental-Policy Plan (NMP) designed to present, in a full and definitive form, the process of internal and external integration of environmental policy at central-government level. The first National Environmental-Policy Plan (NMP-1) (VROM 1989) appeared five years after the IMP-M 1985-1989. The enthusiastic worldwide reception of NMP-1 was largely due to its unique approach based on internal integration in environmental policy.

NMP-1: a Window on the 1990s

The National Environmental-Policy Plan 1989-1993 (NMP-1) was entitled '*Choose or Lose*' ('*Kiezen of Verliezen*') and set out "the strategy for environmental policy in the mid-term" (TK 1989; 7). When the NMP-1 came into effect, it was considered to be the mainstay of environmental policy for the 1990s. The plan retained previous key elements such as the standstill principle, giving priority to at-source measures and the 'polluter-pays principle'. The plan aimed to continue existing policy while introducing a number of new principles, the most influential of which was sustainability (see also § 2.6). By focusing on *sustainable development*, the plan aimed to deal with the consequences of human activity in the long term. The strategy had been "developed in accordance with the wish to solve environmental problems, or render them manageable, within one generation" (TK 1989; 7).

The policy structure based on themes, target groups and areas remained intact, with only a marginal emphasis on areas. The NMP-1 did, however, set out and budget the ROM dedicated-areas policy and a policy for developing a system of integrated environmental zoning. Both these instruments (see § 5.4) are distinct 'integral' forms of area-specific environmental policy. During the 1990s, this policy played an important role in the process of external co-ordination and integration, particularly with the spatial-planning sector, although this process was hardly mentioned in the NMP-1.

The stimuli for external integration were largely missing from the NMP-1, which only made limited reference to the integration of environmental policy with other policy sectors relating to the physical environment (see also Koeman 1989; 50). Neither did it set out the general structure of integration. Integration, insofar as it was mentioned, was only relevant in the context of proposed measures for theme-based policy and target-group policy. It had insufficient influence on other authorities not identified as target groups. To a large extent, the NMP-1 was a repercussion of the internal integration process. In this sense the plan was limited because it was intended as a guideline for the environmental policy of the 1990s: a policy that, following a period of internal integration, was ready to be co-ordinated and integrated with other areas of policy relating to the human living environment.

Most of the strategies and action points in the NMP-1 were elaborated for eight environmental themes and responsibilities were delegated among target groups.[36] Emphasis was placed on at-source measures by making the various target groups accountable for the action points. A reduction in hydrocarbon emissions (action point 32) was achieved partly through an action plan (KWS 2000; VROM 1989b), which was drawn up in co-operation with the parties involved. Agricultural measures were proposed, among others, such as manure storage (action point 183) and keeping records of mineral use (action point 26). Furthermore, 'road pricing' was proposed in action point 193 as an instrument for regulating mobility.

The NMP-1 contained many objectives and goals but did not always clarify how they should be realised. The NMP goals were mainly based on the RIVM report '*Caring for Tomorrow*', which subsequently became the scientific foundation for actions and strategies of national environmental policy. In this report and in the NMP-1, the societal causes of environmental issues were only dealt with very briefly. Koeman therefore concludes that "the planned measures are typical examples of treating only the symptoms and paying too little attention to finding structural solutions" (1989; 51). The NMP-1 was based on a technical-functional approach, despite the process of policy integration, the stimuli for cohesive policy, the intention to pursue a participatory policy and the increasing affinity with the complexity of environmental issues. In this sense there is a parallel with the 1970s, during which policymakers also focused very little on the societal context of environmental issues.

NMP-1 was about realising quantified, concrete goals. In that sense the plan – although this was not the intention[37] – appears to have been in line with the qualifications to blueprint planning and, according to Ringeling (1990; 9), the planning process is somewhat overlooked. Plans were not drawn up *with* target groups but *for* them. Consequently, the status of target groups became that of implementer. The NMP described the goals for target groups, based on '*Caring for Tomorrow*'.[38] Emphasis was placed on covenant policy (see Glasbergen 1998) as an increasingly important method for raising the commitment of target groups to environmental policy. Covenant policy, whereby voluntary agreements were drawn up between government and industry, did not entirely escape discussion from a democratic point of view.[39]

Although Mol does not think the plan should be seen as a blueprint, he points to a central government with a strongly statist and regulatory role geared towards "increasingly refined control of social interaction, insofar as this has consequences for environmental policy" (1989; 35). In its recommendations on the NMP-1, the Central Council for Environmental Protection (CRMH 1989) points to the positiveness of environmental goals and the uncertain contribution made by actors in achieving them. The NMP-1 set out an environmental policy that was primarily oriented as such, with the condition that a strong appeal is made to the commitment of target groups. This inward-looking and hierarchic approach was partly due to the goal-oriented view of the Department of the Environment. In itself, this could be explained as an almost 'natural' reaction to the ever-worsening quality of the environment, which basically required two decades of crisis management.

According to the NMP-1, the 'chaotic state of the environment' required 'dramatic measures'. Glasbergen and Dieperink (1989; 299) asked, for example, whether environmental policy should be adapted to the prevailing political trends of deregulation, privatisation and decentralisation. They argued that revised regulations, 'solidarity organised by central government' and a strong material policy commitment on a national scale would be better approaches. They were not the only ones to hold this view. Other recommendations and comments on NMP-1 point to possibilities for the further development of standards as a central instrument of control, to accelerating the proposed speed of remediation by means of zoning policy, and to extending policy to cover a greater number of pollution sources[40] (VROM 1989c). In its recommendations on the NMP-1 (RAWB 1989), the Science Policy Advisory Council (RAWB) stated that standards were seen as the most important instrument a government could implement in the context of environmental policy. The CRMH (1989) responded that it was: "disappointed that the NMP has not, at the very least, set stricter standards for effect-oriented measures, such as noise barriers and insulation, or made more financial resources available" (1989; 26). Drupsteen (1990) also claims, in reaction to the NMP-1, that standards can best be used to steer policy. The Socio-Economic Council (SER) argued that "consistent application of the polluter-pays principle in the form of price incentives, standards and orders/prohibitions" (1989; 15) was essential for effective environmental policy. There was a societal dialogue on how government should deal with the environment as a collective but undervalued asset. Mol observed a "broad social consensus on the increasing role of the state in controlling environmental problems" (1989; 35). There was consequently a broad social support base for extending the hierarchical framework of environmental standards when the NMP-1 was launched.

Social consensus with regard to environmental policy was also translated into a political context. At the end of the 1980s, 'the environment' was a major political issue because society as a whole was concerned about it.[41] This was a favourable climate for proposals issued by the Department of the Environment.[42] At the beginning of the 1990s there was a clear social, political and administrative consensus with regard to the continued development of a central, framework-setting environmental policy to counteract developments such as deregulation, decentralisation and privatisation. The broad social consensus contrasted sharply with the strong reactions, several years later, against the further development of a hierarchic framework-setting environmental policy in the early 1990s. At the very least, this raises doubts about the supposedly 'sustainable' foundation of environmental policy.

5.4 Area-Specific Environmental Policy: Aversion and Euphoria

Although the underlying reasons for the changes during two decades of structured environmental policy may be easy to understand, it is nevertheless difficult to form a cohesive picture of those changes. From a *decision-oriented* perspective (§ 4.6), a number of changes were designed to make environmental planning more effective and efficient. The compartmental structure, the effect-oriented approach, and the technical-functional cause/effect strategy were replaced as far as possible by an integrated structure based on a thematic and source-directed approach. Environmental problems were no longer seen as straightforward and easily defined issues. More emphasis was placed on the relationships between environmental issues, although this approach did not yet extend much beyond the path between the source of the emission and the location of its impact. Social and contextual causes and effects were not yet part of problem definition and problem selection. Nevertheless, the different constituent parts of environmental issues were brought together in a number of categories: the themes of environmental policy, which were then developed within target-group and area-specific policy.

Viewed from the perspective of *institution-oriented action* (§ 4.7), the most striking aspect of the changes in environmental policy is the inflexible inward-looking attitude of the Department of the Environment. The department guides and directs, and had tried for two decades to steer the behaviour of others according to ministerial and/or political limits. It was a hierarchic, 'unicentric' government policy based on narrowly defined frameworks that had to be developed at local level. This standpoint, given society's concern about the state of the environment that characterised the period, was even encouraged and was also compatible with the way in which relevant target groups in society were involved in environmental policy. The Department of the Environment no longer simply set standards, but also demanded the active involvement of societal actors – and expected them to achieve at least reasonable results.

Considered from the *goal-oriented* perspective (§ 4.5), the most striking aspect is again an unchangeable approach that remained geared towards the further development of the framework-setting system of environmental standards. The development of this system was the most stable part of environmental policy. From the publication of the Priority Policy Document on Environmental Protection up to and including the presentation of the NMP-1, environmental standards were refined and extended (see § 5.2). This development apparently went hand-in-hand with the institution-oriented role of central government with regard to environmental policy.

Nevertheless, given the relationships established in Table 4.1 and Figs. 4.8, 4.9 and 4.10, one would have expected the shift from a compartmental towards a more thematic structure to lead to greater flexibility in the standards system, greater emphasis on processes, decentralisation of policy, and the increased involvement of the contextual environment within which environmental issues arose. This appeared to be only partly the case, at least as far as the NMP-1 was concerned. The NMP-1 was the product of an integration process, but at the same time it confirmed that the nature of environmental policy was hierarchic and framework-setting. But the plan was not a hierarchic *diktat*. It was above all a strategic proposal for central governance based on broad public consensus. A great deal of commitment was expected from society in implementing the plan. However, the plan dealt hardly at all with the social causes and consequences of environmental problems, and paid little attention to effect-oriented and area-specific aspects of policy. The process of external integration was hardly mentioned at all.

When the NMP-1 was published, the internal integration of environmental policy had reached a high point and the next phase in the development of environmental policy – external integration – could begin, at least according to the planning in the IMP-M 1985-1989. In this phase, area-specific environmental policy, which had an intrinsic spatial orientation, would build a bridge between generic environmental policy and spatial planning (see Fig. 5.3). Developments in area-specific environmental policy will be discussed in the following sections. These developments represented a turning point in thinking on governance and on standards in environmental policy. In particular, the government's hierarchic framework-setting role was thoroughly reviewed (see also § 1.2). As a result, there were not only shifts in the decision-led approach, but also within the goal-oriented and institutional approach to planning-based action.

The Third Track

There are a number of reasons why the NMP-1 largely overlooked effect-oriented policy and, by extension, area-specific policy. The environmental policy of the 1970s had been largely effect-oriented and was not considered a success. In addition, at-source policy is by definition more effective in protecting the environment than effect-oriented policy. We must also remember that measures geared towards environmental effects were seen as a dying breed, although it was assumed to be "inevitable that effect-oriented measures will also be necessary in the mid term" (TK 1989; 86). Whatever the case, VROM believed in adopting a conservative approach towards effect-oriented measures, which would only be used in exceptional cases such as decontamination, when structural at-source measures would be ineffective in the short term, when effect-oriented measures would be considerably cheaper, and in the event of a disaster. These conditions apply, for example, to protecting nature reserves and woodlands, when extra efforts at a local or regional level are essential (NMP-1, action point 20). Despite the cautious approach to effect-oriented measures, the Action Plan on Area-Specific Environmental Policy (TK 1990) appeared in 1990 as a follow-up to NMP-1 action point 157. The action plan dealt with issues that were left out of the NMP-1, namely the fact that external integration with spatial planning and other forms of policy relating to the physical environment would have to be based on an area-specific approach.

We have already observed that area-specific environmental policy was not new. Large parts of the Netherlands were covered by some form of environmental zone, or were subject to specific policy aimed at preserving or achieving a particular level of environmental quality. In the Rijnmond area and the area around Schiphol Airport, policy was geared towards the 'grey' environment, while policy on the Veluwe national park and the Waddenzee coastal area focused on the 'green' environment. Various forms of fragmental area-specific policy aimed at defined sectors have used for some time. In the first place, the environmental-standards system, which was based on sectors, was applied on a national scale and established a framework. This system had consequences for spatial planning. The development of the system during the 1970s and 1980s is described in section 5.2. Secondly, "areas that are relatively vulnerable to pollution, impact or disturbance and which, due to their specific function or ecological significance, qualify for a level of protection above and beyond what is generally considered necessary" (TK 1990; 22). Central to this approach are "the quality of the living environment, and the relationship between the condition of the environment and the functions such areas fulfil for society" (TK 1990; 6).

When the 'grey' environment is the only criterion, quiet zones and sanctuaries, groundwater-protection areas, soil-protection areas and phosphate-sensitive areas are relevant area-specific products. This list is manageable, but when the 'green' environment is also taken as a criterion (the Action Plan on Area-Specific Environmental Policy Action Plan is restricted to the grey environment), many more forms of protection are involved: areas mentioned in the Policy Document on Agriculture and Nature Conservation (TK 1975), national parks, protected areas and sites, area-specific REGIWA (regional integral water management) projects, wetlands, the ecological structure of the Randstad conurbation (VROM and LV 1985), valuable cultural landscapes (LNV and VROM 1992; 39), strategic green projects (LNV and VROM 1992; 123), and – of course – the national ecological infrastructure that consists of core areas, nature development areas and buffer zones (TK 1990b; 78). These are mainly areas in which the grey and green environments overlap. For example, 80 per cent of soil-protection areas are included in the national ecological infrastructure – for obvious reasons (Bartelds 1993). In this proliferation of areas, every location was subject to some form of land-use constraint.

When the Environmental Management Act came into effect on 1 March 1993, the protection areas in the grey environment were combined to form environmental protection areas. The legislation, however, still referred to quiet zones and groundwater-protection areas as distinctly separate elements. Soil-protection policy, which was in any case optional and was developed only slowly due to lack of resources, is omitted from the Act (De Roo and Bartelds 1996). Provincial governments had a designating, executive and controlling role in area policy and each province had to shape that role in its own way. Provincial authorities were required to identify areas of special environmental quality, include these in their plans, and draw up provincial byelaws to protect the designated areas. These forms of sectoral area-specific policy are, with the exception of a number of groundwater-protection areas, geared towards rural areas. Designed to protect the ecosystem, these policies all too often transcended local-authority boundaries, once again with the exception of groundwater-protection areas. The majority of these areas offered good recreational facilities and therefore had a double function. These functions involve little or no mutual conflict, and there is a reasonable level of consensus with regard to land-use allocation.

This situation will be somewhat different when an area-specific approach is advocated in situations involving several conflicting functions and various environmental aspects, and when there are transboundary conflicts on several administrative levels involving various actors. In such cases the Action Plan on Area-Specific Environmental Policy advocated an integrated area-specific approach. In this context, the Association of Netherlands Municipalities (VNG) observed that such an approach would make a positive contribution "in complex problem situations involving several environmental aspects and/or different policy areas" (1993; 18, see also Teunisse 1995).

The Action Plan stated that "an area-specific approach to the environment [...], in addition to being based on themes and target groups, [is] a third path to integration in environmental policy" (TK 1990; 1). This was the statement made by Minister Alders (1990-1994) when he informed the Second Chamber that, as far as he was concerned, policy should no longer be predominantly effect-oriented but should be replaced by a policy based on designated-areas and target groups. At first sight, the implications of this statement appear somewhat futile, particularly if the central and framework-setting character of environmental policy were to remain unchanged. If these frameworks remained in place, area-specific policy would essentially be no different from the effect-oriented policy based on standards and zoning that had been followed for many years. It would be a different matter if such an integrated approach were based on regional or local governance rather than central governance and, bearing in mind the principle of subsidiarity, general framework-setting regulations allowed greater local initiative and creativity. Also important was flexibility for a more balanced and fair weighing-up of environmental policy interests and spatial-planning interests at local level. In that case, area-specific policy would represent a completely new approach.

For area policy to be established at local level, location-specific possibilities and constraints should take precedence over generic source policy and effect-oriented policy. A local balance would then be struck between source-based and effect-oriented policy, within the framework of an area-specific strategy. This was not only a question of local governance designed to protect particular environmental qualities by means of environmental-protection areas. It also involved co-ordinating environmental-health norms and values – probably defined at a local or regional level – with developments in the use and management of areas and locations with several different functions, activities and interests.

Many believe that this development "in addition to target-group policy, [should] be the foundation for the integrated implementation of environmental policy" (Kuijpers 1996; 59, see also Biekart 1994). The question arose, however, whether this was actually the intention of the Department of the Environment. The Action Plan on Area-Specific Environmental Policy created a foundation for this, but when it appeared the discussion on this development had not even begun. For the purpose of this discussion, however, it is useful to consider the contribution such an integral policy consideration can make at local level. If that contribution, or 'added value', can be demonstrated, we must ask ourselves what form of governance – national or local – can increase the efficiency and effectiveness of policy at the local level, at the interface between the 'grey' and spatial environments.

The NMP-1 and the Action Plan reflect two developments in area-specific environmental policy that allow us to examine the possibilities and implications of a shift in the administrative level at which decision-making takes place. One of the developments was the extension of the 'classic' standards system through integrated environmental zoning (IMZ). This development follows on from existing goal-oriented policy. Within the Department of the Environment, IMZ was seen as the next step in a policy that had hitherto been reasonably successful and popular. This meant that expectations were high. As an instrument, IMZ can be seen as a confirmation of a hierarchic and framework-setting form of governance. The proposed system for integrated environmental zoning was primarily intended to contribute to the process of internal integration (see Borst et al. 1995). This is probably why the Action Plan on Area-Specific Environmental Policy described IMZ not as an integral, but as a fragmental policy applied to defined sectors. IMZ was not mentioned in national spatial plans, partly because the RPD (National Spatial Planning Agency) saw it as a serious threat to spatial-planning policy based on a balance of interests. By contrast, the ROM designated-areas policy *was* referred to in the VINO (Fourth Policy Document on Physical Planning) and the VINEX (Fourth Policy Document on Physical Planning-Plus). The ROM policy was the second development in area-specific environmental policy and placed emphasis on building local consensus and devising individual solutions geared towards local circumstances. This approach was not based on the goal-oriented facet of planning, but emphasised the institution-oriented facet of planning (see also Driessen 1996, Gijsberts 1995, Glasbergen and Driessen 1993, Glasbergen and Driessen 1994). It created openings for other – mainly spatial – interests, and therefore did not win universal support. The ROM designated-areas policy made considerable allowance for specific local circumstances, stimulated participation by local NGOs and encouraged local and regional authorities to take greater responsibility for solving environmental/spatial conflicts.

IMZ (integrated environmental zoning) can be seen as an exponent of traditional goal-setting environmental policy, while the ROM designated-areas policy was the exponent of the new-style interactive policy. Experience was gained in both methods as a result of the eleven pilot projects. During the process, the evaluation of standard-based policy changed, and the experiences with integrated environmental zoning acquired a negative interpretation. At the same time, the ROM designated-areas policy was increasingly given the benefit of the doubt, despite the fact that it achieved similar results to the zoning project. Both instruments had their pros and cons, but the question is how these innovative policy instruments influenced the institutional and goal-setting character of environmental policy. This question will be dealt with below and relates not only to the environmental policy sector, but also to the co-ordination and integration of environmental policy and spatial planning.

Integrated Environmental Zoning

The aim of NMP (National Environmental Policy Plan) Strategy 39 was described as 'promoting integrated environmental zoning, area-specific policy and effect-oriented policy'. The strategy related to the policy theme 'disturbance'. Because policy relating to quiet zones was to be found elsewhere in the plan (action point 158), Strategy 39 has only one action point, namely point 82 relating to the implementation of the project *Source Aggregation and Integrated Environmental Zoning* (*Cumulatie van bronnen en integrale milieuzonering*), which was officially launched on 3 April 1989 (VROM 1989b). The project involved a proposal for trials with integrated environmental zoning (IMZ) with a view to "improving the quality of the environment around large industrial complexes" (TK 1989; 152). The proposal was intended to lead to "a system for integrated environmental zoning" (VROM 1988; 3) to be included in Section 6 of the Environmental Protection Act.

The system was never legalised, however. The proposal for the zoning system was seen as an indication of the direction in which the environmental department intended to steer the goal-oriented aspect of environmental policy, and as such it was also seen as too great a threat. In 1996, the Ministry of VROM delegated the IMZ project to the Interprovincial Council (IPO), which then set a completely different course for IMZ. In 1996, the IPO published the draft report *Reisgids ROMIO*[43] (IPO 1996). The report put an end to the goal-oriented and prescriptive character of the IMZ project. So great was the administrative aversion to prescriptive standards policy that the term 'standard' was not mentioned anywhere in the main text of the IPO proposal. Standards and goals were no longer expected to provide solutions and were replaced by interaction, participation and communication. In retrospect, it is surprising to consider how the proposal for integrated environmental zoning – heralded in 1989 by project leader Tan as "a natural conclusion to sectoral environmental legislation and a logical step towards integral legislation" (in Van den Nieuwenhof and Bakker 1989; 11) – was viewed at the time, after having been set aside for several years. Although, in the end, the ministry's proposal for integrated environmental zoning[44] was not given a structural place in standard environmental policy, IMZ was an important project, above all because it was the basis for discussion on how environmental/spatial conflicts should be dealt with in the Netherlands.

The concept of integrated environmental zoning was presented at a time when the environment was at the top of the political and administrative agenda. There was also broad public support for far-reaching measures. The government received requests for funding for remediation from several quarters, and there were calls for the further development of the standards system (see § 5.3). The Spatial Planning Council (RARO) was not alone in believing that "it is essential to pursue an integrated policy to prevent the disturbance of living and working environments in urban areas" (1989; 26), and was therefore "in agreement with the plans to define integrated environmental zones for 15 complex industrial locations" (1989; 26). Finally, not 15, but 11 pilot projects completed the zoning programme proposed by the ministry.[45]

In the Netherlands, the feeling was that the scale of environmental pollution had peaked and could only decrease in the future. It was felt that the consequences of zoning would not be too radical. The RARO (1992; 15) reflected this belief when it pointed to (1) the continuous advances in environmental technology, which would improve the environment over time, (2) increasing wealth, which would place increasing demands on the quality of the environment, and (3) the expectation that enforcing environmental policy would also improve the quality of the environment. It was expected that standards could be tightened-up over time. In response to the grassroots support for more stringent standards, the environmental department undertook to "introduce some form of integrated environmental-zoning system as soon as possible" (VROM 1989d; 3) in order to help resolve "the longstanding problems between industry and residential development" (VROM 1989d; 3).

In retrospect, the type of zoning method and system[46] chosen and the way in which it could be incorporated into existing regulations largely determined how the idea of integrated zoning, and the standards system in general, were evaluated. A number of zoning methods were already in use, including that of the VNG (Association of Netherlands Municipalities), the IZ-DSM method (see § 5.2), and methods implemented or being developed by various municipal authorities such as Amsterdam, Arnhem, Maastricht, Oss, Spijkenisse and Zaandam (De Roo and Van der Moolen 1991). Following the Ministry's initiative, the idea of integrated zoning crystallised around one specific system: *VS-IMZ*, a provisional system for integrated environmental zoning (VROM 1990). The discussion on standards was influenced not so much by integrated environmental zoning as a concept but by the basic principles, limiting conditions and possible consequences of the provisional system (VS-IMZ) proposed by VROM.

The VS-IMZ system had a methodological component, a standard-setting component and a procedural component. The methodological component was designed to ascertain, by means of aggregation, the total environmental load. The standard-setting component was intended to evaluate environmental load and the related planning consequences. The procedural component of VS-IMZ aimed to strike a balance between the desired level of environmental quality and the desired spatial-functional structure of a given area (VROM 1990). VS-IMZ could therefore also have administrative consequences that would necessarily result in at-source measures and planning measures, in consultation, negotiation and agreements with industry and other actors, and ultimately in an integrated environmental zone defined in the land-use plan. The system was also valuable in a preventive sense: "the extent to which VS-IMZ allows for the harmonisation of decisions relating to the location, expansion and restriction of environmentally harmful and environmentally sensitive sites is such that it creates a sustainable equilibrium between environmental impact in areas that accommodate environmentally sensitive activities, namely housing, and the development potential of environmentally harmful activities, particularly industry" (VROM 1990). The result was a prescriptive system for the 'sustainable' segregation of residential locations and industry, with a view to achieving a more favourable balance between decisions relating to the location, expansion and restriction of environmentally harmful and environmentally sensitive functions, activities and areas.

The method is based on the proposals of the DSM Task Force (see § 5.2, VROM/Task Force DSM 1987, and Van den Nieuwenhof and Bakker 1989; 9), whose purpose was to draw up an inventory of pollution levels at the DSM site. The Task Force also submitted proposals designed to incorporate workable standards for the DSM site into the existing standards system. An integrated zoning method was then developed for the DSM site, based on actual measurements taken (in contrast, for example, to the VNG method described in section 5.2), and the different levels of pollution were aggregated. The VS-IMZ method comprises standardised levels for industrial noise and odour, for external safety and for toxic and carcinogenic substances (Table 5.1). It also comprises a classification system for balancing the various standards, and an aggregation method that could serve as the basis for integrated zones (see De Roo 1993c; 373-374). Despite the fact that standards varied for the different types of pollution, the VS-IMZ advocated a similar approach for all pollution types and, in the same way as the Noise Abatement Act, it differentiated between new and existing situations. Sectoral classification produced values for each type of pollution. The values, which could be compared with each other, ranged from 'negligible' to 'unacceptable' for industrial noise, odours, risk from industrial installations, and toxic and carcinogenic substances. Local authorities in particular were worried that this classification would lead to stricter standards, particularly with regard to odour and certain forms of air pollution (Borst et al. 1995).

The classification used was not based on the real cumulative effect of pollution. Knowledge and resources available at the time were simplistic and could not produce a single accurate assessment of the dose/effect relationship for all types of pollution taken together (Aiking et al. 1990; 48). However, it was assumed that, at locations where several forms of pollution contributed substantially to the environmental load, the total load would exceed the most prominent individual form of pollution (see also Lammers et al. 1993). The permitted maximum level of total or aggregated environmental load remains dependant on individual (sectoral) environmental standards. The classification method whereby the various load are aggregated was not appropriate as a health and hygiene criterion because it lacked a scientific foundation and wsa arbitrary in character. It could therefore only be used as a policy-based administrative or political decision. This form of aggregation reflected the sentiment that 1 + 1 was greater than 1, which would have been the outcome of a sectoral approach, and was therefore a realistic approach in terms of policy.

The aggregation table could be used to determine the integrated class for a given location, and to predict the consequences for spatial planning (see Fig. 5.4). VS-IMZ comprised six integrated environmental-quality classes.[47] Class I was the 'white area' where pollution was negligible or non-existent. Classes II to V reflected the 'grey area', with increasing planning constraints. Class VI was the highest class, the 'black area' where the level of pollution was unacceptable for residential areas. This class was subject to the greatest constraints in terms of housing development. New housing developments were not permitted in areas placed under 'Class VI for new situations', and homes located in a 'Class VI for existing situations' area would have to be demolished (VROM 1990).

Table 5.1 Categorisation of sectoral environmental load according to VS-IMZ

Sectoral Class[a]	**E**	**D**		**C**	**B**		**A**
Category	Existing	Existing	New	Existing & New	Existing	New	New
Industrial Noise, in dB(A)	> 65	65-60	>60	60-55	< 55	55-50	< 50
Risk from industrial installations, as individual mortality risk per year	$> 10^{-5}$	10^{-5}-10^{-6}	$>10^{-6}$	10^{-6}-10^{-7}	$< 10^{-7}$	10^{-7}-10^{-8}	$< 10^{-8}$
Odour, in odour units per m^3 as 98th percentile	> 10	10-3	>3	3-1	< 1	< 1	< 1[b]
Carcinogenic substances, as individual mortility risk per year[c]	$> 10^{-5}$	10^{-5}-10^{-6}	$> 10^{-6}$	10^{-6}-10^{-7}	$< 10^{-7}$	10^{-7}-10^{-8}	$< 10^{-8}$
Toxic substances, as % of cumulative No Observed Adverse Effect Level (NOAEL)	> 100	100-10	> 10	10-3	< 3	3-1	< 1

a Sectoral Class A does not apply to existing situations and sectoral Class E does not apply to new situations.
b As a 99.5 percentile score.
c The maximum cumulative individual risk for carcinogenic substances is $x.10^{-6}$ (where x = the no. of substances), with a maximum risk of 10^{-5}.

Source: VROM 1990; 14-15
Note: VS-IMZ is the Provisional System for Integrated Environmental Zoning

The results of the eleven VS-IMZ pilot projects provide a clear overview of the areas that were subject to environmental load from extensive and complex environmentally harmful activities[48] (see Chapter 6, Borst et al. 1995 and Borst 1996). Overviews were produced of integral environmental quality levels in areas surrounding the industrial locations studied. Surprisingly, however, these results not only produced detailed overviews and confirmed the relevance of aggregation,[49] but also reflected the perceptions of environmental/spatial conflicts that existed before quantitative information on the situation had been gathered. Perceptions of environmental/spatial conflicts often overestimated or underestimated the situation, but in all cases they differed significantly from measured values (Borst et al. 1995). Given that minor modifications to environmental contours had almost immediate consequences for spatial development, the overview presented of the environmental situation had to be as realistic as possible.

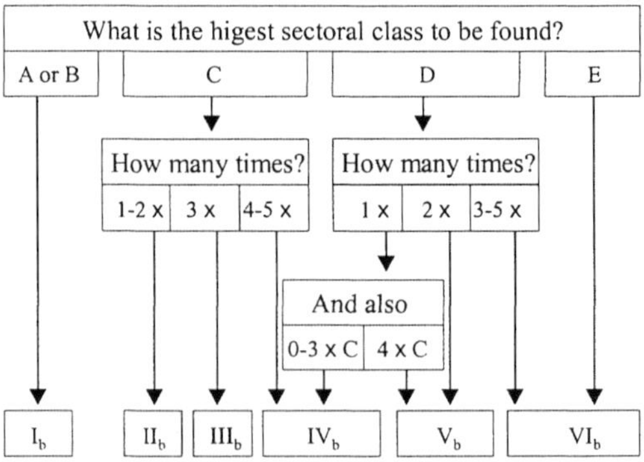

Figure 5.4 Aggregation table for integrating pollution classifications for 'existing situations'

Source: Stuurgroep IMZS-Drechtsteden 1991; 78

Using actual pollution levels instead of indicative or average values had a drawback that was difficult to foresee at the beginning of the pilot projects. A number of projects resulted in the actual demolition of homes. In Hengelo-Zuid it was considered necessary to demolish homes adjacent to the Twentekanaal industrial complex (Flohr and Meijvis 1993; 99). In Burgum, 36 homes were demolished (Borst et al. 1995; 137). In a number of pilot projects, including those in Groningen and Geertruidenberg, demolition was not necessary and in others – Arnhem and the Drecht cities are the best examples – demolition was out of the question because of the cost involved, the clash with the spatial-planning concept of the 'compact city' and – last but not least – the predicted public opposition. The IMZ pilot projects produced the first structured overview of a number of environmental/spatial conflicts involving several types of environmental load. Digital mapping and colour presentations made several conflicts, for which VS-IMZ had disastrous consequences, even more persuasive as examples.

These cases brought unpleasant surprises for many of the parties involved (Boei 1993, Van der Gun and De Roo 1994, De Roo 1992 and 1993c, Voerknecht 1994).

In a number of situations involving considerable aggregated levels of pollution, actors who were likely to be confronted in the future with statutory IMZ were haunted by images of multiple 'black' sites requiring remediation. The most striking example was the black rectangle depicting the results from the VS-IMZ pilot project in Arnhem (Project Organisers IMZA 1991, Boei 1993). The rectangle represented an area of five by five kilometres, at the centre of which lay the Arnhem-Noord industrial complex (Project Organisers IMZA 1991). The nearby city centre of Arnhem was also included in the black area. The detailed maps showing pollution levels within the boundaries of the Drecht cities also provided a clear picture (Stuurgroep IMZS-Drechtsteden 1991). Extensive remediation and demolition was required in an area with a high housing density and considerable economic potential that also had to accommodate a high-speed rail link, the starting point of the Betuwe rail link, and 13,700 homes under VINEX (Fourth Memorandum on Physical Planning-Plus, VROM 1996). It was also an area in which industry was developing fast. What would now happen to the Netherlands if IMZ was implemented in the other 246 industrial sites where heavy industry was located and two or more forms of environmental pollution were present (SCMO-TNO 1992)?

The study *Environmental Zones in Motion* (Borst et al. 1995), commissioned by the Ministry of VROM, examined 11 IMZ pilot projects and a further 25 potential IMZ projects. The study considered the nature and scale of each environmental/spatial conflict. It was found that environmental/spatial conflicts could be usefully categorised according to their complexity (see Chapter 6). The categories were based on the different aspects of environmental pollution, their possible effects, the geographical scale of the pollution, the number of polluting sources, the spatial structure of the area around the source and the spatial dynamics of the area affected by the environmental/spatial conflict.

The first approximate classification into five categories (see Fig. 5.5) was based on the relationship between the spatial structure and the scale of the pollution. Certain environmental/spatial conflicts involved relatively small-scale pollution and/or a relatively small number of homes. In both cases the problems were limited because there was very little overlap between the environmental pollution and environmentally sensitive residential areas. Figure 5.5 shows the distinction made between situations A1 and A2. In situation A1, although a relatively large number of homes directly surrounding the source are affected, the low level of the pollution means that the situation is not characterised as 'complex'. Situation A2 involves extensive pollution but the housing density in the affected area is low, so this is categorised as a 'relatively straightforward' situation. The vast majority of environmental/spatial conflicts fall under the 'A' category. Situation B involves environmental/spatial conflicts in which the spatial structure is such that a relatively large number of environmentally sensitive functions are exposed to several types of pollution. These situations require more integrated co-ordination of limiting environmental conditions and spatial-development potential. The final category comprises 'very complex' environmental/spatial conflicts. The number of conflicts in this category is relatively low but they relate to large areas with a high housing density where industry and homes are usually heavily interspersed. This means that radical environmental and spatial-planning measures were required to resolve the conflict satisfactorily. The distinction between categories C1 and C2 is mainly theoretical, and category C2 is intended to identify the 'hot spots'. The assumption is that this small number of 'hot-spot' conflicts cannot be solved within the hierarchical standard-based framework of environmental policy: "Other approaches are required to achieve the most realistic result" (Borst et al. 1995; 219). The following categories of environmental/spatial conflict resulted from these distinctions: 'straightforward' (A), 'complex' (B) and 'very complex' (C) (compare Fig. 4.9). The categories will be discussed again in Chapter 6 and developed into three decision-making strategies. The choice of strategy will be determined by the level of complexity of the environmental/spatial conflict.

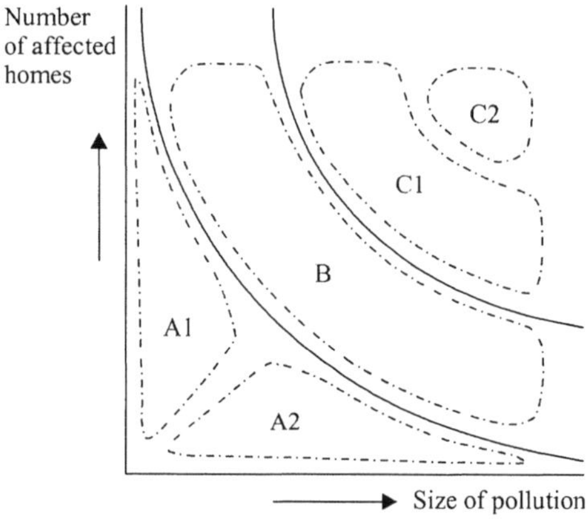

Figure 5.5 Environmental/spatial conflict categories for environmental zoning projects

Source: Borst et al. 1995; 218

Using the *Environmental Zones in Motion* study as the basis, a careful extrapolation was made for the remainder of the 257 'heavier' industrial sites subject to two or more types of pollution. Approximately half of the 257 cases in which the surrounding areas were affected by several forms of pollution from the industrial sites could be categorised as 'straightforward' environmental/spatial conflicts. It will be relatively easy to resolve these conflicts within the framework of standard-based environmental policy (compare Fig. 4.9). The estimated total number of industrial sites in category B was one hundred. These environmental/spatial conflicts could also be solved using standard-based policy, provided this was not done in a cavalier manner. In these situations, environmental standards had to be applied more flexibly and in a location-specific, function-related way (see De Roo 1996). In the Netherlands there were several dozen 'very complex' environmental/spatial conflicts for which standard-based policy was not the obvious method for achieving an acceptable physical and human living environment in an effective and efficient way (see Fig. 4.9). The existence of these 'complex' and 'very complex' environmental/spatial conflicts exerted pressure on the limiting conditions of VS-IMZ and on standards policy in general.

The methodological component of VS-IMZ provided a 'reasonable' to 'good' overview of the environmental situation in a given area (Borst et al. 1995, De Roo 1993b, De Roo and Van der Moolen 1991). This is, therefore, not why VS-IMZ was ultimately scrapped as a system. The method had even more to offer, since it was also used in the pilot projects to update information on environmentally harmful activities and to bring together and manage that information in an integral database for the purpose of issuing permits.

Yet the method was criticised. The Health Council (*Gezondheidsraad*) claimed that it was "not possible to construct a single line of measurement based on empirical science or medically acceptable arguments for measuring nuisance, sickness and mortality and risk" (Gezondheidsraad 1995; 32). The relevance of this criticism is debatable given that the method was designed primarily as a policy tool. Nevertheless, VROM interpreted the comments by the Health Council as a sign of diminishing support for integrated environmental zoning. The practice of restricting the system to only those industrial areas where activities were subject to regulations for large-scale, complex fixed industrial installations was also criticised. In a number of cases, activities on a smaller scale at or close to the industrial site, such as scrapyards, bakeries or paint companies, were also perceived as harmful nuisance by local residents. A number of other sources caused nuisance that affected local residential areas but were not located at the site and had no planning consequences under VS-IMZ. Non-industrial traffic noise was given as an example in the Arnhem-Noord pilot project (De Roo and Van der Moolen 1991; 42). In the Groningen Northeast Flank project, the primary environmental impact was caused by odour emitted from two sugar refineries located a considerable distance away from the project area (Gemeente Groningen 1992). In such cases, residents and industry in the project area do not always recognise or understand the relevance of proposed spatial-planning measures for locations that are not in the immediate vicinity.

Having initially adopted a positive 'wait-and-see' stance, industrial stakeholders soon began to point out the disadvantages of VS-IMZ, including its many inherent uncertainties (VNO-NCW 1991) and in particular the impossibility of 'adding up' levels of noise and odour nuisance. The latter was a frequently voiced objection relating to an aspect that, as such, did not occur in VS-IMZ.

The aim of VS-IMZ was to create uniformity, equity before the law, and clarity where they were needed: in the environmental standards system and in dealing with problems relating to housing and work locations. The belief prevailed that not only the technical details of the system were straightforward, but also the administrative relationship between the environment and spatial planning. At the beginning of the VS-IMZ project, De Boer, the member of the Noord-Holland provincial executive who was responsible for the environment, saw this relationship in terms of responsibilities delegated between the environmental and spatial-planning policy sectors, whereby the *positioning* of zone boundaries was the responsibility of spatial planners and the *nature* of the zone boundary was the responsibility of environmental policy-makers (in VROM 1989b; 9). This approach proved incompatible with the complicated administrative and societal reality of many environmental/spatial conflicts.

Not only the methodological component of VS-IMZ was criticised, but also its standard-setting component. This criticism was more serious. Environmental standards for several types of environmental impact had been developed in line with the Noise Abatement Act (see § 5.2). However, this did not prevent the standards from becoming inconsistent, resulting in irritation among permit-issuing bodies and industry. The inconsistency related among other things to the conceptual framework,[50] the details, the legislative aspects, deviations, the influence on spatial zones, the consequences for spatial planning, and the degree of functional and spatial differentiation. The latter issue was heavily debated.[51] The general character of standards systems is such that all actors, regardless of location, are aware of their position. This should result in a large degree of equity before the law. Each individual can therefore reasonably expect the same degree of protection. In determining the degree of protection, a large population group is taken as representative of Dutch society. This evaluation is not necessarily in line with the local perception of nuisance by individual residents or groups of residents. Cleij, Director of the Amsterdam Environmental Department, questioned whether it was practical to reduce noise levels at the city's Rembrandtplein in line with legislation (in Bakker 1994; 5). The RARO also stated that it was desirable to "deal more explicitly in day-to-day policy practice with the dynamic, time-related and place-related aspect of environmental zoning" (1992; 10).

The level of odour standards was also heavily criticised (§ 5.5, De Roo and Van der Moolen 1991). Perceived odour nuisance was not in line with either the level of impact expressed in standards or the extent of measures taken. In addition, there were environmental aspects for which standards had been developed only in part or not at all, while foreign guidelines (mainly German) were applied to other aspects such as dust and vibration. Integrated environmental zoning appeared to be the logical solution to these 'blank spots' in the standards system. However, VS-IMZ incorporated only two functions (living and work) and five aspects of environmental impact, while dust, vibration, traffic noise and soil contamination are also forms of pollution that result from industrial activity and can seriously affect the quality of the local environment (Borst et al. 1995). More importantly, the initiative could not develop fully due to the labyrinth of sectoral specifications and related interests and issues. Each environmental standard has its own history of development and adaptation that are partly the result of attempts to avoid spatial-planning consequences and prevent costs from escalating. The inconsistency in the environmental standards system, and its varying statutory, recommended and interim values, guidelines and circulars, also influenced the degree of support for VS-IMZ as a legal instrument. In practice, most of the guidelines and recommendations were statutory to a certain extent (RARO 1992; 13) but this did not prevent critics from labelling them as 'weak'. Only the standards for industrial noise had a clear legal status, which meant that executive bodies were confronted with legal constraints when applying integrated environmental zoning. Industry did not therefore consider itself bound to comply with the possible consequences. In addition, the development of the standards system and the underpinning of VS-IMZ were increasingly influenced by the consequences of target-group policy, which was being followed at the same time as effect-oriented environmental policy. Thanks to these developments, VS-IMZ sometimes caught up with the standard-formulating process, although agreements were sometimes made that resulted in the standard-formulating process failing to keep pace with proposals for VS-IMZ. These internal and external developments relating to the standards system not only affected the consistency of the integral approach, but also made it less effective.

Nevertheless, a hierarchical top-down standards system had its advantages. Local authorities, for example, did not need to define, consider or discuss the level of environmental quality that should be acceptable for each individual environmental/spatial conflict (see also Rondinelli, Middleton and Verspoor 1989). National standards also provided support for local authorities when they held industrial actors accountable for achieving targets relating to a local environmental/spatial conflict. The discussion on the proposed methodology of VS-IMZ shows that "in general, an integrated approach to environmental/spatial problems is viewed in a positive light" (Borst et al. 1995; 79). However, these advantages could not swing the balance in favour of VS-IMZ.

In addition to standard-setting and methodological components, VS-IMZ also comprised a procedural component. Procedures linked the existing/permitted level of pollution, translated into spatial zones, and the consequences of those zones for spatial planning. This procedural structure resulted in qualifications to the prescriptive nature of VS-IMZ as an instrument. After an overview of the environmental impact had been obtained, it was measured against the environmentally sensitive area in order to ascertain the scale of the environmental/spatial conflict. This information did not immediately lead to the relocation of factories or the demolition of homes. In the event of an environmental/spatial conflict, the first step was remediation, whereby at-source measures took priority. VROM had expected that environmental zoning, as an effect-oriented instrument, would be effective in pressing for at-source remediation (Van den Nieuwenhof and Bakker 1989). The procedure was intended to define and initiate the necessary remediation, although it was not always possible to force a company to take satisfactory measures. Permits might have expired, the company may have failed to comply with the permit conditions, or may not even have one. The chicken-and-egg question (i.e. which had been at the site longest – industry or housing?) might be relevant, as might the extent to which housing developments had used up the space that would otherwise have been available to the company.

In this sense the municipal and provincial authorities, which issued building permits and environmental permits, were equally responsible for the problems they now faced. This places one of the central principles of environmental policy – the Polluter-Pays principle – in a somewhat different light. Although regional and local authorities were responsible for safeguarding the quality of the environment at local level, spatial development was also an important criterion in their negotiations with industry. New developments were needed and the open space between an industrial location and the local housing was suitable for this purpose. However, because companies generally aim to continue expanding, they seek a larger 'allocation' of space than is required for current activities.

It was not easy to reach agreements on the scale of remediation required or cost allocation. The deadline for completing the remediation was also a subject for negotiation. VROM expected this term to be approximately one year, after which the 'provisional' integrated environmental zone would be incorporated into the land-use plan (VROM 1990). In cases of extensive environmental impact, it was often impossible for companies to implement measures in the short term without incurring excessive costs. They preferred measures that allowed for gradual replacement investment. It was not surprising that the point at which the implementation of VS-IMZ at a particular site was announced to industry, and the point at which industry was first actively involved in the project, were subjects of debate and became more important as the zoning project progressed.

It was not only industry that required greater flexibility. The incorporation of the integrated environmental zones into land-use plans brought objections that stood in the way of flexibility. Land-use plans offer legal certainty because they are legally binding, but "the lack of flexibility in spatial planning and the impossibility of incorporating modifications to environmental standards immediately into the land-use plan" (RARO 1992; 19) do not sufficiently address spatial urban dynamics and the consequences of increasing knowledge of and insight into environmental impact.[52] If, in addition to this, the integrated classes for new and existing situations are translated into different integrated environmental zones, the result is a complicated jumble of contours and a list of consequences for spatial planning. All in all, this does not make for a transparent and workable situation.

The pilot projects revealed that the stringent application of VS-IMZ was not always possible. The environmental/spatial conflicts in which VS-IMZ was used were generally more complex than they at first appeared (Twijnstra Gudde 1994; 39). Cases of supra-local environmental impact can have far-reaching consequences (Borst et al. 1995). "It is precisely this complexity that creates the impression that, in a number of cases, VS-IMZ is ineffective as a method and procedure for realising the sustainable segregation of environmentally harmful activities and environmentally sensitive functions. Defining integrated environmental zones is more difficult than the relative straightforwardness of VS-IMZ would lead one to expect" (Borst et al. 1995; 78-79). Given these tensions, incremental standards could not be considered realistic. However, there was discussion about the tendency on the part of stakeholders to use restrictions to the full, in other words they would allow emissions etc. to reach the upper limit, rather than aiming to keep levels as low as possible. We should note here that this problem did not apply exclusively to environmental zoning. The stringent implementation of sectoral standards would have led to similar problems (De Roo 1992).

The resistance to VS-IMZ was largely due to the method of 'aggregation' of consequences and the transparency with which they were presented. The nature of the consequences of the system initially led to doubts as to how realistic it was (Borst and De Roo 1993, Ten Cate 1992, 1993, Weertman and Nauta 1992). The purpose of environmental zoning was to achieve a given level of environmental quality by means of spatial segregation. Therefore, particularly in complex situations, environmental policy was prescriptive in terms of spatial-planning policy. "Environmental policy dictates the consequences for spatial planning, so that spatial planning can no longer weigh up alternatives" (Borst et al. 1995; 77). This was unfortunate because different aspects of spatial policy could contribute to improved liveability too. Environmental zoning could therefore create a link between environmental policy and spatial policy, but it could not camouflage the essential differences between the two policy areas.

In the VS-IMZ instrument, the government's definition of environmental/spatial conflicts was based on technical arguments that did not necessarily need to match local perceptions of environmental/spatial conflicts. The relationship between the problem and the extent of the consequences was set out in arbitrary terms and did not always find agreement at the local level. Standards and spatial-planning consequences imposed by central government restricted local involvement in finding solutions to problems that had also been defined at a higher level. Local and regional governments encountered administrative problems once it became apparent in local discussions that the conflicts were not the responsibility of industry alone. Past local developments and the need for local authorities to realise a level of environmental quality compatible with housing development meant that two sets of interests were involved. It became clear that an acceptable environmental quality level could not be attained without consultation and co-operation between actors. The fact that problems were defined by a higher level of government, which also defined the framework for seeking solutions, limited the possibilities for shared governance at a lower level. This resulted in friction. Local authorities required detailed information regarding what was or was not possible or permitted. This need increased in proportion to the degree of overlap between environmentally harmful and environmentally sensitive areas, and in proportion to the number of actors involved (Borst et al. 1995, see also Chapter 6). At the same time, central government was unable to harmonise its generic regulations with the detailed information required by local authorities. National standard-based policy could not remain inconsistent in the face of specific and very diverse local situations, and the uncertainties surrounding government policy increasingly became an obstacle.

Moreover, it was difficult to predict to what extent other forms of government policy would influence environmental/spatial conflicts. Local authorities came under a great deal of pressure from compact-city policy (see Chapter 3). Space was a scarce resource in 'compact' cities and it was conventionally allocated by striking a balance between the needs of local actors. Scarce space came under even more pressure due to the implementation of environmental standards and zones. The 'ABC location' policy (see Chapter 3) also resulted in dilemmas that made it difficult to prevent environmental/spatial conflicts and deal with them in the most effective way. The aim of locating labour-intensive workplaces close to railway stations could not always be achieved because office development was not permitted due to the risk contours resulting from the presence of harmful substances at shunting yards. In such cases it was almost impossible to strike a balance between the various interests involved. Environmental quality was presented as an absolute that was not negotiable beyond narrow margins.[53] By contrast, the process of weighing up all the interests involved produced a given spatial quality over which environmental quality would take priority.

This was increasingly seen as unacceptable, partly because of the large amount of space required for urban development in the Netherlands (see Chapter 3).

It was not surprising that some actors came to fear the consequences that environmental zoning would have on local spatial planning, local administrative relationships, the relationships between interests and parties at local level, and the relationship between local, regional and national government. These reservations were discussed at the Nunspeet Conference (see Chapter 1) in 1994. At the conference, various authorities raised the subject of how prescriptive environmental quality should be after it appeared that, in a number of cases, the costs exceeded the benefits when standards were stringently applied. These were cases in which strict compliance with environmental-quality standards had a disproportionately negative impact on spatial quality.

For a long time, it was taken for granted that environmental pollution would decrease. It was also assumed that the relationship between spatial planning and environmental policy could be managed simply by maintaining a 'safe' distance. In a number of cases, however, environmental/spatial conflicts proved to be extremely complex. Partly for this reason, the basic principles of environmental policy came up for discussion, including the 'Polluter-Pays' principle, the standstill principle and incremental standards. It became clear that, in these complex environmental/spatial conflicts, central government only had a limited overview of how effective its regulations were at a local level. This in turn meant that insufficient allowance was made for local – and often unique – circumstances. The conclusion of the Nunspeet conference was therefore that, in principle, local environmental problems should be resolved at the level at which they arose (§ 1.2, VROM 1995). It also became clear that this did not necessarily apply to all environmental/spatial conflicts. The majority of conflicts could be resolved effectively and efficiently by means of centralised prescriptive policy. Moreover, it was not simply a matter of resolving the conflicts. In more complex situations, it was also a question of local authorities and other non-governmental organisations (NGOs) sharing responsibility for the whole spectrum of tasks relating to problem definition, solution strategies and the resulting planning consequences.

The following chapters will examine the possibilities for shared governance and for placing more responsibility with local government with regard to developing and implementing operational and strategic environmental policy. In the past, a consequence of policy-implementation methods was that local government was willing to take a greater share of the responsibility but lacked the resources and experience (see also CEA 1998). After all, hierarchic government regulations did little to motivate local authorities to place local environmental/spatial conflicts in a broader context. This means that there was very little interest at the local level in strategic, area-specific and participatory policy.

In 1995, the environment department at the Ministry of VROM was still seriously considering defining integrated environmental zones for the 50 most appropriate sites of the 257 sites subjected to two or more forms of environmental impact. It proposed to do this by means of a governmental decree (AmvB). In 1997, however, VROM announced that it considered statutory integrated environmental zones to be 'a bridge too far', and that it no longer supported that goal (VROM 1997). It believed that VS-IMZ should be seen only as "an instrument intended to support complex local policy considerations relating to the spatial and environmental demands on areas surrounding industrial locations" (Baaijen 1997).

The ROM Designated-Areas Policy

When examining several different forms of environmental impact at the same time, it is more practical to use an approach that considers the forms of impact together and translates them into an environmental zone, rather than one that is fragmented or sectoral. However, we also observed above that the 'added value' of such a method, and therefore also the integrating function of environmental zoning, decrease as environmental/spatial conflicts become more complex (i.e. as the level of impact increases, as functions become more interspersed, and as the number of actors increases). In such cases integrated environmental zoning – at least in the form proposed by VROM – would have a negative rather than a positive effect on the relationship between environmental policy and spatial-planning policy and could even result in a decrease in spatial quality and the integral quality of the immediate environment. The VNG observed that: "The remediation, preservation and protection of special areas and the resolution of complex environmental problems demand an unorthodox form of result-oriented co-operation based on projects" (1993; 9). The ROM designated-areas[54] policy was also an integrated area-specific approach to spatial planning and the environment, and it seemed to be an appropriate solution. The ROM designated-areas policy represented a shift away from policy that was largely restricted to setting limiting conditions (TK 1990; 14). It was a move towards an 'unorthodox' participative area-specific approach.

The ROM designated areas were experimental and implemented at a time when prescriptive environmental policy was becoming increasingly unpopular and enthusiasm for participative decision-making was growing. The question was whether the ROM approach was efficient and effective enough to produce the added value that a prescriptive instrument could not. The ROM policy was more explicit than IMZ in stating that "the goal of the experimental ROM designated-areas policy [...] is also to integrate spatial-planning policy and environmental policy" (RARO 1992; 10). However, the ROM policy had very little else in common with integrated environmental zoning. The ROM approach was based on external integration, which would be achieved by striking a balance between various important aspects within a pre-defined geographical problem area. This involved geographic differentiation and the integration of administration and policy-making (see Bakker 1989). It was, therefore, not simply a question of environmental policy exerting unilateral influence on spatial planning, but of creating a level playing field for all areas of policy, particularly for environmental policy and spatial planning.

ROM policy brought about a shift away from command-and-control governance – which had been so evident in integrated environmental zoning – towards direct participation by actors.[55] In other words, its approach was based on developing 'self-governance' in addition to direct regulation (Van Tatenhove 1993; 23) and was in line with the network-based approach (see § 4.7). The policy was based on the assumption that the actors were all dependent on each other to the same extent, which meant that a balance would have to be struck when dealing with commonly perceived problems. The designated-areas policy was typically a product of its time. It was expected that co-operation between the various actors (*participation*) within the framework of local possibilities and constraints (*tailor-made solutions*) would generate added value. Herein lay the fundamental difference with standards-based policy. The ROM method was not only the opposite of environmental zoning, but also differed considerably from existing top-down policy.

This does not mean that the ROM approach ignored the basic principles of national environmental and spatial-planning policy (see also Glasbergen and Driessen 1993; 135), rather that there was a shift of focus towards institutional relationships, interactive processes and an area-specific approach. The goal-oriented aspect of planning was therefore less concrete and apparent than in integrated environmental zoning. Nevertheless, the ROM policy aims for a general and centrally formulated objective, namely to realise a 'general' level of environmental quality in areas where this is lacking and to maintain or improve any 'special' qualities in areas that are considered to be exceptional from an environmental point of view (TK 1989; 177-178).

Part 1 of VINO (Fourth Policy Document on Spatial Planning; VROM 1988c) named six areas that were 'polluted' but also had high economic potential. The idea was that, during the decision-making process for the spatial development of these areas, spatial-planning measures would be used to underpin environmental policy. One of the principles of VINO was therefore to make the most of area-specific opportunities. Schiphol and Rijnmond, two areas designated as national *mainports,* were included in the list of 'polluted' areas.

Schiphol Airport in particular was seen as a test case that would have to show "to what extent environmental interests count with regard to 'the most powerful driver of the Dutch economy'" (Van Peperstraten 1989; 8). On 21 September 1989 an agreement was signed that was a stimulus for the development of an *Action Plan for Schiphol and Environs* (Steering Group, *Actieplan voor Schiphol en Omgeving* 1990). The nature of the relationship between Schiphol and its surrounding spatial and natural environments had been a subject of discussion for many years (see § 5.2). Schiphol must, could and would be expanded, and space was required for that expansion. However, the expansion clashed with other interests. For example, the regional plan for the Amsterdam-North Sea Canal area, drawn up by the province Noord-Holland (1987), included the construction of 80,000 homes in the area between Amsterdam-Diemen and Haarlemmermeer-Uithoorn. This was in line with the national policy set out in VINO and confirmed in the subsequent VINEX policy document.

In the Rijnmond area too, the proposed expansion of the port area, which was necessary to accommodate economic growth, clashed with environmental protection interests and posed a risk to nature conservation areas and liveability in residential areas (Stuurgroep ROM-Rijnmond 1992). Problems loomed large: a shortage of port capacity, increasing congestion, and declining environmental quality (Glasbergen and Driessen 1993). These physical problems, together with the administrative structure of the Rijnmond area, gave rise to many complex environmental/spatial issues. The transboundary nature of the problems was not matched by consensus between the local authorities or the will to co-operate when the ROM project was launched (Beerkens 1998). The different authorities believed that they could solve the problems independently and they therefore rejected almost every initiative proposed (Aart et al. 1993). This attitude eventually changed and actors acknowledged the many different – and often conflicting – interests involved. It was also acknowledged that the 'costs and benefits' were not shared equally among the actors (Stuurgroep ROM-Rijnmond 1992; 1). This fact presented them with a 'considerable challenge' (Stuurgroep ROM-Rijnmond 1992; 1).

The Zeeuws-Vlaanderen Canal Zone was also a ROM designated area. At least, a project had already been launched,[56] which was then taken up in the national initiative. In 1987 in the memorandum *An Area-specific Approach to the Environment in the Canal Zone*, the province of Zeeland had already referred to the possibility of a joint administration project within which local authorities and private-sector stakeholders would co-operate to solve complex regional problems using an area-specific approach. The Zeeland memorandum served as an initial agreement, which – despite initial misgivings on both sides (Beerkens 1998; 37) – was supported by the various government authorities as well as industrial stakeholders. The aim of the project was to improve the environment and the economic structure of the region, focusing on the "fair distribution of the costs and benefits to industry" (Beerkens 1998; 37).

The Gelderse Vallei area was exposed to environmental impact caused by the excessive use of fertilisers for agricultural activities. The structure of the economy was also weak and the extensive agricultural activities threatened the ecosystem. The province of Gelderland's environmental-policy plan for 1987-1991 designated the Gelderse Vallei as an experimental area for the restructuring of intensive livestock farming. In VINO the reorganisation of the regional spatial structure was added to this. The problems were widely recognised throughout the region (Kusiak 1989). The situation was similar to that in other ROM areas. The Gelderse Vallei, De Peel and Central Brabant belonged to the 'polluted' category and priority was given to measures for improving the 'grey' environment.

In the national plans VINO and in the NMP-1 (action point 95), four 'relatively clean' locations were designated as ROM areas in addition to the six 'polluted' areas.[57] In the 'relatively clean' areas, priority was given to conserving the 'green' environment. These areas were the Gooi/IJmeer, the 'Green Heart' of the western Randstad conurbation, the Limburg hills and the peat grasslands of Friesland. Initially, an area of the Drenthe Plateau adjoining the Friesian peat grasslands was part of this fourth 'clean' ROM area. The provincial authorities of Drenthe, however, wanted to designate another region: the Aa river valley / Elperstroom. This resulted in five ROM areas in which certain environmental-quality aspects were higher than the nationally accepted level. It was predicted that these unique features would be lost if measures were not taken to protect them.

The ROM projects had a number of features in common, namely environmental/spatial conflicts involving many diverse economic, social and area-specific aspects and activities. These projects were also characterised by a large number of actors representing a diversity of interests. The nature of the environmental/spatial issues therefore ranged from 'relatively complex' to 'very complex'. As we pointed out in Chapter 4, this has consequences for the choice of approach. The issues were specific and interwoven with the regional context to such an extent that a policy that was generally formulated and centrally directed would have been ineffective.

Each of the eleven ROM projects involved supra-municipal issues. The nature of the problems was such that "the environmental impact transcends municipal boundaries, but policy is constrained by them" (Bouwer 1996; 49). It was not possible to identify a single responsible authority or find an appropriate administrative level for dealing with the issues. Administrative relationships would therefore determine the path to be followed. Conventional policy, which was based on traditional levels of administration and clearly defined administrative frameworks, would – it was expected – provide insufficient possibilities for reaching a coherent vision and related measures (VROM 1994c, VROM 1998b). It was acknowledged that authorities still worked too much within existing administrative boundaries, basing policy documents and plans on their own perceptions of a problem (see e.g. Glasbergen and Driessen 1993, Gijsberts and Van Geleuken 1996). This resulted in a large number of plans, but they failed to provide a solution for supra-local issues. The VNG commented that "the accumulation of environmental problems requires a combination of solutions from different policy areas" (1993; 15). Policy should not therefore "be restricted to a single layer of government and requires intensive co-operation between the various administrative levels and the joint implementation of instruments over a period of several years" (1993; 15). Actors should "be prepared to look 'outside their own boundaries'" (Kuijpers 1996; 64), which would involve a partial departure from their usual perspective (Menninga 1993). The ROM designated-areas policy was therefore presented as a decentral instrument to *supplement* regular policy with a view to "implementing environmental policy to accelerate sustainable development" (TK 1990; 14). It was an *accelerated and decentral intervention* designed to solve regional issues relating to the environment, spatial planning and the economy. The ROM approach had a 'rejuvenating' effect on policy content and procedure (Driessen 1996). Above all, ROM presented an *administrative integration framework* for dealing with the above problems. Within that framework, existing policy can be better co-ordinated, and more labour and knowledge can be made available, and more effective results can be achieved through joint efforts: "In the ROM areas policy, parties who formerly opposed each other or waited to see what happened are encouraged to take action" (Driessen 1996; 79).

Glasbergen and Driessen (1993; 147) argued that, owing to administrative integration and the participative character of the ROM approach "current policy cannot serve as a 'hard' limiting condition in this type of project". On the other hand, the aim of a policy that was fully geared towards a specific problem was to gain increased acceptance (legitimacy), thereby leading to greater efficiency. Measures would be more likely to be implemented and complied with (see Gijsberts and Van Geleuken 1996). This is an important aspect because the consequences of the ROM designated-areas policy were not necessarily less far-reaching than the consequences of the

Provision System for Integrated Environmental Zoning (VS-IMZ). In the Canal Zone, therefore, various residential areas were demolished, including the village of Boerengat/Hoogedijk, and industrial activities were relocated. Now, with growing local support for policy on the immediate environment, it was hoped that there would be greater consensus and acceptance of the measures.

The aim of the ROM designated-areas policy was "to generate as much public support as possible for a policy on the local environment by involving all target groups in the planning process" (TK 1990; 14). This theory of participation was based on the assumption that problems and local initiatives should not necessarily be dealt with by the government only. Any individual actor could submit proposals. The emphasis should be placed on impulses – i.e. rational and emotional standpoints, ideas and needs – of all parties involved. Equally understandable was the assumption that the government would not propose a unilateral solution, neither would it unilaterally determine what measures were required.

In practice, the ROM areas policy was less about *all* actors than about the support base formed by stakeholders directly involved in dealing with a regional environmental/spatial conflict. The question was, however, how to identify directly involved parties. In each case, the actor taking the initiative would identify the other actors to be involved and specify how they could contribute to solving a problem perceived by the community. The actors were 'selectively motivated' on the basis of their possible role in decision-making, planning and implementation. This meant that certain actors would be asked to become directly and actively involved. Other actors were asked to give advice, 'sounded-out' indirectly, invited to give their views, or were completely excluded. The ROM projects varied in the extent to which their organisational structure was 'closed' or 'open'. In the Schiphol and Rijnmond projects, direct involvement was limited mainly to the relevant authorities and economic actors. In the Canal Zone project, local environmental and residents' associations were also involved. In the *Gelderse Vallei* project, the Rabobank, agribusinesses and the Rural Women's Organisation for Gelderland & Utrecht participated in the *Gelderse Vallei* Committee. In the Gooi/IJmeer and Green Heart projects, involvement was largely restricted to the various government bodies. Environmental and residents' associations were excluded from the steering group and project group. This was also the case in the Rijnmond project (Glasbergen and Driessen 1993, VROM 1994c).

In general, the best decision-making structure for this specific policy was considered to be one that was not completely open. This raised questions about the democratic nature of the approach, given that decision-making in the context of the ROM designated-areas policy was no longer supposed to be a matter for the government alone (see Glasbergen and Driessen 1993; 137). A selected group of actors, including economic actors with specific interests, would help to formulate the solution strategy.

Because the various actors representing various interests "face each other as negotiating partners and will influence each other to a large extent" (Glasbergen and Driessen 1993; 147), existing policy frameworks came under pressure. Furthermore, the extent to which actors were involved was not always in proportion to their interest. Although the *Schiphol and Environs* project saw the environment as part of the 'double objective', the parties directly involved in the negotiations were economic actors and spatial-planning actors with an interest in developing Schiphol. Residents and their organisations who were experiencing environmental nuisance – or the local authorities that could have represented them – were not invited to participate in the planning talks. Conversely, certain actor interests were sometimes disproportionately represented to such an extent that a level playing field was virtually out of the question. Furthermore, actors were not always well organised and did not always have sufficient grassroots support. The question remained, for example, to what extent individual farmers would 'voluntarily' participate in the business relocations in the *Gelderse Vallei*. When groups feel excluded, they may be less willing to co-operate. According to Kuijpers (1996; 62), the emphasis on building consensus among actors could mean that the chosen solution to an environmental problem was not necessarily the most workable solution for all concerned.

In some cases, interest groups would come forward or be formed much later while the project was actually under way, usually as a reaction to proposed measures for dealing with an environmental/spatial conflict. They voice their dissatisfaction even at such a late stage and attempt to halt, stagnate or redirect the process. One example was the action that led to a referendum on the IJburg development project (see Fig. 7.1). These actions take place mostly at the elaboration stage. Groups of actors are also formed after a project has been completed. These have little alternative but to accept the consequences of the ROM process. Such belatedly formed groups could include residents from new housing developments that were partly the result of a ROM project.

The participative nature of the ROM approach means that the most important stage is the first phase in the decision-making process (see Fig. 5.6). During this phase, actors are selected to become actively involved, the project structure was determined,[58] a joint problem definition is formulated and implementation agreements are signed. This is necessary in order to involve and retain the commitment of relevant actors as early in the project as possible. The most important questions in the ROM designated-areas policy are therefore: which actors would be involved, who will select them and how could they contribute to formulating a solution strategy.

The aim of the first phase of the ROM procedure was to formulate a joint definition of the problem. A number of projects had a *double objective*: economic development *and* improved environmental quality. The aim of the ROM integration framework was to achieve "a well-considered balance [...] between economic and ecological aspects with a view to achieving sustainable development that is compatible with the functional requirements of the region" (VROM 1993). In contrast to the IMZ pilot projects, which more or less excluded local economic development from the problem equation, the ROM projects sought to strike a balance between economic and other interests, on the basis of an area-specific approach. To a certain extent this involved achieving a spatial equilibrium between the environment and the economy, with spatial planning as the foundation.[59]

In the following planning phase area-specific plans of action are formulated. Here, the question of *selection* arises again,[60] although this time it is not a matter of selecting project participants but of defining the area, functions, activities, the extent of the problem and the appropriate policy fields.[61] The resulting plans propose measures for dealing with the problem and specify the resources and procedure to be followed. Here, according to Kuijpers (1996; 64), it is necessary to prioritise measures, because the available financial resources have to be weighed against the urgency of the need for intervention.

ROM policy was not based on a procedure or an administrative level, but on the scale of the problem and the stakeholders involved. Geographic delineation and differentiation were essential in determining which actors were involved, in determining objectives, and in arriving at a joint strategy for dealing with the problem. The ROM strategy was therefore strongly implementation-oriented, geared towards "placing the solution to environmental/spatial problems in the broader context of the development potential of an entire region" (Glasbergen and Driessen 1993; 147). The quality of the environment should no longer form the reference framework for this process, although one of the objectives was "environmental policy-makers would be actively involved in setting conditions in the planning process for a given area" (TK 1990; 14).

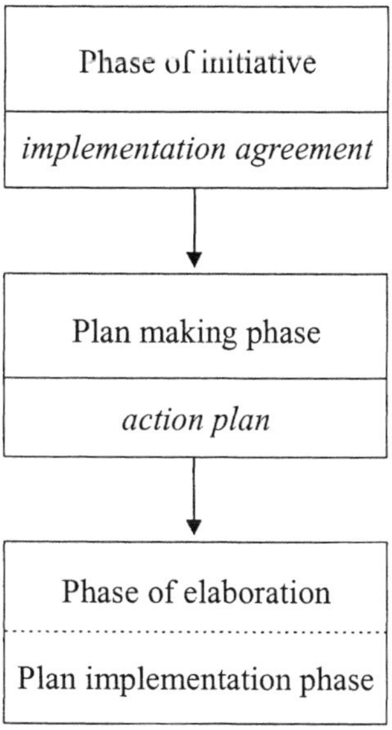

Figure 5.6 Phases in the planning process for ROM projects
Source: Glasbergen and Driessen 1993; 144

The above results in the redefinition of environmental/spatial conflicts: "The basic question is not what resources will be deployed in a region to deal effectively with environmental problems, but rather how environmental impact could be reduced and other socio-economic problems solved simultaneously " (Kuijpers 1996; 62, see also Glasbergen and Driessen 1993; 147). Benefits could be gained through geographic differentiation (Ten Heuvelhof and Termeer 1991) and also by broadening the scope of themes, thereby creating more opportunities through the interaction of a complex supra-local environmental/spatial conflict with its administrative, social and physical contexts. Kuijpers also argues that "providing solutions to indirect environment-related problems also secures greater acceptance for measures" (1996; 62). Here, it is important to co-ordinate those aspects of environmental protection, spatial planning, the economy, nature management and the infrastructure that were considered relevant to the environmental/spatial problems on which the ROM project would focus. Environmental/spatial conflicts are therefore no longer assessed in terms of cause and effect. The physical, social and administrative contexts are seen as parts of a conflict and are taken into account when the solution strategy is formulated.

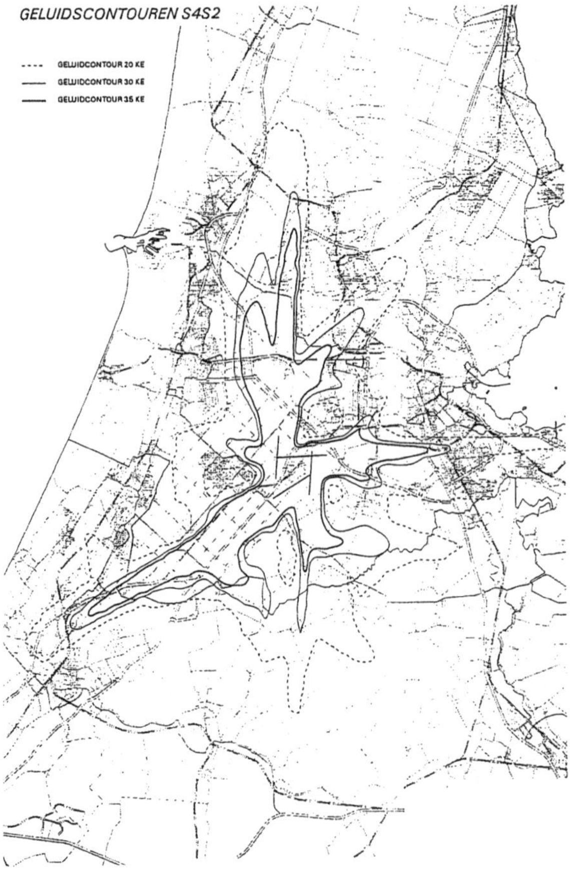

Figure 5.7 Prescriptive noise contours for Schiphol Airport, 1 November 1996
Source: VW and VROM 1996; Appendix, map E3

ROM designated-areas policy is essentially a results-oriented policy. It is primarily geared towards the administrative consequences of decisions taken jointly. The emphasis is on consensus, which means that the administrative path must be free of obstacles and become more efficient. The central questions are: who is to reach consensus and on what? Equally important is *how* this consensus would be reached. It can also happen that all the actors are satisfied with a level of consensus that is ineffective in terms of solving spatial, environmental or ecological problems. This relates to the *effectiveness* of decision-making. Driessen also argues, therefore, that the question should not be "whether consensus has been reached, but [...] whether it has produced an effective method for dealing with the problem" (1996; 81).

In addition, it is not always immediately obvious which of the stakeholders will fulfil their commitments. Public-private participation under the ROM policy often generates the will – after a great deal of effort – by industry to "make a financial and material contribution to reducing environmental impact but, 'in return', the government has to be prepared to invest in economic development in that region" (Glasbergen and Driessen 1993; 136). Glasbergen and Driessen (1993) conclude that this approach ensures that the polluter pays, but this is within a participative process of give-and-take.

The effects of measures must also be monitored and actors kept informed about deviating or unexpected results. Environmental/spatial conflicts can be driven by powerful dynamics, which means that policy must be continually amended[62] (Driessen 1996; 82). The ROM approach, however, assumes a temporary project structure. The ROM approach is carried by an administrative/organisational structure that remains in place until the problem had been solved. This *temporary nature* is a feature of network strategies. Once the processes have been completed in the manner proposed by the various parties, the organisational structure is dismantled. This could mean that, over time, it becomes less clear who is responsible for policy amendments and how they are to be translated into definitive measures.

The possibilities offered by the ROM approach were therefore limited. In this light it is interesting to take the *Schiphol & Environs* project as an example. The aim of this project was to realise the "sustainable development of the mainport, based on the principle of a controlled, directed expansion" (Project Mainport & Milieu Schiphol 1991; 9). Accessibility was another important aspect of the Schiphol project. It was also considered necessary to reduce the number of homes within the 35 KE zone (see Fig. 5.7 and footnote 13 of this chapter) from 16,000 to 10,000. Schiphol has a clear impact on the region in terms of infrastructure, spatial planning, the environment, employment, and attracting businesses (Project Mainport & Milieu Schiphol 1991; 93). If Schiphol and its flight paths are regarded as a single source, it is the largest single source of environmental impact in the Netherlands. The zone is bounded by the 35 KE contour referred to above and is located in one of the most densely populated regions of the Netherlands. This clearly regional context meant that Schiphol fulfilled the requirements for a ROM project. However, the economic, political and administrative interests of the airport extended far beyond the regional boundaries of the project. In the case of Schiphol, national actors – including four ministries – and international actors were directly involved and intervened in a process that should have been dealt with by balancing only local and regional interests. Individuals in the region were therefore forced to wait on the sidelines while supra-regional actors reached compromises. Residents in the immediate vicinity were not directly represented. Haarlemmermeer and Amsterdam, the municipalities involved in the project, had economic interests and a say in the development of Schiphol, but they could only consider residents' interests to a limited extent. Residents' organisations were represented in the sounding-board group, but left it in 1993 at approximately the same time as the environmental organisations because they were dissatisfied with the state of affairs. Municipalities that were not directly involved also had reservations. A striking example is the municipality of Leiden, which took part in the *Bulderbos* project, a tree-planting protest organised by Friends of the Earth Netherlands to obstruct the building of a fifth runway at Schiphol. The actors selected for this ROM project were therefore restricted to economic actors and actors who had control over the spatial development of Schiphol. Actors exposed to the negative environmental impact of Schiphol's activities were not selected to participate in the project, but they were able to take action – albeit indirectly – to obstruct the decision-making process. The actors who were directly involved steadfastly defended their own interests. This was especially apparent in the way in which information was exchanged. So poor was the provision of information regarding the Schiphol developments and their environmental consequences that Enthoven, who had long been a national player as Director-General of Environmental Management (Ministry of VROM), accused the 'other' authorities of "beating about the bush for twenty years by repeatedly postponing the introduction of statutory noise zones" (in Van Kasteren 1989; 8). Equally revealing is the remark made in the political current-affairs programme *Buitenhof* on 15 February 1998 by Wijers, (Minister of Economic Affairs, 1994-98) to the effect that it was about time Schiphol produced 'hard' figures. Schiphol was accused of being above the law (Van Peperstraten 1989; 12). The roles of national actors, and the parameters within which they were expected to fulfil those roles, were nothing if not vague. But this was not the only problem. Although it is usually difficult to establish a cause-and-effect relationship in environmental problems, this was certainly not the case for Schiphol.

However, it was difficult to define goals and work towards them because up-to-date information was either lacking, withheld or difficult to interpret. It is difficult to give realistic growth forecasts for air travel. Glasbergen and Driessen correctly observed that "a lack of clear information about the extent to which regions or target group should be aiming to reduce environmental impact, [...] will result in a great deal of leeway at project level" (1993; 135). In the case of Schiphol there proved to be a large gulf between technical expertise and information on the one hand, and actors and the consensus-building process on the other. Gijsberts and Van Geleuken therefore point out that "the legitimacy of area-specific policy is undermined because it is often difficult to define the relationship between pollution generated in a given area and its actual effects on the environment" (1996; 7). Here, national actors play an important role. The role of the central government – insofar as we can refer to it as a single body – was multifaceted and far from transparent. Although the decentralisation of environmental policy was supposed to be the issue, national economic interests called for direct government intervention in order to ensure that these interests were weighed against other interests. Apart from this, national actors provided no clear information on the parameters within which Schiphol could manoeuvre and within which local and regional interests could be considered. The recommendations of the SER[63] regarding the NMP-2 were perhaps the most clear statement on this: "the government is responsible for defining a framework of standards on the basis of clear political choices regarding the approach to environmental problems [...]. Without such a framework there can be no self-regulation because responsibilities and tasks are not clearly defined" (SER 1994; 33). Where clarity exists, it is almost always temporary because the parameters shift, their value for the future is often minimal and uncertain, or non-compliance is tolerated. That is to say, in the words of the Advisory Council for the Environment (CRMH): "problems are either not dealt with [...], or they are postponed, or they are placed in a context (international, European) for which others are responsible" (1991; 35). Given that it was supposedly impossible to provide hard figures on the growth in air traffic, some stakeholders were surprisingly creative in promoting their interests. The government, for example, appears to have adopted two conflicting and interrelated standpoints with regard to Schiphol. On the one hand, it regarded 'Schiphol' as a complex environmental/spatial conflict requiring a participatory approach, but on the other hand 'Schiphol' was regarded as an economic issue that could be dealt with by imposing relatively straightforward top-down regulations. The economic context was, however, inextricably linked to the complexity of the environmental/spatial conflict. Intervention was necessary because of the considerable economic interests involved, so it is difficult to establish whether the Schiphol & Environs ROM project complied with the 'rules' of the network approach. This was partly why the CRMH (CRMH 1991; 46) concluded that *Schiphol & Environs* was not a suitable ROM project. In other words, the discussions about Schiphol and its immediate surroundings would have proceeded in more or less the same way if there had not been a ROM policy at all.

The ROM designated-areas policy has the general characteristics of participatory decision-making and network strategy. It is therefore used to tackle complex supra-local issues involving many interrelated aspects, many actors and many diverse interests. The project-based character of the ROM policy also means that it is an *ad hoc* instrument. However, this does not mean that we have returned to the environmental policy of the 1960s, which was also *ad hoc* in nature. The ROM designated-areas policy is intended to support regular policy. Above all it was designed as an integral approach emphasising co-operation between the public and private sectors. Shared governance has to lead to a joint approach geared specifically to local circumstances. ROM policy is based on the principle that no two situations are the same. This means that each situation requires a specific problem definition, decision-making process, solution and implementation process.

The lack of explicit regulations may also partly explain the popularity of the ROM policy. When applying the policy to a given situation the assumption is that the situation is complex and unique, which means that prescriptive standard regulations are inappropriate.

However, we can conclude from the above that network-based and participatory policy strategies for dealing with complex environmental/spatial conflicts should comply with a number of rules, albeit implicit ones. This relates above all to balanced interdependence, shared responsibility and precise processes for defining a problem and selecting the actors to be directly involved in dealing with it. These implicit rules will have to be taken into consideration if a network-based approach or participatory policy is to be effective. On reflection, these implicit rules have benefited environmental policy, with the result that the ROM designated-areas policy has not been implemented only to supplement regular policy, but has also evolved in a certain way from a marginal to a mainstream approach (VROM 1998b).

From 'Technically Sound to Consensus-Based ... '

If we compare the performance of integrated environmental zoning (IMZ) with the ROM designated-areas policy from the goal-oriented perspective and institutional perspective of planning, we can see that both instruments have advantages and disadvantages. If the two instruments are considered from a *goal-oriented perspective*, the most noticeable contrast is between the centralised programming and the more or less fixed regulations of IMZ and the flexibility of the ROM policy. The integrated environmental zoning programme proposed by the ministry of VROM had a number of explicit principles that were the main determining factors for defining problems. These principles established a directive framework for possible solutions and predetermined the measures to be taken, thereby narrowing the spectrum of possible solutions. This method for dealing with relatively small-scale and relatively straightforward environmental/spatial conflicts achieves its goal without extensive policy efforts. That goal is the satisfactory segregation of environmentally sensitive and environmentally harmful functions and activities. In such cases the outcome of the IMZ method is reasonably predictable and a centrally defined routine approach is sufficient. The rules of the integrated environmental zoning system therefore have definitive advantages, although they cannot always be harmonised with policy developments at the interface between the environment and spatial planning. The main disadvantage of IMZ is that its effectiveness diminishes as environmental/spatial conflicts become more complex and dynamic.

Despite the lack of parameters and the enthusiasm for self-governance and network-based strategies, there are a number of rules with which the ROM approach must comply if it is to be efficient and effective. In contrast to the rules for IMZ, these are mainly implicit and informal rules relating to institutional circumstances and relationships, and they have implications for goal formulation and effectuation. Environmental quality will no longer be realised within centrally defined frameworks, but by achieving a balance between the various interests within a given area. The aim is no longer to maximise goals but to optimise the planning process with a view to achieving a desired integral level of environmental quality. The fact that this desired outcome will not always be achieved should be seen as 'part of the game'.

If we compare the IMZ system and the ROM approach from the *institution-oriented planning perspective*, the term 'consensus' (see Table 5.2) requires some elaboration. The first National Environmental Policy Plan (NMP-1) set out a strategic policy that rested on broad public consensus for a prescriptive and effective policy that allowed for radical interventions if required. This was a matter of winning extra – but not essential – public support for centralised policy. There was undoubtedly broad public support for the VS-IMZ system, but support was lacking at the local and administrative levels. The integral zoning system was a technical instrument that produced largely predictable results, which meant that local actors were confronted with a *fait accompli*. The actors involved in implementing measures were not always prepared to take account of public concern for the environment. There was a feeling that standards-based environmental policy failed to take sufficient account of the local actors' interests. This feeling was so strong that it became an obstacle to implementing zoning policy and delayed the development of the environmental standards system.

Table 5.2 Comparison between the IMZ system and the ROM designated-areas policy

		IMZ	ROM
Problem	(how)	Direct cause-and-effect relationship.	Participatory approach.
	(what)	Comply with generic standards.	Environment in conjunction with other aspects of physical surroundings.
	(who)*	Defined by the government.	Local, regional and national authorities.
Solution		Proposed by government. Implemented at local level.	Formulated and jointly implemented by actors selected to be directly involved.
Consequences		End result predictable within predefined parameters.	Determined during the process, with limited predictability for end result.
Consensus		Broad public consensus needed for national prescriptive environmental policy.	Consensus needed between directly involved actors.

* see § 4.2.

However, the lack of consensus among actors in the implementation of standards policy was a stimulus for the ROM approach. In the ROM designated-areas policy, consensus-building was a priority, and was to be achieved by involving all the relevant target groups in the planning process. This was not simply a question of gaining broad public acceptance but of *directly involving* selected actors who could make a positive contribution to the decision-making process and the final implementation of joint measures. The emphasis was not on winning public support but on the participation of actors involved in the problem to which the decision-making was geared and who were essential to the success of the policy. With regard to the efficiency and effectiveness of decision-making, the aim was to achieve continuous interaction and direct involvement with actors with a view to gaining acceptance for decisions and building the political will to jointly implement decisions taken. At the same time, the authorities involved in decision-making fulfilled different roles. On the one hand they remained responsible for protecting the public interest and on the other hand they had to be willing to accept shared governance in order to strike a balance between their own specific interests and those of the other actors experiencing the problem in question. In this context Healey (1997) refers to 'institutional consensus'. In such cases, non-governmental actors are categorised by government bodies as 'societal (f)actors'. The consensus then serves as the foundation for a policy with public support, which suggest democratic legitimacy. However, this suggestion is not strictly accurate. The degree of consensus is not the same as that achieved for the NMP-1 or initially on the IMZ system.

The two instruments differ in the extent to which they can delineate an environmental/spatial conflict within its social and physical contexts, the extent to which transparent regulations exist with regard to the conflict, and the predictability of the end result prior to decision-making. If a conflict is difficult to define, if regulations are not transparent or even conflicting, and if the progress and outcome of decision-making are difficult to predict, the conflict is categorised as 'complex' or 'very complex'. In such cases, one central actor can no longer judge all aspects of the conflict in the same way, but various actors must consider its complexity in relation to the unique local and contextual environments in which it has arisen. This conclusion is elaborated in Chapter 6.

The framework for planning-oriented action described in Chapter 4 can also be used to represent the goal-oriented and institution-oriented development in the IMZ system and ROM policy (see Figs. 4.8, 4.9 and 4.10). In Fig. 5.8, the two extremes of the goal spectrum (shown on the vertical axis) of the framework are 'standards' (the 'single fixed goal' in Fig. 4.9) and 'multiple composite and dependent goals' (§ 4.5). Comprehensive goals have to be dealt with and realised in an ongoing process, depending on the circumstances, timing and possibilities of the planning process. In the proposed framework for planning-oriented action (Fig. 4.9), the spectrum of relationships is represented on the horizontal axis, with 'central governance' and 'shared governance' as the two extremes. These terms, if applied to area-specific policy, do not necessarily need further clarification, but are referred to in previous studies as 'hierarchic' versus 'consultative' or 'consensus-based'[64] (see Bartelds and De Roo 1995, Borst et al. 1995, De Roo 1995 and 1996b, De Roo and Miller 1997).

In this framework for planning-oriented action, the 'central prescriptive policy' in the upper left quadrant is represented by the integral environmental-zoning system and the 'local policy' in the lower right quadrant is represented by the ROM approach (see Figs. 5.8 and 5.9). In terms of systems and complexity (see Chapter 4), IMZ has to function within a stable system. It can do so as long as it is not heavily influenced by external factors. In other words, it is an instrument based on the idea that, when the starting point is known and the principles are given, the outcome is largely predictable. It is an instrument that is subject to *universal* patterns. By contrast, the ROM approach is designed to take account of inevitable external influences and results within '*self-organising*' administrative systems that have specific rather than universal applications. These self-organising systems "allow for adjustment to local circumstances" (Biebracher, Nicolis and Schuster 1995). However, a fair degree of stability can be achieved with the ROM approach, provided that a number of conditions are met (see Chapter 4). According to Emery and Trist (§ 4.5), IMZ should be implemented in a 'placid, randomized environment', while the ROM policy requires a 'disturbed reactive environment'. IMZ and ROM therefore have definite and somewhat extreme positions in the policy spectrum.

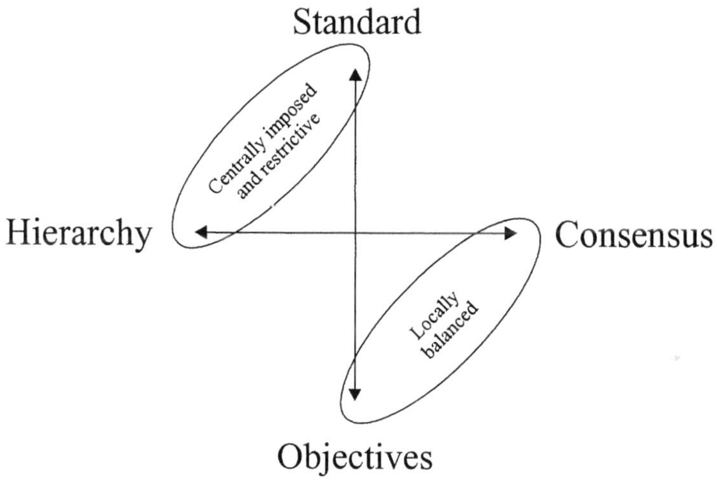

Figure 5.8 Four approaches to environmental policy translated into policy characteristics

Note: After Fig. 4.8

The pros and cons of the VS-IMZ initiative and the ROM approach provide a clear picture. In the case of more complex environmental/spatial conflicts, the zoning system needs to be more flexible in terms of problem definition and should allow greater scope for balancing interests and for collaboration and negotiation. Conversely, the ROM designated-areas policy is less effective when implicit rules are not complied with, particularly with regard to participative decision-making. A problem may be relatively too straightforward, unilateral interests may be involved that are determining factors, the parameters for participation and balancing interests may be unclear to actors, or a problem may be so complex as to transcend jointly agreed parameters.

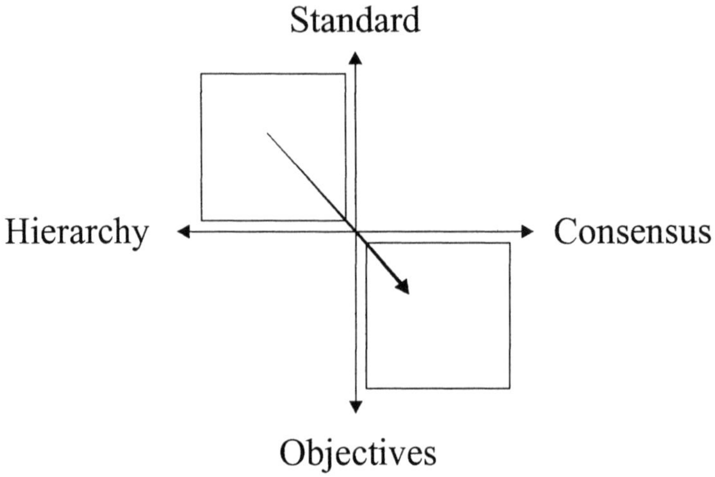

Figure 5.9 The shift of emphasis in environmental policy

This examination of the IMZ system and the ROM approach allow us qualify the assertion that *a policy based on social consensus is more effective than a technically sound policy*. This was the conclusion of the inter-authority conference at Nunspeet (NL) and the message underlying the rhetoric often used to explain developments in environmental policy (see § 1.3, VROM 1995). 'Technically sound' refers to policy that can be generally applied to a clearly defined problem, based on a fixed process that leads to a predictable result. Such policy is ineffective in complex conflicts, but is a successful approach to relatively straightforward issues because limiting conditions do not need to be formulated each time. In such cases, national values and standards can be useful in simplifying environmental/spatial conflicts. By contrast, the aim of policy based on 'social census' is to remove as many barriers to decision-making as possible and achieve goals on the basis of joint efforts by directly involved actors. At local level this is often referred to as 'administrative renewal'.[65] This form of policy is particularly effective for 'complex' or 'very complex' environmental/spatial conflicts.

5.5 Hierarchic Governance versus Local Initiative

The debate during the 1990s on the direction environmental policy should take crystallised around the IMZ system and ROM pilot projects. The debate centred on the question of whether to continue down the same path or strike out on a new path towards a more flexible, interactive approach. The IMZ and ROM projects also acted as early-warning signals for the Ministry of VROM. One signal was clear: "a transition has to be made from 'command-and-control' approach to 'policy delegation'" (VROM 1995). Environmental/spatial conflicts have to be solved at the level on which they arose, in accordance with the principle of subsidiarity. In that case, central governance will have to be replaced by policy development at local and regional level.

Will standards-based policy be abandoned in favour of an alternative path with recognisable elements of the ROM designated-areas policy? Is this choice in favour of participative decision-making a knee-jerk reaction to all that is 'wrong' with traditional standards-based policy? In the conclusion to section 5.4, this argument is qualified with the complexity factor. Or will the decision be taken to continue with conventional policy and develop it further to incorporate – depending on the problem – innovative elements? This is the question underlying this section, which examines the developments in area-specific environmental policy that took place after the ROM and IMZ projects.

The First Steps towards a Structural, Decentralised and Flexible Environmental Policy

Reactions to ministerial proposals on the further development of standards-based policy were not confined to those who were directly affected by them. Within the Ministry of VROM, too, policy based on standards and zoning came under pressure. This was partly as a result of various developments initiated by the central government that did not always prove to be compatible. The clearest example was the conflict of interest between the expansion of Schiphol and the extent of the noise-nuisance problem (§ 5.4). There was also a conflict of policy in other major government projects, such as the Betuwe Line and the high-speed rail link.[66] This led to an internal conflict at VROM between the 'purists' – the steersmen of traditional environmental policy – and the 'moderates' who looked beyond the boundaries of environmental interests (VROM 1998b; 124). The debate centred on policy consistency. Most of the 'purists' had a technical and analytical background, or were legal specialists who wanted to ensure that instruments and related regulations were used consistently throughout the ministry. The 'moderates', including economists and spatial planners, fought for broader frameworks that would allow for economic and spatial considerations. Two factors eventually blunted the conflict and led to greater understanding between the two ideological 'camps': the exchange of staff between the National Spatial Planning Agency (RPD) and the environmental department, and the relocation of both departments to shared premises. The further development of goal-oriented policy based on quantitative standards also became a less desirable aim following the heavy criticism directed at the first National Environmental Policy Plan (NMP-1) at the end of 1993. The emphasis shifted from defining new goals to policy implementation itself. There were tensions "in the available financial and human resources and in the consensus of the public and industry" (TK 1993; 33). Around 1994 and 1995, a number of factors conspired to diminish the ministry's enthusiasm for discussing the further development of environmental standards: criticism of standards-based policy – and the VS-IMZ system in particular – the demonstrable and considerable consequences of environmental policy that was too stringent, and the toning-down of proposals for odour policy. By contrast, a surprising amount of interest was shown in alternative approaches, especially *compensation*.[67]

The environmental department saw the need to amend its goal-directed policy and increase the emphasis on policy implementation. It also wanted to share responsibility for the environment with other authorities. It no longer wished to be held accountable for failing to meet its own targets. Consequently, the '*Shift of emphasis from top-down governance to self-governance within parameters*' became a principle of the NMP-2[68] (TK 1993; 42). This had been a government-wide aim since the 1980s, but was largely ignored in standards-based environmental policy. Now this development was being taken up again, it also involved a degree of decentralisation.

Local and regional governments promoted the shift from '*top-down governance to self-governance within parameters*' through innovative incentives aimed at the interface between spatial planning and the environment. It was not only the ROM designated-areas policy that showed how deregulation and decentralisation could have a mutually reinforcing effect and at the same time accelerate the process of external integration. There was more – much more (Oosterhoff et al. 2001). A number of provinces introduced an integrated plan for the local environment (Buysman 1997, De Roo and Schwartz 2001, De Roo and Schwartz 2001b, Schwartz 1998, Wissink 2000, Wissink and Lingbeek 1995). The purpose of these plans was to improve coherence with regional plans, water-management plans and environmental policy plans. The strategic and intermediary role of the province – Voogd calls this a 'hinge function' (1996; 194) – is an important motivating factor in the aim for greater coherence in policy on the physical environment. This need for coherence increased when the provincial environmental policy plan was introduced as a third strategic strand of planning for the physical environment in Section 4 of the Environmental Protection (General Provisions) Act (WABM).[69] Moreover, provinces were urged to draw up a traffic and transport plan (EK 1998; 2, TK 1997b). Briefly it appeared that the provinces would have to draw up nature-management plans too but, to date, this has not been done (TK 1996b). All existing legal strategic plan for the physical environment have specific objectives and themes, and their content overlaps to a certain extent. For provincial authorities and other actors, it was confusing to have to co-ordinate and implement several of these plans simultaneously (Buysman 1997; 20, Oosterhoff et al. 2001). Furthermore, provincial authorities saw the development of an integrated plan as a positive foundation for resolving internal organisational issues (Paauw and De Roo 1996; 207). Equally important is the fact that the provinces were to become more involved in issues that required an integrated and area-specific approach, including infrastructure projections and finding new locations for residential and work functions (Buysman 1997; 21, De Roo and Schwartz 2001).[70]

This last approach – the area-specific approach – has become even more popular than the integrated and comprehensive plan for the physical environment. Nationwide, approximately 150 different projects have been implemented in which environmental, water, nature and spatial-planning policy have been bundled within a specific approach geared towards a given location[71] in order to solve rural issues that are specific and complex. It is no longer the government or provincial authority alone that makes use of this approach. Municipal authorities, too, have become increasingly aware of the benefits of this approach (Oosterhoff et al. 2001).

While provincial initiatives were geared largely towards the integration of statutory plans for the physical environment and the management of actual complex issues that transcended municipal boundaries, the initiatives of municipal authorities were mainly geared towards active day-to-day implementation policy and dealing with actual environmental/spatial conflicts that usually arise in urban areas. Although there were substantial differences, municipal authorities and municipal partnerships were able to benefit from a number of financial incentives provided by the central government as a stimulus for local environmental policy (Commissie Ringeling 1993). The BUGM, FUN and VOGM[72] grant schemes set up specifically for this purpose were terminated, so municipal authorities now had to finance their own environmental policy as they saw fit. Municipal environmental services continued to employ a surprisingly large number of staff to perform statutory duties such as licensing and enforcement of regulations on noise, soil and waste, while the commitment to other aspects such as developing strategic policy was comparatively limited. In this respect, too, the municipal government apparatus proved to be a vehicle of prescriptive central-government policy.

This emphasis on implementation could change in the future, thanks to the Investment Budget for Urban Renewal (ISV; *Investeringsbudget Stedelijke Vernieuwing*). ISV was an initiative of the VROM, EZ and LNV ministries. The budget was made available under the Urban Renewal Act, which came into effect at the beginning of 2000 and was intended to make government funding for municipalities less compartmentalised and replace: "the large number of subsidy schemes [...] relating to living, working, the environment, the physical conditions for economic activities, and green urban spaces" (VROM 2000; 7). Municipal authorities were therefore given greater responsibility with regard to spending government subsidies on issues and requirements relating to the urban physical environment. This incentive applied to a number of aspects of environmental policy, including soil remediation and noise measures. The municipal authorities themselves decide how to allocate funds to the various policy areas. 'Tailor-made', 'integration' and 'interaction' are again keywords. However, environmental policy has also become more negotiable and will have to consider a wide range of interests. This is an exciting development given that, in the past, the municipal services implementing environmental policy at a local level had considerably less political influence than departments such as Spatial Planning, Town Planning and Economic Affairs. Because environmental policy at local level will become much more flexible, local environmental policy will now have to fight its corner with other forms of strategic municipal policy. In the past, considering the strategic aspects of municipal environmental policy rarely advanced beyond the contents of the Municipal Environmental Policy Plan, if there was one. Most of these plans were structured in terms of fulfilling the standard responsibilities such as waste and noise. In any case, the consequences of ISV policy for the environment will not go unnoticed, although they will not manifest themselves until sometime in the future.

The acceptance of environmental policy as an extra responsibility for municipal authorities was not a smooth process (Van den Berg 1993, VNG 1993). One of the main problems was the interrelation with spatial planning. During the 1990s, responsibility for the 'grey' environment increased as a structural part of local-authority apparatus. However: "the institutionalisation of environmental policy is not complete and is engaged in a growth process" (Teunisse 1995; 37). Two of the main elements that helped to structure the broadening and intensification of local environmental policy were the NMP Framework Plan of Action for local authorities and the Framework Plan for municipal environmental policy, which were introduced in 1991 and 1995 by the VNG (VNG 1995).

Many local authorities proved to be a source of inspiration and creativity in dealing with local environmental/spatial issues (VNG 1993, VNG and VROM 1990, VROM 1996b). The more active authorities were particularly creative when it came to day-to-day operational policy. A number of environmental assessment methods were developed for ascertaining the relationship between environmental quality and spatial development (see Humblet and De Roo 1995), and this number is still increasing. The new methods reflect the shift that has taken place in policy: from a quantitative, prescriptive and directive approach[73] to a comparative and participatory approach,[74] whereby the method is based on specific local circumstances to a greater extent than in the past. A large number of local authorities have also committed themselves to Local Agenda 21[75] and translated it into concrete policy measures (VNG 1996). These efforts go beyond formal compliance with statutory plans and programmes.

The big cities in the Netherlands, headed by the environmental departments of Amsterdam and Rotterdam, held up a mirror to the national government by pointing out the problems caused by centralised prescriptive policy and by proposing strategies for solving them (for Amsterdam, see Chapter 7). The most talked-about strategy was the Bubble Concept (*Stolpmethodiek*), devised by the Amsterdam Environmental Department and developed by the Institute for Environmental Studies (IVM) of the Vrije Universiteit Amsterdam[76] (Rosdorff et al. 1993) The method was introduced in the Amsterdam Policy Document on Spatial Planning and the Environment (BROM) (Gemeente Amsterdam 1994, Meijburg 1997) as 'a possible application for integrated area-specific policy'. According to the policy document, the Bubble Concept made it possible to "achieve a balance between environmental and spatial components in decision-making with a view to devising the best solution for the environment *and* the economy" (Gemeente Amsterdam 1994; 9). These words echo the objectives of the ROM designated-area approach.

The bubble idea was not completely new. It was referred to as a possibility in the PIM (see § 5.3). In the United States, the Bubble Concept is used to delineate the boundaries of 'environmental utility space' at industrial sites.[77] With its attempt to realise the Bubble Concept for Dutch cities, Amsterdam took a stand in favour of goal-oriented alternatives in environmental planning. The method did not have a direct practical value – it was never actually implemented – but was nevertheless important in the discussion on possible changes to national environmental policy. The proposal expressed ideas that were particularly well-received in the context of the discussion on environmental standards policy.

The Bubble Concept was not a participatory approach, but a more or less quantitative goal-oriented method that used indicators and data on various scales to determine the environmental quality of a given area – in this case the city of Amsterdam. Contrary to zoning methods, the principle of the Bubble Concept was not to use activity-related environmental quality as the criterion for economic and spatial development. The emphasis is not on the relationship between the source and its surroundings (externalised approach), but on the quality of a predefined area (internalised approach). The aim was, within a given period of time, to reduce the total environmental impact within that area to an integrated target level that was considered acceptable and practicable. Unlike the standards system, the method was not geared towards a single goal but towards a multifaceted result.

The Amsterdam local authority stated that "the establishment of an urban bubble will not require the same contribution from all parts of the city" (Gemeente Amsterdam 1994; 10). That contribution would be based on the possibilities and constraints within the various sectors of the bubble. This would mean that "it is possible to trade-off the various forms of environmental impact" (Gemeente Amsterdam 1994; 11). In the bubble-concept report (Rosdorff et al. 1993), the fixed maximum level under the environmental-standards system was seen as an obstacle to finding efficient solutions to the 'paradox of the compact city'. Making the various forms of environmental pollution 'exchangeable' could make that system more flexible. The substitution option reflected the fact that in some cases major but costly efforts achieved only a minimal but legally required reduction in a particular form of environmental impact, while for other forms the same degree of effort could achieve much more. In that case, "if a spatial or environmental measure in one area is ineffective (from an economic as well as an environmental point of view), another solution must be found to improve the area or the situation must be compensated elsewhere with extra measures (but within the total bubble)" (Gemeente Amsterdam 1994; 10). The aim was therefore not to achieve a maximum goal formulated by central government for each environmental/spatial conflict, but to work towards an optimum result by adopting an area-specific approach that guarantees local environmental quality by means of compensation. This method was a proposal for area-specific standards, substitution and compensation designed to achieve more area-specific 'tailor-made' solutions in the planning process (Meijburg and De Knegt 1994).

The Bubble Concept found favour with many actors and was well-received. Expectations were high – with hindsight perhaps too high. It was assumed that "it would no longer be necessary to devise an individual solution for each environmental problem" (Meijburg and De Knegt 1994; 10). With "area-specific environmental zoning it is also possible to find an optimal mix of spatial and environmental measures for each situation" (Meijburg and De Knegt 1994; 10). In addition, "returns on available funding are maximised, pollution levels are reduced throughout the city and the quality of the spatial environment is improved" (Meijburg and De Knegt 1994; 10). In response to the criticism of zoning instruments, this proposed new method would introduce flexibility, help to achieve policy integration and devolve responsibility for environmental policy to local authorities. These objectives were music to the ears of many people in the Netherlands, and were given an impetus of their own. Local authorities frequently claimed 'we will be implementing the bubble method too'. In Amsterdam itself, however, the policy has never been implemented and is not likely to be in the foreseeable future.

It did not prove easy to translate the principles of the Bubble Concept into practical applications based on indicators, combined indicators[78] and delineated areas and area levels. The aim of the Bubble Concept was to reduce total environmental impact as efficiently as possible, more or less independent of the nature and scale of the sensitive area. The bubble model did not focus on the relationship between the pollution source and the quality of the immediate environment. This could mean that the aim of the source-directed approach was no longer the individual protection of local residents, contrary to standards-based policy that is designed to guarantee the same basic level of protection for all individuals in the Netherlands. Although making the standards system more flexible through an area-specific approach and pollution trade-off can both increase financial and policy performance, it can also limit the level of individual protection. As it is proposed, the environmental impact at micro-level (the individual citizen) is becoming part of environmental impact on the macro-level (e.g. the city).

In 1996, the IVM[79] reviewed the Bubble Concept and examined research from the discipline of welfare theory in order to substantiate it (Boer et al. 1996). As a result the concept became even further removed from day-to-day practice. The compensation aspect made it clear that it was not easy to put the Bubble Concept into practise. Compensation remained a national policy theme when the bubble euphoria had ended, but translating it into definite measures for use in urban environments remained such a complicated and sensitive matter (De Roo 1996c, Wiersinga, Ronken and Ten Holt 1996) that a feasible formula has still not been found for the Bubble Concept.

The Bubble Concept presented a completely new administrative perspective on goal-oriented action in environmental planning. On the one hand it was the extreme opposite of operational and goal-directed zoning instruments and, on the other hand, it was an area-specific strategy that was a useful supplement to zoning instruments. The bubble model can be placed opposite the standard-setting approach on the goal axis in the framework for planning-oriented action (Figs 4.8, 5.8 and 5.9). It was, after all, designed to achieve an integral goal – a given level of environmental quality – within a specified period of time, thereby contributing to a multi-faceted result based on cohesive policy for the local environment (see also § 4.5).

The Bubble Concept also made a valuable contribution to the discussion on harmonising environmental health and hygiene policy, spatial planning, economic feasibility and economic development. Although it did not prove its worth as a concrete policy instrument, it gave an impulse to the discussion on compensation, was a welcome new element in the discussion on environmental standards, and delivered an 'escape route' – the 'third step'– in the City & Environment Policy, which is discussed in the following section.

City & Environment: Standards, Compensation and Local Creativity

The NMP-2 (TK 1993) refers to a 'paradox of the compact city' (see Chapter 3), regarding which "the Ministry of VROM [...] in collaboration with other authorities [will] launch a project to develop environmental policy for urban areas (action point N90). The project will explore the possibilities for ecological urban concepts, the urban bubble model (Amsterdam), urban traffic and environment and the differentiated implementation of environmental standards" (TK 1993; 203). This action point was implemented as the City & Environment (Stad & Milieu) project in the autumn of 1993 by VROM, the IPO, the VNG and a number of the larger local authorities. It was a step towards a structural solution for specific urban problems that were primarily due to policy barriers at the interface between spatial planning and the environment. The project focused on "problems in urban areas relating to achieving environmental and spatial planning objectives" (Stad & Milieu 1994; 1). A further aim of the City & Environment initiative was to gain insight into how integrated policy could optimise the use of space and at the same time improve liveability (Kuijpers and Aquarius 1998).

The City & Environment project was based on the problem definition that while urban environmental standards were often exceeded and environmental quality was poor, meeting targets in the short term would 'paralyse' life in the cities (Stad & Milieu 1994b; 1). This bold statement – that reflected the aversion to the standards system – soon required modification after it became apparent that "local administrators generally cope extremely well with the paradox of the compact city" (Stad & Milieu 1995; 6, Stad & Milieu 1995b). In a number of cases, however, it was claimed, administrators encountered "deficiencies in the instruments that could turn 'win-win' situations into 'either-or' situations" (Stad & Milieu 1995; 6). This was a somewhat characteristic way of indicating that policy initiatives were not always mutually reinforcing, but could sometimes counteract each other and produce unintended results. A provisional aim of the City & Environment project was to produce a policy document setting out methodical, organisational and administrative/legal solutions for these urban problems. The document explained that, because the problems were local, the solutions – taking the ROM policy into account – would be based on the principle of local governance. An area-specific, integral policy approach was therefore advocated. In addition – in line with the Bubble Concept – reductions in environmental quality would have to be compensated in order to achieve a measurable and acceptable improvement in liveability (Stad & Milieu 1994b; 2).

Actors struggled for more than a year to agree on a result. On 22 December 1995, the report '*Where there are many wills, there is a way*' (Stad & Milieu 1995) was submitted to the Lower House. The report contained a proposal designed to remove the policy barriers contributing to the dilemma of the compact city. In the spring of 1997, 25 pilot projects were launched to test the practicability of the proposal: "These inner-city locations are almost always required to concentrate on restructuring, changing uses, remediation and infill development. They encounter various environmental problems simultaneously, such as noise nuisance, air quality, public safety and soil contamination" (Kuijpers and Aquarius 1998; 31, see also § 3.4).

The City & Environment project was and is based on a three-stage approach that local authorities were supposed to follow in the event of environmental/spatial conflicts. The first two steps emphasised existing policy. Step 1 involves implementing as many source-directed measures as possible and involving environmental aspects in the spatial planning process at as early a stage as possible. This last element is particularly important because, often, environmental data is not gathered until it is actually required at a particular point in the planning process, e.g. for issuing permits.

The proposals and working method for most of the 25 City & Environment pilot projects[80] do not show a great deal of drive or creativity on this point. Most of the projects did not map the environmental health and hygiene situation or the confrontation between this situation and environmentally sensitive areas and desired spatial development. If the ministry's proposed integrated environmental-zoning system has taught us anything, then it is the importance of having such maps at an early stage to provide information on the condition of the environment (§ 5.4 and Chapter 6). We have already observed that it was not sensible to base policy on unsubstantiated estimates of local environmental quality.[81] However, this observation is at odds with the reasons for promoting change in soil and noise policy, which assume that local authorities are sufficiently internalised to assume responsibility for those policies (VROM, IPO and VNG 1995, MIG 1998; 8).

If step 1 fails to produce the desired result, creative solutions have to be sought through step 2 within the boundaries of existing legislation. Maximum use has to be made of the scope offered by the existing standards system. The emphasis on this step meant the bankruptcy of the conviction that existing standards could be reduced in the future due to concrete policy efforts, agreements with industry and technological developments. This conviction had found favour when the first steps were taken towards a quantitative standards policy at the beginning of the 1970s. Step 2 puts pressure on the 'stand-still' principle. In this step of the City & Environment approach, the existing level of environmental quality is no longer regarded as the target level that should be preserved. Instead, the target level is now the maximum level permitted under environmental legislation. This tendency, which was still being dismissed as a risk in the 1970s, is now seen as part of the process for preventing limit values being exceeded.

The integrated-zoning pilot projects showed, among other things, that there were environmental/spatial conflicts for which even the results of steps 1 and 2 of the City & Environment strategy were unsatisfactory. These were conflicts for which it was exceptionally difficult to co-ordinate spatial, environmental and economic possibilities and constraints in order to achieve a multifaceted goal through the maximum application of environmental standards. It also applied to environmental/spatial conflicts that arose because legislation was not appropriate for the specific possibilities of a particular situation that might be incompatible and result in procedural barriers (Kuijpers and Aquarius 1998). For such cases the City & Environment strategy has a third step that allowed for deviation from legislation.[82] This step is an implicit acknowledgement of the fact that centralised legislation does not always take sufficient account of local circumstances. It is also recognised that local authorities are responsible for aiming for optimum results for a given location or area. Step 3 enables local authorities to deviate from existing standards if there are sound reasons for doing so. Reduced environmental quality would then have to be compensated "in terms of liveability"[83] (Stad & Milieu 1995; 11). Initially, this option was made available only to the 25 designated City & Environment pilot projects under the special City & Environment Experiment Act (TK 1998).

The Ministry of VROM took a deliberate decision not to draw up compensation legislation, since this could the options open to local authorities. The ministry hoped that a great deal could be achieved through step 1 and step 2 of the City & Environment approach, thereby minimising the need for compensation.

In short, standards remain the point of departure for the City & Environment projects. However, area-specific differentiation is possible when this can be shown to have a positive effect, on condition that compensatory measures are taken. Policy continues to be goal-oriented, albeit on the basis of shared governance. While the existing standards system was accepted as a framework for the ROM designated-areas policy, a new step has now been taken to allow for specific local circumstances requiring greater local responsibility in the development of policy for environmental protection and local surroundings. The decision to implement this step is now the responsibility of local government, not the central government.

At the request of the Upper and Lower Houses of the Dutch parliament, the City & Environment Experiment Act included a provision on evaluation: "Within two and six years of the implementation of this Act, the minister shall submit to both Houses of the States General a report on the practical effectiveness and results of this Act" (Stb. 684 1998; Art. 14). A City & Environment Evaluation Committee was established for this purpose on 1 August 2000. The task of the committee was not only to present the results of the two evaluations to the minister, but also to advise the minister when it considered this necessary (Ministers of VROM and BZ 2000). The committee indicated that it did not wish to restrict its evaluation to Step-3 decisions: "The advice we give the minister with regard to policy renewal will [...] relate to all the steps of the City & Environment approach" (Evaluatiecommissie S&M 2000; 8). The evaluation would also include developments in related aspects such as the renewal of policy on noise and soil protection, the consequences of the introduction of the Urban Renewal Investment Budget and the influence of urban policy. The committee thereby indicated that the principles of the City & Environment approach had to be linked to "the large-scale dynamics of the urban environment and urban policy" (Evaluatiecommissie S&M 2000; 9). It was recognised that City & Environment should not be seen in isolation, that its principles may be outdated by the time the evaluation is completed in 2004 and also that, given the context of the City & Environment policy, useful lessons could be learned about *local creativity*, integration, strategy development and the force of urban environmental policy. The City & Environment project could therefore provide insight into future relationships and developments relating to environmental interests within the total context of policy for the urban physical environment.

The position taken by the City & Environment Evaluation Committee was an interesting one and probably a necessary one, since the original objectives of the City & Environment initiative – step 2 as well as step 3 – are to a certain extent being overtaken by developments in environmental policy, which is geared towards individual forms of environmental impact. Largely as a result of these developments, local authorities involved in the projects have questioned the value of continued contribution to the City & Environment prpject. As a result of the most recent developments, which will be discussed below, it has been suggested that local authorities can – if they wish – more or less formulate their own strategy and environmental regulations in the near future. A number of municipalities are therefore questioning whether they need to continue devising 'difficult constructions' that require substantiation, while it is becoming ever clearer that they will soon be able to 'make all their own decisions'. The approach of those local authorities is reflected in the motto 'Leave it to us!'. This applies particularly to sectoral regulations for odour, soil contamination and noise.

Odour Policy: Varied According to Complexity

The most relevant developments in odour policy had already taken effect before the City & Environment three-step approach was introduced. Nevertheless, odour policy will be discussed here because developments in that area are similar to changes in soil protection policy and noise policy following the City & Environment initiative. The developments in odour policy are also relevant because complexity is a specific criterion for action. In the Policy Document on Odour Nuisance (TK 1992) – an elaboration of NMP action point 75 – it was proposed that odour-nuisance problems in the Netherlands could be dealt with by drawing comparisons between similar activities and by adopting a complexity-related approach. The policy document opted for an approach that was *differentiated according to complexity* (see Dönszelmann 1993). Three types of company (i.e. odour sources) were distinguished: (1) Companies belonging to a homogeneous industrial sector, within which companies emit similar levels of odour. It was usually unnecessary to measure emissions in this category, and odour could be dealt with using a standard approach; (2) Large companies, of which there are only a small number in the Netherlands. Odour problems in this category would be dealt with by means of an Odour Abatement Plan; (3) Very complex companies with a large number of odour-emitting sources. A plan of action would be required for this category, too, and would have to be drawn up in conjunction with a liveability survey among local residents (TK 1992).

The first proposal of the Policy Document on Odour Nuisance was still based on quantitative standards. The policy document and standards had the status of a directive. This status, combined with a participatory approach, led Dönszelmann to assume that "the odour concentration standard is becoming increasingly accepted" (1993; 25). However, this proved not to be the case (Te Raa 1995). The many types of odour, the scale of odour emissions and the subjective nature of odour perception made it difficult to express odour problems in quantitative terms. The objections made by the joint Provincial and Municipal Executive of Groningen were the clearest example of this. The proposed odour standards, it was stated in a joint letter to the Minister of VROM "are causing problems in Groningen West with regard to reaching an optimum decision on the best way to separate and intersperse functions in that area" (GS & BW Groningen 1995). The letter was written in response to the requirements placed on the two sugar refineries in Groningen, which had to reduce odour emissions to the maximum level permitted by the odour standards, which were also an obstacle to housing development plans in Groningen. The first modification of official odour policy in 1993 represented a considerable shift in standard-setting odour policy (TK 1993b). The second modification heralded the end of the standard-setting approach in odour policy (VROM 1995b). Odour units were no longer maintained as an upper limit. Instead, abatement of new odour emissions was based on the ALARA principle in accordance with Article 8.11, paragraph 3 of the Environmental Protection Act. The NMP target was, however, retained as the basis for national environmental policy: odour nuisance should be perceived by no more than 12 per cent of the national population by the year 2000 and the level of serious odour nuisance should reach nil by the year 2010.

Soil: from a Multifunctional to a Function-Oriented Approach

If we look at staff numbers, the average municipal environmental-health department concentrates on issues relating to soil protection, noise and waste treatment. Despite the fact that local authorities are responsible for these matters anyway, and despite the fact that they are the most frequently occurring issues, the focus on these themes is proportionately very large. The main reason for this is legislation, particularly that for noise and soil. This has led to a situation in which local authorities see soil protection, noise nuisance and waste treatment as their main environmental problems. The approach to these problems is more or less specified in legislation and regulations, thereby reducing the role of municipal authorities to that of implementer. This role is changing dramatically, however, and this can be seen most clearly in soil remediation policy.

Soil remediation policy in the Netherlands was formulated more or less out of necessity. It was a direct response to the exponential increase in the number of cases of soil contamination in the early 1980s (see § 5.2). For a long time, soil remediation policy was based on the Soil Clean-up (Interim) Act (IBS). It was clear from the discussions in 1994 and 1995 on the inclusion of this interim act as a paragraph on remediation in the Soil Protection Act (WBB) that there were problems with "translating strategic policy principles into practice and devising solutions to operational problems" (*Werkgroep bodemsanering* 1993; 5, see also VNG 1992).

There were two general objections to soil-remediation policy: it was too expensive and it caused spatial development to stagnate. In principle, after remediation, the soil should be free of contamination and should be multifunctional,[84] i.e. able to support many functions.[85] Until these criteria were met, planning permission could not be granted under Art. 8, paragraphs 2 and 3 of the Housing Act,[86] and spatial development could not take place (see also Moet 1995). This type of remediation measure produced clean, multifunctional 'patches' in urban areas, where there is usually an unnatural background concentration resulting from many years of diffuse contamination.

The renewal of soil-remediation policy was initiated by the appointment in September 1992 of the Soil Remediation Working Group (*Werkgroep bodemsanering*), later known as the *Commissie Welschen I*), whose brief was to solve the problems described above. Earlier, in 1989, the Steering Group for the Ten-Year Soil-Remediation Scenario (*Stuurgroep Tien Jaren-Scenario Bodemsanering*) had recommended that the polluters and users of contaminated locations be required to share the cost of remediation. This recommendation was taken up by the Soil Remediation Working Group and placed within the framework of "a decentralised approach to soil contamination [...], whereby responsibility is placed with the various target groups" (*Werkgroep bodemsanering* 1993; 16). Following on from this, the working group pointed out the advantages of further developing an 'active soil management' approach[87] that would be required to provide sufficient up-to-date information on the condition of the soil prior to the spatial planning decision-making process (IPO, PGBO, VNG and VROM 1996). Remediation measures often had disappointing results because of the lack of advance information on soil quality, and this in turn obstructed building and other activities.

The standards system for soil remediation (see § 5.2), which was designed to return the soil to its multifunctional state, remained intact at this stage. Its scope was increased, however, with a second remediation option,[88] whereby contaminated soil was not removed but contained and controlled as a definitive solution (known in the Netherlands as the 'IBC' approach). The aim was "to implement the IBC approach so as to remove a number of constraints on land use" (*Werkgroep bodemsanering* 1993; 36).

As the remediation paragraph of the Soil Protection Act was introduced in three phases in the period 1994-1995, the old 'ABC' standards system (see § 5.2) was abolished. At the same time a process was set in motion whereby powers and tasks were devolved to the provincial authorities and the four main cities. Further privatisation of soil remediation activities was also proposed. The principle was no longer 'the government will foot the bill, unless...' but 'the government is not responsible for clean-up measures, unless ...' (Moet 1995; 13). Responsibility for soil remediation was therefore placed with those who had caused the pollution and with those who owned or used seriously contaminated sites. A system was developed for assessing the urgency of remediation. Based on actual risk, it enabled responsible parties to assess the level of soil contamination. The system determined which cases of serious contamination required urgent remediation and which did not. 'Serious soil contamination' and 'urgency' are key terms in the remediation provisions of the WBB. The necessity for remediation was not determined by the actual nature of the pollution, but by the *seriousness* of the pollution[89] (VROM 1994). The degree of urgency determined *when* the remediation measures should be taken[90] (VROM 1994b). In very urgent cases – in any event those involving risk to human health – remediation was required within four years. For other urgent cases, a distinction was made between ecological risk (remediation required within ten years) and the risk of the pollution spreading, in which case remediation had to commence within 25 years 'of the date of the decision'. Under the new principle ('the government will not foot the bill, unless...'), urgency was the only criterion that could be used to exert pressure on the parties responsible to clean up polluted soil.[91] If the responsible actors did not fulfil their obligations, the soil remediation paragraph of the WBB contained provisions that could be used to enforce this requirement. This development was a first step towards what is known as the 'communalisation' of soil remediation policy (VROM, IPO and VNG 1995).

Despite these developments and recommendations, signals were received from 'the field' that "soil remediation and other factors are obstructing strategic projects at the centres of urban networks" (Welschen 1996; 1). A second working group on soil remediation (known as 'Welschen II') was set up to advise on this issue. It concluded that the legal framework "[offers] more or less sufficient scope to harmonise soil remediation policy with practical situations" (*Werkgroep bodemsanering* 1996; 5). Contaminated locations that had potential for spatial development but were not categorised as urgent cases should no longer prevent local authorities from formulating development policy. In this respect, there was no longer a distinction between urgent and non-urgent cases. It was therefore down to the local authority to formulate an area-specific integral policy that, as necessary, specified an order of priority for cleaning up contaminated sites.

The soil-remediation working group (1996) nevertheless observed that policy was not sufficiently integrated, despite the fact that "soil remediation cases [are] dealt with less and less as separate clean-up projects. Instead, the 'soil' is a factor that is increasingly considered in broad societal processes, such as spatial planning" (VROM, IPO and VNG 1995; 3, see also Hofstra 1996). The Target Scenario for Soil Remediation Policy (*Streefbeeld bodemsaneringsbeleid*; VROM, IPO and VNG 1995) was based on an inter-administrative approach to soil remediation policy and attributed the lack of integration to factors such as 'dense regulation', and consequently to an inaccessible remediation policy.

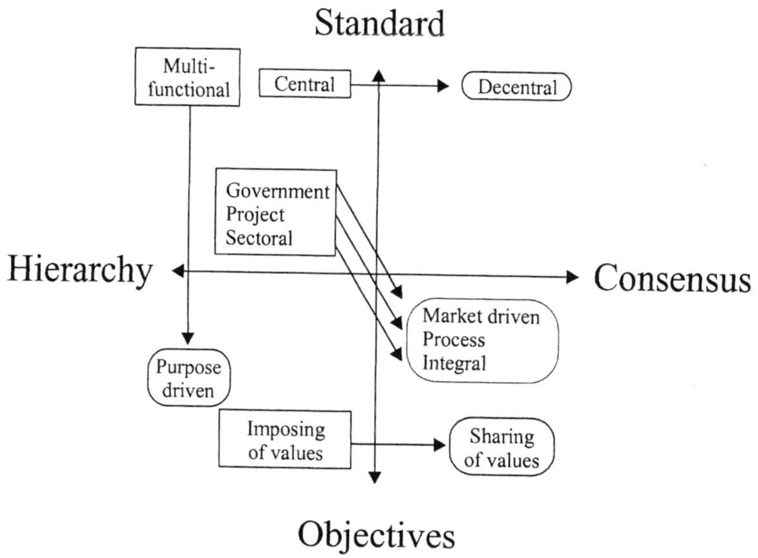

Figure 5.10 Trends from the target scenario placed within the framework for planning-oriented action
Source: VROM, IPO and VNG 1997; 17

Hofstra (1996) also showed that regulation could lead to complex conflicts in cases of inner-city soil contamination. As a result, locations with potential remain unused and other locations are chosen for development. Hofstra used ten case studies to show that, if good information about the nature and extent of the soil contamination is available in advance, the conventional processes leading from remediation to development can be partly integrated. This can be the first step towards improving the development planning for a location. It is essential to have information about the environmental situation of a site and to make agreements with non-governmental actors on procedures. The shift in the nature of the planning process from 'consecutive' to 'co-ordinated' can accelerate this integration process.[92]

The Target Scenario for soil remediation therefore proposed the further decentralisation of powers, thereby involving soil remediation in a wider balancing process, and allowed for "location-specific soil remediation for a specific function"[93] (VROM, IPO and VNG 1995; 6). Using a number of key terms, the Target Scenario described the shifts on which consensus had been reached. The shifts were the basis for administrative renewal in remediation policy:

- from sectoral to integrated;
- from multifunctional to function-oriented remediation;
- from a project-based to a process-based approach;
- from centralised to decentralised;
- from government dynamic to market dynamic;
- from imposed values to shared values.

Using the consensus as a starting point, the BEVER[94] consultative platform on soil-policy renewal was established. Within this structure, working groups developed a number of 'strands' to be included in the new process of 'communalisation' of soil-remediation policy. One of the strands was designed to lead to short-term solutions and sought, for example, to shorten implementation procedures and clearly set out the legal framework. A second strand involved developing a long-term vision. Here, the intention was not to bind competent authorities to strict regulations, but to base local soil policy on multifaceted objectives. A third strand was designed to alter relationships between stakeholders and achieve greater participation by market players. This would be achieved through absolute limit values to be set by the government and differentiated according to function. The government would also specify a general standard of environmental quality – it was not clear how – from which local and regional authorities could deviate if they had good reasons for doing so (BEVER-werkgroep 1997). These developments were supposed to lead to a situation in which the spatial function of a location was no longer determined by the environmental quality of the soil. Instead, the desired quality of the spatial environment would determine the quality requirements for the soil[95] (Roeters 1997; 29). The reasoning was that a function-oriented and area-specific approach would reduce the cost of remediation and allow the market mechanism more room to operate (VROM 1997b).

This 'communalisation' of soil-remediation policy, together with the trends in the Target Scenario, were summarised by the BEVER working group 'Typology, Bandwidth and Decision Support' using the framework for planning-oriented action (see Chapter 4 and Fig. 5.10; VROM, IPO and VNG 1997; 17). The principle was to decentralise responsibilities and powers to 'the most local and relevant level of government possible'. These are familiar words. The argument for this 'integrated decision-making in a social context' was supported by the conclusion that soil remediation could be seen as a regional problem and therefore required a regional approach (VROM 1997b).

A number of local and regional authorities had problems following the developments in soil policy, but others were quick to respond to the changes. Roeters (1997) has shown that a number of provinces were already following an 'anticipatory' soil protection policy. Provincial authorities are reducing their involvement as actors in implementation policy and focusing instead on "motivating other actors, combining interests and consensus-building" (Roeters 1997; xv). Some municipal authorities also tried to set an example. The Amsterdam Municipal Executive was the first to state that "it [is] only acceptable to use public money for soil remediation measures if these can be shown to have a beneficial effect on soil quality in terms of designated urban land use" (Gemeente Amsterdam 1996; 11). The council of Groningen, among others, following the example of Amsterdam, also saw the need to elaborate its soil remediation policy. It pointed out that "there is contamination radiating out from the centre towards the development districts [and that] the soil in a large area of the city centre has not been 'clean' for many years" (Gemeente Groningen 1997; 4). The councils of Dordrecht and Rotterdam made similar claims (IPO, VNG and VROM 1997).

Multifunctional remediation proved too costly and yielded too few environmental benefits. Multifunctionality was now no longer a starting point but a desired goal to be reached at some point in the future (TCB 1996), and was replaced by function-oriented and area-specific remediation. Function-oriented remediation was based on the principle that "a decision taken on the goal of remediation is based not only on general principles (that the soil should, in principle, be clean), but also on what the local community wants and is able to achieve, and what is justified from the point of view of environmental protection" (IPO, VNG and VROM 1997; 38). Function-oriented remediation meant that the physical remediation of the site in question was no longer the most important factor. The spatial constellation would be given priority in the sense that new functions must not be put at risk (in terms of 'fitness for purpose') by soil contamination.

On 19 June 1997, the government announced its position on this new soil policy. It wanted "the remediation method in individual cases to be determined by the intended use of the land in question and the prevention, as far as possible, of the occurrence and spread of contamination" (*Kabinetsstandpunt* 1997; 7). This position is based on two variants of function-oriented remediation. In "the efficiency-gain variant it is matter for the originator(s) of the remediation to decide whether their clean-up measures will go beyond the minimum requirements in terms of environmental protection" (*Kabinetsstandpunt* 1997; 7). In addition, there was an 'environmental-gain variant', whereby "the result of remediation measures is of central importance from the point of view of the environment and is used by the appropriate authorities to assess whether the level of risk should be reduced below what is required or what the originator considers desirable or necessary" (*Kabinetsstandpunt* 1997; 7).

Irrespective of the variant chosen as the foundation for the legal framework for soil remediation, the appropriate authorities would assess whether minimum requirements had been met. The competent authority and the originator(s) of the remediation would jointly define these requirements in order to "arrive at a customised solution, for each individual case, chosen from a range of remediation variants" (Kabinetsstandpunt 1997; 7). This policy shift was expected to result in a larger number of remediation projects and reduce the stagnating effect of contamination on 'societal' developments. There would be more residual contamination following clean-ups, but this was seen as acceptable, provided that "the safety of human beings and the ecosystem is guaranteed in all functions" (Kabinetsstandpunt 1997; 7).

Meanwhile, in 1999, the BEVER working group published the report '*From funnel to sieve*' (Kooper 1999), in which the approach to remediation was partly based on the degree of complexity of the case in question and the level of influence from the contextual environment. Three alternative 'routes' were proposed:

- a standard approach for each case or cluster of cases. This approach can simplify the decision-making process;
- a 'tailor-made' approach for each case or cluster of cases. This is the logical solution if a standard approach is ineffective;
- a 'tailor-made' approach per region. This variant is for exceptional cases. The goal of remediation is based on the specific characteristics of an area (Kooper 1999; 9).

In other words, the policy was based on the principle: "a standard approach where possible, a customised approach where necessary" (Kooper 1999; 9). This was reaffirmed in the final report by Bever/UPR (Eindrapport, Bever/UPR 2000). The approach was designed to resolve the stagnation in soil remediation noted by the cabinet in 1997 and lead to "optimum integration of soil remediation with other societal activities and dynamics" (Bever/UPR 2000; 1). These fine words were intended not only to bring an end to the stagnation, but also to generate cost savings of 35 to 50%. The cost saving would be achieved through the function-oriented remediation of immovable contamination,[96] provisionally based on four function categories. The official message was also that "further cost savings will be achieved by planned integration with development activities" (Bever/UPR 2000; 1, see also Hofstra and De Roo 1997).

The government wished to refine these measures further by designating not only the polluter as the cost sharer, but also other actors. The government considered itself to be one of the actors and would also make a financial contribution under the 'participatory system' that would remain in force until 2023 in order to accelerate remediation work as far as possible (Bever/UPR 2000; 21 and 26). Financial resources would be integrated in the Investment Budget for Urban Renewal (ISV). The ISV identified 30 local authorities that qualified for direct government grants on the basis of urban development scenarios submitted by them. As yet, this scenario is nothing more than a five-year development programme setting out how the authority will spend the grants. It is hoped that, in the longer term, the scenario will result in an integrated approach to remediation and development (Bever/UPR 2000; 32). Other municipal authorities obtain funding from the provincial government.

This proposed funding system involves the re-designation of competent authorities. In the past, the competent authorities were the 12 provinces and the four largest municipalities. Under the new system the competent authorities include not only the four largest municipalities, but also the other 'direct' local governments, if they so wish. Two of the 30 'direct' authorities have not yet taken this step, which means that there are at least 40 competent authorities for soil remediation. This will lead, for example, to limiting conditions and to geographical differentiation in policy. The provincial authorities of Flevoland and Zeeland have already indicated that they do not wish to make soil remediation part of their general policy because their goal is to achieve soil that is as clean as possible. Policy has become more diffuse.[97] A further result of this is that the authorities themselves determine *when* they assess the urgency and seriousness of a case of contamination and formalise the result in a decision. The deadline for remediation is not specified until the decision is announced. If the decision is postponed/delayed, so are the measures. This form of 'playing for time', often introduced to allow extra time for arranging finance for the remediation, can – if one so wishes – be seen as a blurring of standards. The Lower House has not yet pronounced on the body of proposals.

These developments in soil-remediation policy have shifted the emphasis from soil quality to the functional allocation of land uses. They also represent a shift from central governance to shared governance, which must allow scope for local area-specific policy whereby integrating the various policy 'strands' can lead to added value. The shift from central to shared governance by no means implies that it is now easier to formulate policy for dealing with soil contamination. The principle of 'no development until the soil is clean' has been been replaced by the point of view that a minimum level of soil quality should be jointly agreed as part of the spatial planning process. Local authorities are taking greater responsibility for soil policy, and its parameters are partly based on local consensus. It is up to local authorities to motivate the various actors to clean up the soil above the minimum risk requirements, although it is not clear how this can be achieved. This 'blurring' of the policy parameters can have the following consequences: substantial contamination may remain below a certain level in the soil (BEVER-werkgroep 1998), there is an increased likelihood of standards being regarded – and used – as maximum permitted levels, economic interests may prevail, and limited clean-ups may be justified by claims that they are within the law. As a result of the principle 'remediation follows functions', when a change in land-use is planned and the new function is not compatible with the residual contamination, further remediation measures will be required – or the new function will simply not be implemented. On the other hand, an area-specific approach is conducive to active soil protection, which can provide the necessary information for allowing equal consideration for the environment in the complex process of balancing interests. It is not yet clear whether these developments will make a positive contribution to relatively straightforward cases of soil remediation – apart from the re-evaluation of the standard 'clean' to that of the background level. But for more complex environmental/spatial conflicts between spatial planning requirements and soil contamination, there is scope for developing area-specific strategies whereby uncertainties relating to the environmental situation can be minimised and environmental/spatial development encouraged. Nevertheless, the regulations on soil contamination have not become more accessible or transparent as a result of the decentralisation of soil policy. It is noticeable that, in contrast to the proposals in the stepped City & Environment approach, environmental targets have been lowered in advance to create leeway for multifaceted goals.

Noise Policy: Directional, Locally Justified and Area-Specific

For a long time, the common elements of national noise abatement policy and soil policy were their complex procedural regulations and their centralised and detailed nature. Following the discussion on soil protection policy, noise regulations were also debated. The centralised detailed legislation of the Noise Abatement Act (WGH) was often seen as unnecessarily complicated by its implementers. Moreover, enforcing the WGH is a labour-intensive task. In the early 1970s, only a small number of officials dealt with noise abatement policy at the Department of the Environment of the Ministry of VROM (§ 5.2). Now, however, noise abatement officials are one of the largest groups in municipal environmental health and hygiene departments. The provisions of noise-abatement regulations are not always compatible with wishes and requirements at the local level. Moreover, there are reservations regarding the delegation of responsibilities for the 'higher-values' procedure[98] and "experience with implementing environmental policy on a local scale show that it is increasing impracticable to deal with noise abatement as a separate element of policy" (MIG 1998; 8).

The discussion on the review of noise abatement policy is not only the result of local dissatisfaction with the delegation of responsibilities and complex legislation that does not dovetail with local circumstances and needs. The review also resulted from initiatives taken outside the sphere of environmental policy. The coalition agreement of the first cabinet of Prime Minister Wim Kok introduced the Competition, Deregulation and Legislative Quality (MDW) initiative 'to encourage markets to function more effectively'. The MDW plan of action was submitted to the Lower House on 19 December 1994 by the Minister of Justice and the Minister of Economic Affairs. This policy document pointed out that regulations in the Netherlands do not always tie in closely with social realities. Overly complex and detailed regulations and the lack of transparency are an obstacle to local responsibilities and creativity. The aim of the MDW operation is to remove unnecessary central rules, to provide better laws, and to ensure that regulation and competition go hand in hand (TK 1994). The first step was to categorise specific activities for which general regulations could be formulated. In the area of environmental legislation, for example, this involved the developments in odour abatement policy and retail licensing, which require extensive preparation and are experienced as excessively 'heavy' instruments.[99] Under the MDW operation, the BUGM/VOGM regulations and the Environmental Protection Act (WM) have also been reviewed. As was the case with the Housing Act, the WM also identified three categories: "developments requiring planning permission, developments requiring notice, and 'free' developments requiring neither of these. The last category is subject to safety-net regulations" (Van Geest 1996; 120).

However, the "review of noise abatement policy instruments and the redistribution of powers among the levels of government" (TK 1996; 1) is also a subject of the MDW project. The aim is to determine whether existing regulations were "still an optimum mix of instruments for attaining noise abatement goals" (TK 1996; 1) and whether "the burden of regulation can be reduced for industry while retaining satisfactory acoustic protection for local residents" (TK 1996; 1). The MDW recommendations for amending noise-abatement instruments point to the local character of noise nuisance, which justifies a decentralised approach. Local authorities are regarded as being in a position "to prevent noise nuisance as far as possible or to reduce it through spatial planning measures that prevent or encourage particular combinations of functions in a location" (MDW-Werkgroep 1996; 12). The national government must determine "the noise levels that constitute a risk to public health and specify these as statutory target values", the purpose being to indicate a target value at which, from the point of view of environmental protection, no further noise-abatement measures are required. Local authorities may, if there are sound reasons for doing so, deviate from the statutory noise standards. The reasons must relate as far as possible to a municipal noise-abatement plan and objectives based on local circumstances. In the view of the MDW Working Group, such "area-specific standards offer satisfactory opportunities for 'customised' solutions" and it recommends that an area-specific approach be made compulsory for local authorities (MDW-Werkgroep 1996; 14).

The MDW recommendations are the starting point for a follow-up project under the title Modernisation of Noise-Policy Instruments (MIG). The aim of the MIG project is to realise "a robust system that includes incentives for reducing noise nuisance and is compatible with the responsibilities of actors" (MIG 1998; 5). The MIG project will also have to produce a noise-abatement bill. The bill is intended to give local authorities the power to draw up their own noise-abatement policy, preferably within the integral framework of the municipal environmental policy plan. Local authorities will have the power to deviate from statutory targets by setting their own *area-specific limit values* as part of an area-specific approach designed to produce local solutions. The proposals (MIG 1998) expressly state that the local authority must, if necessary, be able to deviate from its own limit values in specific circumstances. According to the explanatory notes: "'plan-specific flexibility' prevents a limit value being set at the highest conceivable level of noise nuisance that only occurs in exceptional circumstances" (Bouman 1998; 1). In short, limit values must not be seen as a maximum that cannot be exceeded. When there are no area-specific limit values, targets set by the government serve as the statutory limit values. This target value indicates the noise level that must be aimed for or maintained as far as possible. By setting target values as opposed to fixed standards, the government is choosing to create the right conditions for setting up a local framework, rather than setting prescriptive frameworks.

The Lower House agreed with this but felt that there should be an absolute limit, i.e. a maximum ceiling of 70 dB(A). This was an unfortunate choice because thirty years of standards policy have shown that setting limit values only encouraged the perception that they are an allowable maximum to which both nuisance and pollution levels could be allowed to rise. As a result of the ceiling set by the Lower House, in the worst-case scenario, the achievements of traditional noise-abatement policy could be very effectively dismantled.

A number of local authorities in the Netherlands are anticipating the new regulations and related responsibilities. Five municipalities – Amsterdam, the Drecht cities, Ede, Emmen and Nijmegen – are working under the supervision of the ministry of VROM on a MIG project to come forward with proposals for dealing with local noise problems. Two approaches are already emerging on which municipalities will base their new policy. One approach is solution-oriented and the other would lead to an integrated policy strategy (Eikenaar et al. 2001).

Amsterdam and Emmen have opted for a solution-oriented strategy to deal with conflicts between noise nuisance and spatial development policy. Amsterdam has drawn up a policy document on noise nuisance for the Bijlmer area (Van Breemen 1999) consisting of a framework for spatial planning, an integrated framework for noise-nuisance measures and a basis for simplifying procedures. The main objective is to reduce the number of people experiencing noise nuisance in the area, namely by 15 per cent in the period 2000-2010. The new-style policy document on noise nuisance sets out the current and desired situation in the Bijlmer neighbourhood. It also sets out development plans and the consequences for noise-abatement policy. The document discusses not only road/rail traffic and industry as sources of noise, but also air traffic and disturbance of the peace. It includes a zoning map that is designed to serve as a framework for assessing other policy aspects relevant to the area. The provisional conclusion is that results can only be achieved in the Bijlmer in cases involving physical sources of traffic noise and disturbance of the peace. These are precisely the aspects covered by local policy.

The MIG project for the Drecht Cities is developing 'modern' noise abatement policy as part of a broad process designed to improve the quality of the local environment. This 'liveability' project links various policy themes, including the development of a spatial vision for the Drecht Cities. The MIG project is one of the 'liveability' implementation projects. An interdisciplinary project team has been appointed for each participating municipality and the team leaders are also members of a regional MIG team. This organisational structure is designed to ensure that the interests of the various actors are represented at regional and municipal level, and that local and regional problems are identified. The requirements for noise abatement policy are then elaborated by the government and the provincial authority. The aim is to realise an official noise abatement policy based on a broad consensus within a couple of years. The MIG project for the Drecht Cities has identified physical problem areas but the emphasis is on the larger whole: formulating a strategy for an integrated noise-abatement policy (Eikenaar et al. 2001).

On the afternoon of Saturday 13 May 2000, an environmental disaster took place in north Enschede on a scale hitherto unknown in the Netherlands (see also § 1.1). Television reports clearly showed how the 'grey' environment 'demanded' space. Everything within hundreds of metres of the SE Fireworks factory was destroyed. The disaster temporarily interrupted the discussion on the decentralisation of environmental policy. It was as if the engine driving an apparently irreversible process had stalled as a result of the tragic firework explosions. Now that the dust has settled, the effect of 'Enschede' on the decentralisation process is becoming clear. Regulations on fireworks were amended radically and in a somewhat *ad hoc* way, which was apparently sufficient for the decentralisation process to resume in January 2001, the Ministry of VROM submitted the draft bill on noise abatement to the other departments involved.

A shift away from general statutory regulations to a more plan-based participatory approach will make environmental policy more flexible and 'softer'. On the one hand this could have a positive result at local level but, on the other hand, it could leave a loophole for allowing pollution and nuisance to rise to the maximum allowable level or exceeding standards in specific cases. One of the effects of this will be a shift in the technical-analytical relationship between noise nuisance and those who perceive it. With the emphasis on local multifaceted objectives rather than generic and sectoral standards, there will be less scope for holding local authorities accountable for their performance. Policy will be set out in qualitative rather than quantitative terms (see also De Roo 1997).

The transfer of responsibility to local authorities could involve a dilemma in terms of assessment. This dilemma is inherent to the principle of subsidiarity: the central government sees the local authorities – because of their knowledge of local circumstances – as the appropriate bodies for solving local environmental/spatial conflicts and developing area-specific policy. However, this means that the government will be less justified in holding local governments accountable for unsatisfactory performance. The City & Environment pilot projects, among other things, will show the extent of this problem. It is possible that the shift towards local governance is based on a strong belief in the powers of local democracy: the success of municipal policy will be reflected in local voting patterns. However, a future scenario in which elected environmental health and hygiene officials are 'punished' for unsatisfactory noise-abatement policy by voters, resulting in policy amendments, may simply be a pipe dream.

Municipalities are required to draw up not only area-specific limit values, but also strategic objectives as a framework for local noise-abatement policy. These objectives must also specify how the authority proposes to reduce local noise nuisance in the long term. According to the proposals, the provincial authorities and the government are also required to formulate strategic goals. Under this structure, national goals will set the course for provincial and municipal noise abatement policy based on joint consultation (MIG 1998; 12-13, 18). In addition to these proposals regarding cohesion at a strategic level, there has also been a shift towards operational goals. The preference is no longer for 'strict, but realistic goals with feasible options for less strict requirements,' but – in compliance with the second alternative set out in the PIM (VROM 1983; see also § 5.3) – for a 'new' system of standards that will be 'less stringent, but will allow for tightening-up'. This new framework is expected to emphasise the process whereby local authorities translate national objectives – set jointly by various administrative levels – into integrated policy and environmental plans (VROM 1998). The emphasis is also expected to be on the formulation of local frameworks for policy on the environment, spatial planning and local liveability, and on the development of function-related and area-specific policy designed to ensure that local spatial and economic development is not realised at the expense of local assets – including the environment.

5.6 Conclusion

The changes in policy on noise nuisance and soil, and the proposals in the context of the City & Environment project, are characteristic of a notable development in environmental policy, namely decentralisation following a period of several decades in which environmental policy was shaped by government initiatives. This appears to have brought an end to centralised policy implemented by the government in order to put the environment on the administrative agenda and put an end to undesirable developments in the physical environment as quickly as possible. In this chapter we have discussed the evolution of environmental policy in the Netherlands. This concluding look at area-specific environmental policy considers the developments in the context of the three planning perspectives discussed in Chapter 4, which provide insight into the continual friction between goal-oriented, institution-oriented and decision-led planning with a view to continuing the decentralisation process.

Environmental interventions by the government in the early 1970s can be seen as *decision-led*. They determined the way in which Dutch environmental policy developed in recent decades (see A in Fig. 5.11). The aim of that policy was to achieve short-term concrete results by placing constraints on human activity in order to tackle its negative effects. The decision was taken to develop a standards system that provided quantitative and generic answers to questions such as 'What problems exist in the environment?' and 'What do we want to achieve with environmental policy?' This approach reached a high point with the introduction of the Noise Abatement Act in 1979. This act focuses on one particular sector of environmental policy (noise) that will be enforced on other policy sectors. This principle served as an example for regulations formulated for other forms of environmental pollution. The goal-setting nature of policy at this time is particularly striking. Once set by the government, fixed standards were imposed on the process of balancing local interests.

In the 1980s, policy implications were drawn from the fact that environmental issues could no longer be dealt with in isolation. These issues were related to each other and to the physical environment in which they arose. Consequently, a decision-led approach was adopted that would lead to integrated policy. Meanwhile, the sectoral standards system was gradually elaborated, resulting in a complex range of proposed standards, e.g. for noise and odour nuisance, soil quality, radiation and risk. Environmental policy based on standards was consistently implemented according to a *goal-oriented* approach. This method was compatible with a hierarchic and prescriptive governance strategy. While the first step was taken towards the decision-led restructuring (i.e. integration) of environmental policy, the principles of goal-oriented and institution-oriented planning remained intact (see B in Fig. 5.11). Local authorities and various target groups are still governed within a narrowly defined framework by a ministerial environmental department that wants to keep control of measures in the fight for a cleaner environment.

The introduction of the IMZ zoning system and ROM designated areas – at the same time as the publication of the first National Environmental Policy Plan – threatened to destroy the cohesion between the goal-oriented, institution-oriented and decision-led perspectives of environmental policy. Despite the broad societal consensus reached by the end of the 1980s, the implementation of ministerial proposals for integrated environmental zoning led to unforeseen policy conclusions and spatial effects. It proved almost impossible to harmonise the system of sectoral standards using an integrated approach. Instead of benefits in the form of a consistent and transparent standards-based policy, attempts at integration had far-reaching spatial and financial consequences. Moreover, the 'sustainable' nature of integrated zones as an instrument for separating environmentally sensitive and environmentally harmful functions was not always compatible with the dynamics of urban development. Stringent environmental parameters led to tensions with local spatial planning policy, so opportunities for achieving positive 'customised' results with local projects were under-utilised. A number of environmental/spatial conflicts had far-reaching spatial consequences, so zoning was almost impossible. Proposals for standards-based odour nuisance policy were abandoned along with integrated environmental zoning. The broad consensus for these initiatives could not compensate for/conceal the fact that institutional consensus was lacking.

The fact that, by contrast, the faith in the ROM designated-areas policy grew rapidly over a short period of time is partly due to the fact that it placed greater emphasis on the question 'Who should implement the policy?', which proved to be more relevant than was at first thought. The participatory nature of the ROM approach soon came to be seen as a valuable and effective innovation in dealing with complex transboundary environmental/spatial issues, despite the broad public consensus on strict standards policy (see C in Fig. 5.11). There was a growing awareness that centralised prescriptive policy was not satisfactory for dealing with complex local problems. The ROM approach embraced this awareness and was geared towards building consensus among actors involved in implementing policy. This new area-specific approach made it clear that environmental/spatial conflicts could also be solved on the basis of a broad problem definition and overall objectives.

The discussion on harmonising and integrating environmental policy and spatial planning policy at regional and local level became a structural one when the hierarchical and prescriptive zoning policy reached its limits in terms of efficiency and effectiveness, when the ROM designated-areas opened the door to shared governance, when integrated area-specific policy broadened the focus beyond the environment and when, in the first half of the 1990s, municipal and provincial authorities came forward with their own proposals – which the government could not ignore. The diverse motives of actors led to a discussion on the principles of delegating environmental responsibilities among the various authorities. The context of location-specific environmental/spatial conflicts was emphasised, including the unique character of local circumstances and friction between local policies. Greater emphasis was placed on economic issues, liveability and the need for spatial development. These motives were underpinned not only by abstract arguments (such as complexity and uncertainty regarding policy consistency). Emotional and subjective perceptions (e.g. power struggles, aversion and fear) also played a role. Once remediation policy had reached maturity and preventive policy on the physical environments began to take shape, the effects of this policy became fully noticeable. As a result, at its height, 'command-and-control' planning became less desirable and other forms of governance were considered more effective and efficient.

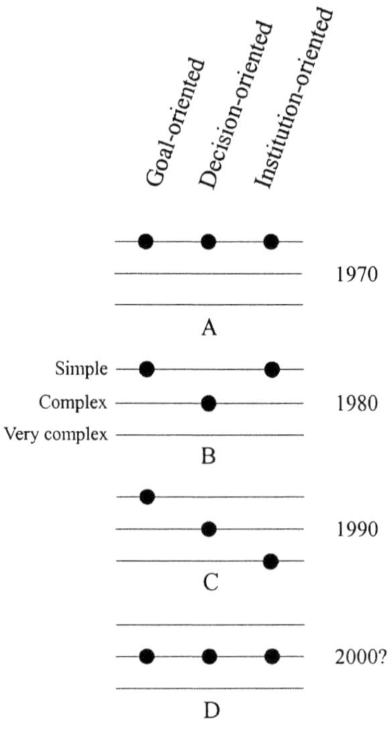

Figure 5.11 Shifts in the planning-based perspective of environmental policy over time, measured against the degrees of complexity that can be attributed to environmental issues

Note: For A, B, C and D, see text

By the mid-1990s, the traditional goal-oriented approach appeared to have reached a dead-end. The relevant themes in *institution-oriented* policy appeared to confirm this. However, existing policy and regulations cannot always be simply pushed aside. This results in a major imbalance between the goal-oriented and institution-oriented approaches to planning. The decision-led alternative to retain a system of standards, and at the same time allow deviation from it in certain circumstances if justified, was an interesting compromise whereby the primacy of decision-making was shifted to a decentralised level of government. The goal-oriented and institution-oriented contours of environmental policy then have to adapt to the decision-led alternative (see D in Fig. 5.11).

The decision-led approach did not simply dismiss standards policy and replace it with area-oriented tailor-made objectives and network-based participatory strategies. The government retained central responsibility, with a degree of decentralisation. The decentralisation of policy responsibility also resulted in a shift in the role of standards, which were no longer prescriptive, but served more as guidelines from which local authorities may deviate if there are sound reasons for doing so.

A shift was initiated from policy aimed at relatively straightforward issues towards a policy that steers a middle course between straightforward issues on the one hand, and very complex issues on the other. This policy perspective offered scope for dealing with a diversity of issues and conflicts that varied in complexity, whereby the local authority decides what level of complexity to attribute to the issue or conflict. The authority can then choose between (1) the target values formulated by the government (2) an area-specific approach based on a detailed evaluation of standards and a flexible stance in relation to other forms of policy, or (3) a more strategic overall approach that hooks into other forms of policy and whereby the direct involvement of non-governmental actors is seen as essential. If local authorities hold the view that the effort involved in taking on greater responsibility for environmental policy does not measure up to the advantages of a policy that is area-specific, integrated and strategic, they can resort to the government's target values, which will then serve as frameworks for local policy – a sort of 'default' policy. These proposals and developments brought the goal-oriented, institution-oriented and decision-led approaches reasonably into line with each other again.

This partially solves the issue of the restrictive frameworks that environmental policy imposed on other forms of policy relating to the physical environment. The effect was to lower the threshold for co-ordinating and integrating environmental policy with other forms of policy concerning the physical environment. The proposals and developments brought goal-orientation and the institutional structure more into line with the process for balancing interests in spatial planning. The door has therefore been opened to situation-specific and area-related planning that evolves towards shared governance.

But this is not the end of the story. The developments in soil policy (function-oriented remediation) and noise abatement policy (area-specific approach and tailor-made solutions) show that spatial planners increasingly determine the timing of decisions on environmental measures. Developments in the environmental-standards system also show that it has not only become more guiding rather than prescriptive in character, but also that the level of environmental quality it represents has been lowered. It is therefore naïve to expect that a decentralised environmental policy, which aims to deliver comprehensive, situation-specific solutions and strike a balance between the interests of actors, will have no negative consequences for the desired level of environmental quality. In Chapter 4 we discussed the consequences of the shift away from a policy that focuses on the elements of an issue towards a policy that considers the context in which the issue has arisen. In this chapter, the discussion on the development of environmental policy appears to confirm these conclusions. The path it follows reminds us of the critique by McDonald (§ 5.1). In any case, the changes in environmental policy are an implicit acknowledgement of the fact that it is simply not possible to achieve a 'clean' environment in the Netherlands.

Now that this has been acknowledged, complex and very complex environmental/spatial conflicts can be tackled more effectively and efficiently, thanks to an approach based on situation-related comprehensive solutions and interaction with actors. It also means that environmental policy no longer focuses on protecting the individual but on the 'public interest', which is represented by the municipal authorities. Time will tell whether this confidence in the ability of local democracy is justified.

The chosen approach is not based on 'limits to growth' (§ 5.2) but on 'limits to standards' – a choice that does not impose constraints on societal developments. The role of environmental standards will change as a result. At the institutional level this will largely involve a redistribution of responsibility. This change – despite the principle 'from technically sound solutions to solutions based on societal consensus' – is, however, not radical enough to be described as participatory policy with a pluriform character. Nonetheless, strict policy parameters have been replaced by what can be described as *optional* policy: based on the complexity of an environmental/spatial conflict, local authorities determine what degree of responsibility they wish to take for developing and implementing local environmental policy, and with whom they wish to share that responsibility. The 'grey' environment has become an element of municipal policy that must be balanced against other elements. It will be interesting to see if environmental policy is able to maintain a high profile in the local policy arena. For a long time, municipal environmental policy was underpinned by standards set by the national government. Will local environmental health and hygiene departments have enough power, or will they be overridden by Spatial Planning, Development or Economic Affairs departments? The question is to what extent the 'grey' environment has become a structural policy element.

From a theoretical perspective, the decentralisation of environmental policy and the dismantling of its frameworks (see Chapter 4) mean that, for a given initial situation, it is no longer possible to obtain 'complete' certainty with regard to the end situation. The intention to do so in the past made Dutch Environmental policy '*too good to be true*'. Policy is increasingly formulated on the basis of *causa remota* (§ 1.4). Instead of a clear and predictable result, increasing distinctions will have to be made between the ways in which local authorities shape their environmental policy. There will be definite success stories, but just as many failures. This is an inevitable consequence of decentralisation and is often underemphasized.

If local authorities wish to take control of the development of policy for the 'grey' and physical environments, they will have to develop strategic policy (see Fig. 5.12). Although this step is almost inevitable, local authorities do not appear to be fully aware of it. It requires an approach that is fundamentally different to municipal policy in the goal-oriented, interactive and strategic sense. Chapter 6 proposes a method for local-authority decision-making on formulating policy for environmental/spatial conflicts. Chapter 7 examines the considerable efforts made by the local authority of Amsterdam in order to implement an acceptable strategic environmental policy. In line with the principles that apply to environmental policy in general, local strategic policy on the 'grey' and physical environments must achieve a balance between decision-led, goal-oriented and institution-oriented measures.

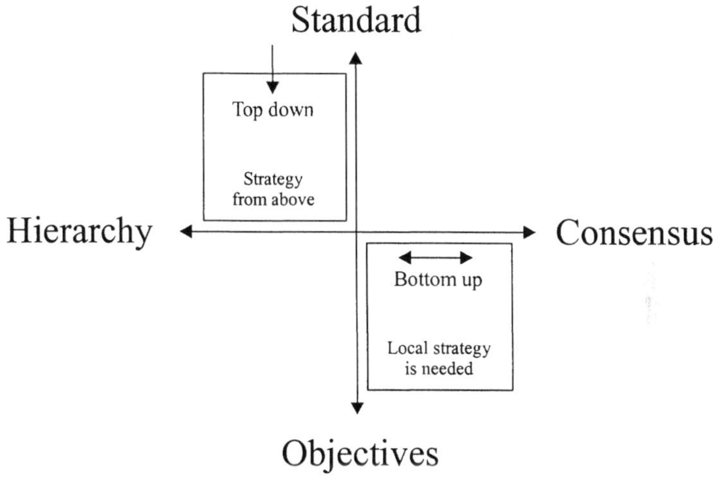

Figure 5.12 The need for strategic policy at the local level

Note: This will depend to a large extent on the decision-led measures and related goal-oriented and institution-oriented measures relevant to the policy issue.

Notes

1 In a certain sense, the report by the Club of Rome was the climax of a long discussion about the finite capacity of the natural environment and the physical environment in which man exists. In 1962, a similar stir was caused by Rachel Carson's book *Silent Spring*, which described the consequences for man and the ecosystem of overusing pesticides. In 1968, Garrett Hardin published his article 'The Tragedy of the Commons', which warned that the world's resources are finite and cannot support the unlimited growth that results from the perpetual fulfilment of individual needs. Disaster is almost inevitable if we continue to fulfil our own needs without considering the wider context and the finite nature of the ecosystem. In Hardin's opinion, pollution should be subject to regulation that also places restrictions on our freedom to produce and consume. In 1970, Barry Commoner published a much-discussed book called 'Science and Survival', in which he established a connection between the physical, chemical and biological aspects of the ecosystem and societal, political, economic and philosophical factors. He concluded that the effects of environmental problems are expressed in physical, chemical and biological terms and their causes lie in socio-economic structures. This is why, he argued, political choices are ultimately inevitable (see also Chapter 2 and De Koning 1994; 24).

2 Although the concept of 'equilibrium' was extensively used at the beginning of the 1970s, the importance of 'sustainable development', which did not become popular until the end of the 1980s, began to be recognised.

3 Environmental Protection was transferred to the Department of Public Health (*Volksgezondheid*), one of the reasons being an equal distribution of ministerial posts for the Biesheuvel cabinet (1971-1973). Stuyt was appointed to this department and became the Netherlands's first Minister for the Environment (see also De Koning and Elgersma 1990).

4 The Priority programme is set out in the second and final part of the Policy Document, which described a large number of measures whereby existing environmental pollution would be dealt with during the following five to ten years. The significance of the Policy Document for the development of environmental policy lies in the first part, which discussed the role of environmental protection as an essential part of the societal process in general, and of the entire range of environmental problems in particular. The Priority Policy Document was a key document that set the course for the development of environmental policy in the years that followed.

5 "It shall be the concern of the authorities to keep the country habitable and to protect and improve the environment." This is how the responsibility of the government is described in Article 21 of the Dutch Constitution (see also Neuerburg and Verfaille 1991). Over the years, this responsibility has been defined by the changing needs of society. Traditionally, this responsibility has been interpreted by the government primarily in terms of public health. The increasing threat to public health led to differentiation in healthcare, and environmental protection became a fixed policy area.

6 This despite a shift in the principles of spatial planning. The revised Housing Act and the new Spatial Planning Act, which both came into effect on 1 August 1965, meant that spatial planning was no longer based on functionality but also on quality (Faludi and Van der Valk 1994; 124). The main aim was now "to allocate space to all relevant functions in society" (Winsemius 1986; 25) with the emphasis on "the quality of the environment in general and the occasional balancing of interests that conflict in the short term" (Winsemius 1986; 35-36). This new legislation and regulation for spatial planning was designed to lead to decentralised decision-making, a development about which people were very optimistic (Faludi and Van der Valk 1994; 123).

7 The publication of the Priority Policy Document is referred to somewhat arbitrarily here as the beginning of environmental planning. Despite the framework it established for the development of a separate environmental policy, there are also, if we choose to look for them, other points in time that can be identified as milestones in environmental planning. Dutch environmental policy was first developed via the Occupational Health & Safety Inspectorate (*Arbeidsinspectie*): "Increasingly, labour problems and environmental problems had the same underlying causes" (De Koning 1994; 26). In 1962, the Public Health Inspectorate (*Inspectie voor de Volksgezondheid*) was set up to monitor environmental quality (De Koning 1994; 23). Van Tatenhove regards this as the "beginning of the institutionalisation process [...], the establishment in 1968 of the Environmental Protection Department under the Directorate of Food and Environmental Health of the Directorate-General for Public Health of the Ministry of Social Affairs and Public Health" (1993; 13).

8 The OECD 'Polluter-Pays Principle' (1975) was adopted by the Dutch government. The principle rests on the idea that those who cause environmental problems should

also finance the measures required to resolve them. However, it is not always easy to identify the polluter, who could be bankrupt or deceased, or have relocated. Furthermore, the environmental cause-and-effect chain described in Chapter 2 also shows that environmentally harmful activities are the result of a complex amalgam of social and individual developments and needs.

9 This programme appeared in the autumn of 1973 (EG 1973), and provided "the basis for establishing environmental standards in a communal context" (VM 1976; 4).

10 Legislation against nuisance dates back to the Factories Act of 1875. From 1896, shortly after the introduction of the Safety Act (1895), it was called the Nuisance Act (De Koning and Elgersma 1990). Brussaard et al. (1989; 95) point out that the concern about the effects of industrialisation on society was the underlying reason for this legislation. In the pre-war period, the upper classes in the Netherlands became concerned about the natural environment, which partly prompted protective legislation on the 'green' environment (Van der Windt 1995). The Nuclear Energy Act came into effect in the 1960s. It was designed to protect workers and the environment from radiation. Other pre-1970s legislation was also partly oriented towards environmental protection, for example the Commodities Act and the Rendering Act, but was primarily designed for other themes, in particular worker protection and public health (see Brussaard et al. 1989, Neuerburg and Verfaille 1991, De Koning 1994; 40, Van Tatenhove 1993).

11 In March 1993 the Chemical Waste Act and the Waste Substances Act were incorporated into the Environmental Management Act.

12 Despite that fact that this was recognised as a national poblem, 1973 only one technical specialist and one policy official at the Department of the Environment were working on the subject of noise (Tan and Waller 1989).

13 The main measures for preventing and controlling aircraft noise were set out in Section IV of the Aviation Act. The measures are threefold: (1) at-source measures, including technical requirements for aircraft, (2) regulations for airport use, and (3) spatial-planning policy based on environmental zoning around airports. The zones are based on 'KEs' ('Kosten Units') named after the Kosten Committee, which, at the end of the 1960s, stated that 40 KE was an acceptable maximum for noise nuisance. The Health Council for the Netherlands (*Gezondheidsraad*), however, claimed that serious nuisance began at a level of 15-20 KE (*Gezondheidsraad* 1971, 1994). This view was shared by the provisional Central Council for Environmental Protection (*Centrale Raad voor de Milieuhygiëne*, 1994). The accepted maximum is set out in the Major Airports Noise-Nuisance Decree, and was finally set at 35 KE, a limit which is more flexible that the 50 dB(A) limit specified in the Noise Abatement Act as a preferred limit value for noise from traffic and industry. Twenty-five per cent of residents experience nuisance at 35 KE, compared to 10 per cent at 50 dB(A) (CRMH 1994; 27).

14 Germany preceded the Netherlands in translating environmental standards into spatial contours designed to separate environmentally sensitive functions and activities from those that are environmentally harmful (see e.g. AGS 1974, Rheinisch-Westfälischer TÜV 1974).

15 The Noise Abatement Act did not regulate the full spectrum of noise nuisance. Aviation-related noise falls under the Ministry of Transport, Public Works and Water Management and, as previously mentioned, is regulated by the Aviation Act. Noise in the workplace is regulated by the Working Conditions Act (*ARBO-wet*) and falls under

the Ministry of Social Affairs. Minor sources of noise nuisance are regulated in the Nuisance Act, which was incorporated into the Environmental Management Act in 1993. A major issue is neighbourhood noise. There is virtually no legislation for this problem, except perhaps the provisions of the Buildings Decree and a number of byelaws.

16 There were a number of objections to the Act. Environmental zoning in existing locations was opposed on the grounds that it could lead to under-valuation of land and property in comparison to non-zoned areas. Moreover, certain land-use designations were excluded from zoned areas, or had to comply with certain conditions. It was also feared that zoned areas would have a stigma attached to them (Lurvink 1988). This criticism was made again later when the Ministry of Housing, Spatial Planning and the Environment (VROM) proposed implementing integral environmental zoning (§ 5.4). However, this criticism did not hold water. There was a direct relationship between the problem (excessive noise nuisance), the consequences for planning (remediation during pollution transfer or at the recipient location) and the financial resources for this measure (construction of noise barriers). Because the allocation process for available resources was inflexible, it was not easy to redirect financial resources. Furthermore, noise barriers are not welcome in all locations. In some cases they are perceived as a visual nuisance. Regulations offer no options for re-allocating funds reserved for noise barriers to insulation, for example.

17 The absence of specific soil legislation was not due simply to the lack of emphasis on the consequences of soil contamination. Soil protection should have been regulated for the most part through existing legislation (Lambers 1989; 169).

18 WBB = Soil Protection Act.

19 The method for assessing soil contamination was amended when the Soil Clean-up (Interim) Act was incorporated into the Soil Protection Act (the first phase came into effect on 15 May 1994). Intervention values replaced 'C' values as an instrument for determining the level of contamination. The intervention values indicated the "the level of concentration of pollutants above which the functional properties of the soil for man, flora and fauna are seriously reduced or threatened" (*Leidraad Bodembescherming* 1995; 45). The values therefore related to serious contamination, the urgency of which would determine whether a deadline would be set for remediation. The level of urgency depended on the actual risk to humans and the ecosystem, and the risk of spread. Under the new Act, a decontamination order could no longer be issued for all cases of serious soil pollution (see also § 5.5).

20 An odour unit per m^3 is used to indicate the concentration of an odour that is distinguished from 'clean' air by 50 per cent of an assessment panel. The percentile indicates the period of time within which the local community does not smell the odour (Inspection Working Group on Odour Nuisance 1983). Air samples (e.g. from the odour source) are diluted by an olfactometer until 50 per cent of the assessment panel can no longer smell the odour.

21 In the policy document *Dealing with Risks*, risk was described as "the probability of unwanted events occurring, in relation to the extent of the possible consequences for the surrounding area" (TK 1985; 173).

22 A risk standard has also been proposed for ecosystems. It is based on the "provisional assumption that the functioning of the ecosystem is protected if 95 per cent of species experience no detrimental effects" (TK 1989; 2).

23 ALARA stands for As Low As Reasonably Achievable, and is a principle of source-oriented policy geared towards preventing unnecessary pollution at source. According to the ALARA principle, at-source emissions should be kept as low as reasonably possible (Meijden 1991). The principle derives from European legislation on radiation (Van den Hof 1988), and is also incorporated into the Environmental Protection Act (Art. 8.11, paragraph 3).

24 The unwanted side-effect of this was that farmers were reluctant to plant or conserve woodlands on their land because it might then be classified as 'acidification-sensitive'.

25 In contrast to the method for integrated environmental zoning proposed later by VROM, (§□5.4) this method does not calculate the various sectoral levels of pollution cumulatively. The sectoral component and maximum load are taken together as the basis for calculating an acceptable distance to the source.

26 Although there are several ways to deflect environmental problems, deflection between the different compartments in particular has led to the insight that a new approach to environmental policy is needed. Pollution is not only deflected to other environmental compartments, but also to other physical areas (spatial deflection; 'the solution to pollution is dilution'), chronological deflection to future generations (see e.g. De Roo 1998) and financial deflection (see e.g. SER 1989).

27 As a result of the threat of polarisation in the discourse on the environment, changing political priorities, and a stream of sectoral legisaltion and regulations, the foundation for the WABM was laid during Minister Vorrink's term of office (1973-1977). In the mid-1980s, there was a series of relevant environmental legislation with specific licensing procedures, appeal procedures and public consultation procedures. The purpose of the WABM and other Acts was to contribute to the co-ordination, cohesion and harmonisation of the regulatory process. The WABM was incorporated into the Environmental Management Act in 1993.

28 The PIM pointed to the purpose and possibility of including standards in spatial development plans. The nuances in this line of thinking can be seen in the various decrees of the 1980s, among other things on incorporating environmental standards in land-use planning. The most striking example is the Royal Decree on the Port of Rotterdam industrial complex (*KB Maasvlakte*, 10 March 1979, no. 66), which claimed that the Spatial Planning Act (WRO) provided no basis for "the inclusion in land-use plans of standards that are not directly aimed at the use of land and buildings, or realisation of works and activities". The appendices by Otten (1980) De Boer and Van Bolhuis (1980), Bos et al. (1980) and Hunfeld and Schreiner (1981) make it clear that this statement could be interpreted in different ways (1980). The general interpretation was that standards for emissions, immissions and other environmental aspects could not be incorporated into land-use plans (see Van Zundert 1993).

29 Winsemius (1986; 78) points to three forms of regulation in environmental policy: (1) self-regulation (popular during the 1990s), whereby target groups take responsibility for their activities. (2) indirect regulation, whereby the conduct of target groups is influenced e.g. by financial incentives, and (3) direct regulation, which involves coercive legislation and regulation. The regulation chain refers to the last form of regulation.

30 Progressive standards were based on the idea that standards could gradually be tightened up over time in order to allow the economy and society to adapt to stricter future standards. The standards would progress at regular intervals from a guideline, to

a limit value, to a target value. The limit value indicates a maximum permitted level of pollution, while the target value indicates a level of pollution at which there is no material impact. The target value is usually equal to one-hundredth of the limit value. In practice it has proved difficult to comply with this. The Noise Abatement Act therefore refers only to a guideline and a limit value. Progressive standards are mainly applied to emissions. Progressive standards for effect-oriented policy were never actually implemented, one reason being that they involve spatial situations that have consequences for the longer term (see De Roo 1996; 113).

31 In the United States the 'bubble concept' allows companies to recoup environment-related investments by engaging in emissions trading, i.e. selling emission permits to other companies (Van den Nieuwenhof and Groen 1988).

32 The bubble concept did not begin to cause a furore until the early 1990s, when policymakers were looking for alternatives to the standards system, which was perceived as inflexible (Rosdorff et al. 1993). It was not the government, but the Amsterdam municipal authorities that took the initiative (Gemeente Amsterdam 1994, Meijburg and De Knegt 1994, see also Humblet and De Roo 1995, De Roo 1996).

33 Plan integration was not yet considered necessary in 1984, but "given the need for co-ordination, there is much to be said for the joint preparation and recording of environmental policy plans and regional plans" (VROM 1984a; 48). It was acknowledged, however, that organisational factors could make this difficult. In 1984 it was recommended to "make it compulsory in the relevant legislation to indicate how environmental policy plans or regional plans still to be drawn up will relate to current environmental or regional plans". This recommendation was not followed up until ten years later, with amendments to Art. 4.9, paragraph 5 of the Environmental Management Act and Article 4a, paragraph 1 of the Spatial Planning Act – the so-called 'leapfrog' amendments. Still not everyone was satisfied. In April 1994, the provincial authority of Groningen was the first in the Netherlands simultaneously to set out its regional plan, environmental-policy plan and water-management plan. During the 1990s, other provincial authorities were also working to improve the cohesion of plans relating to the physical environment. The first integral Provincial Environment Plan was introduced on 1 March 1999 by the Province of Drenthe.

34 Initially the themes were (1) Acidification, (2) Eutrophication, (3) Toxic/hazardous substances, (4) Waste Disposal, (5) Disturbance and (6) Improvement of environmental-management tools. The RIVM report '*Caring for Tomorrow*' ('*Zorgen voor Morgen*', RIVM 1988) categorised these themes and elaborated on them according to the geographical scale on which they usually occur. The NMP-1 1989-1993 (VROM 1989) eventually defined eight themes that were developed in environmental policy by linking them with strategies and concrete proposals for action. The themes were: Climate change, Acidification, Eutrophication, Geographical spread, Removal, Disturbance, Dehydration and Waste.

35 These unwanted 'negative effects' or 'externalities' (see Chapter 2) are "the deterioration of the conditions of existence and the health of humans, animals and plants, and the deterioration in the condition of material objects" (VROM 1984b; 56).

36 NMP-1 distinguished between five geographical scales (local, regional, fluvial, continental and global) for the purpose of placing environmental issues in an administrative and policy context. The distinctions are based on the relationship between the ecological consequences of human activity and the scale on which these

consequences manifest themselves. The IMP-M 1986-1990 had already referred to the interaction between the different geographical scales on which environmental problems arise. The RIVM report '*Caring for Tomorrow*' ('*Zorgen voor morgen*', 1988) used the scales to categorise environmental problems. In the NMP-1, the categories were applied to geographical areas (see also De Roo 1996; 29-30). In Chapter 2, it was mentioned that this relationship is not simple to apply in all cases. The impact of acidification is continent-wide, due to the emission of SO_2 and NOx, but can also have a local impact due to the emission of NH_3. Mol also points to this problem: "Traffic as a source of acid rain", he rightly argues, "can be seen as a global source with global impact, just as it is a source or effect on a much smaller geographical scale" (1989; 32).

37 The NMP-1 clearly stated that the 'blueprint' method was not compatible with current thinking on the role of government regarding social developments (TK 1989; 44). Biezeveld (1990; 28-29) also argues that the NMP-1 was not simply intended to present desired situations. The main purpose of the plan was to put the environment on the political and interdepartmental agendas. This second goal, he claims, has been successful. NMP-1 was thus drawn up jointly by four ministries: Housing, Spatial Planning & the Environment (VROM), Public Works & Water Management (V&W), Agriculture, Nature Management & Fisheries (LNV) and Economic Affairs (EZ). The Ministry of VROM was responsible for initiation and co-ordination (De Jongh 1989; 12).

38 De Jongh describes the method as follows: "(i) precisely identify the effects that we perceive as negative, and which involve unacceptable risks, (ii) draw up environmental quality requirements, (iii) on the basis of these requirements, calculate the total emission reduction required and, finally, (iv) distribute the reduction targets between the sources" (1989; 15).

39 In this case, we refer to 'management by negotiation'.

40 With regard to noise nuisance, the NMP-1 focused on aviation-related noise ("the second-largest source of noise nuisance" (VROM 1989c; 45), noise nuisance caused by the military, and neighbourhood noise.

41 The fall of the second Lubbers cabinet was due to a financial aspect (i.e. the standard tax-deductible travel allowance) of the NMP-1, not to its objectives.

42 The importance of the environment as a political theme was evident from the environmental targets of the NMP+, which had been adjusted upwards. The NMP+ appeared shortly after the NMP-1 (TK 1990).

43 ROMIO is the Dutch acronym for Spatial Planning and the Environment in Industry and the Immediate Environment. ROMIO was in fact a project-management model that presented a more or less universally applicable organisational framework for dealing with problems in the form of a structure for processes and participation. "The keywords are integration, flexibility, practise-based, area-specific and the equal treatment of stakeholders" (IPO 1996; 7). This reflected the philosophy of the time. ROMIO's general character was such that its specific aim (i.e. the successful co-existence of housing and industrial activities) was not immediately obvious without knowledge of the specific intentions of the project.

44 It is interesting to note that the Dutch VS-IMZ system served as an example for the development of a 'Baseline Aggregate Environmental Loadings (BAEL) Profile' by the New York City Department of Environmental Protection. BAEL evolved into a method similar to VS-IMZ, but was primarily intended to provide an overview of the condition

of the environment. The overview does not automatically have consequences for planning, but is designed to contribute to 'communicative action' by local residents and action groups who are working to improve liveability in the city. The method was applied in a neighbourhood of Brooklyn. Eventually the BAEL Profile will be developed for the entire city of New York (Anderson, Hanhardt and Pasher 1997, Blanco 1999, Miller and De Roo 1999, Osleeb et al. 1997).

45 This relates to the pilot projects for Central Amersfoort, IMZS Drechtsteden, the Arnhem-Noord industrial complex, DSM Geleen, the Theodorushaven industrial complex in Bergen op Zoom, Geertruidenberg, the Burgum/Sumar industrial complex, PISA Maastricht, the Twentekanaal industrial complex in Hengelo, IJmond, and the Northeast Flank of Groningen.

46 The Institute of Environmental Studies (IVM) of the Vrije Universiteit Amsterdam (VU) was asked to submit proposals based on three aggregation methods, one of which should reflect as far as possible the real effects of aggregate environmental load in environmentally sensitive areas. In contrast to this complicated method, which was also difficult to underpin technically and statistically, the second method was based on the existing system of standards. The third method had to act as a bridge between the first two (Aiking et al. 1990).

47 Evaluations indicated that six integrated classes was considered excessive (Twijnstra Gudde 1992 and 1994). In terms of information, the benefits of having six classes were only marginal, and the measures relating to each class differed only slightly.

48 In a number of cases the spread of pollution was not mapped on the basis of source and effect measurements, but according to estimates derived from the number of permits and dispensations, etc.

49 In relation to the sectoral approach, this method resulted in only a marginal increase in constraints for planning (Borst et al. 1995).

50 One such example is limit values. In the Noise Abatement Act the limit value is a lower limit, but in risk standards it is an upper limit.

51 Spatial and functional differentiation is appropriate primarily in nuisance issues. For risk standards there is less reason to deviate from general regulations (see also Chapter 2).

52 This inflexibility is a good reason for amending spatial-planning policy. In order to create greater scope for a project-based approach, the government replaced outlined planning permission under Article 19 of the Spatial Planning Act (WRO) with an 'independent project procedure'. According to the planning-permission rules, applications for planning permission (Art. 46 paragraph 8 and Art. 50 paragraph 8 of the Housing Act) that did not comply with the land-use plan might be granted exemption on submission of a preliminary planning decision or draft zoning plan, thus anticipating a future plan with which the applications *would* comply. This condition was abolished and replaced by the requirement that applications had to be 'well-substantiated in spatial terms'. That substantiation was largely based on existing spatial-planning policy approved by the municipal council. In this way, projects could be carried out in addition to and separate from the zoning plan (Hofstra and De Roo 1997, TK 1997).

53 The Netherlands is not the only country to experience unforeseen policy dilemmas as a result of implementing national standards. Kaiser et al. (1995) give the example of "the Los Angeles metropolitan area [where] the EPA has imposed sanctions for the failure to

meet federal air quality standards, particularly for ozone, and has suggested that new sewage treatment facilities not be federally funded because they induce growth. At the same time, Los Angeles is attempting to meet a goal of 'no discharge' into Santa Monica Bay by sewage treatment plant improvements that will reduce the level of toxins in sewage effluent" (1995; 11). One of the reactions to this dilemma stated that: "the conflict between these environmental mandates ('clean the Bay' versus 'clean the air') is a clear example of intermedia tradeoffs that are difficult to make in the current regulatory system" (Los Angeles 2000 Committee 1988; 31).

54 Designated spatial planning and 'grey' environment areas (*ROM-gebieden*).

55 In 1989 the concept of target groups had not yet entered the discussion, mainly because central government based its policy in the first instance on various government organisations.

56 The Zeeuws-Vlaanderen Canal Zone project was initially submitted by the provincial authority as an integrated environmental zoning project (TK 1989). When the noise zones around certain industrial locations were defined, it appeared that a large number of homes in Sluiskil-Oost fell within the area covered by the 65 dB(A) noise limit (Van den Nieuwenhof and Bakker 1989; 11). The government was prepared to grant EUR 1.1 million for the demolition of 114 homes, on condition that a zoning action plan was drawn up for the entire Sluiskil area and on condition that the demolition was part of a pilot project for integrated environmental zoning (Nijpels 1988). Local administrators and industry had doubts about solving local problems in this way and the method was eventually rejected.

57 The Swildens-Rozendaal/Van Noord motion (TK 1990c) requested that the government implement spatial-planning and environmental instruments not only for ROM projects, but also for all relevant projects throughout the country. This was read as a request to the Ministry of VROM to promote co-operation between the various administrative bodies and other stakeholders in environmental/spatial cases that transcended local boundaries. The following projects are examples of initiatives outside the scope of the ROM designated-areas policy: Northeast-Twente, the 'agrarian enclave' in the northeast of the Veluwe nature reserve, the Grensmaas Valley in Limburg, the Chaam Creeks in Noord-Brabant, and the Kop van Schouwen in Zeeland.

58 In almost all the projects – including the VS-IMZ pilot projects – a structure was chosen with a steering group and project group. The latter was responsible for day-to-day project management, delegating specific tasks and preparatory work to a number of special working groups, and reporting to the steering group. Final decisions were made within the steering group, which also formalised plans and signed agreements. The steering group received most of its information from a 'sounding-board' group, which consisted mainly of stakeholders who did not play a central role in implementation, but nevertheless had a voice, if only in an advisory capacity. Interest groups were often also asked to join the advisory group, the aim being not only to exchange information and views but also to increase local support from the point of view of acceptance of the project.

59 Bouwer has pointed out that "the co-ordination and integration of environmental policy and spatial planning can be no guarantee whatsoever for coming closer to achieving environmental objectives so long as spatial-planning policy, which is directed on the basis of 'the economy', is not geared to that end but is often in conflict with it." (1996; 48).

60 Selectivity was relevant in every case. Selectivity meant determining which aspects

must be considered, and which need not. Because the ROM designated-areas policy was based on a participative approach, selection was an important aspect of every project. As a theme, selectivity was therefore more important to the ROM policy than to integrated environmental zoning. In the ROM policy the context of an issue (see Chapter 4) was more important in defining the problem than for zoning. The extent to which the contextual environment was considered, and which of its aspects were relevant, were determined through the selection process. Selectivity had to be taken seriously in order to set out an environmental/spatial conflict as clearly and concretely as possible in terms of problem definition, problem selection and problem delineation. The conflict was therefore rendered more manageable, stakeholders had a clearer picture of the situation and aspects of the issue could be dealt with by means of processes that were already under way.

61 The process of geographic delineation is not a simple one when it comes to localising causes and effects within clear boundaries (Bouwer 1996). Selecting functions and activities proves equally difficult. Bouwer and Van Geleuken (1994) point out that defining geographic boundaries for an area where a specific policy has been implemented automatically means that actors, qualities and problems are excluded. Gijsberts and Van Geleuken (1996) argue that "as the chosen areas and its boundaries become more and more difficult to justify, [...] area-specific policy in that location will generally be seen as less appropriate" (1996; 8). They also refer to the resistance that can be encountered to the supplementary character of area-specific policy. Proposed measures could also have consequences for socially-related activities outside a ROM designated area (Driessen 1996; 81). In this context, Bouwer (1996) refers to inequality before the law and the deflection of spatial problems.

62 Integrated environmental zoning is intended to provide a 'sustainable' (i.e. long-term) solution long after the project organisational structure has been dismantled. In this sense it also differs from ROM designated-areas policy.

63 Socio-Economic Council, a national advisory board for economical and societal issues. The council comprises representatives from employer and employee organisations and the government.

64 In previous studies, the four approaches to environmental policy are referred to as the 'standards-based', 'hierarchic', 'multiple objective' and 'consultative' approaches.

65 'Administrative renewal' means that the direct involvement of local authorities is encouraged not only during elections, but also during policy development and decision-making processes (Nelissen, Ikink and Van de Ven 1996, and Nelissen, Godfroij and De Goede 1996).

66 In such cases the CRMH referred to 'activities that lay claim to usable environmental space', mainly in relation to "major infrastructural works, large-scale industrial activities and waste processing. The concentration of such activities in certain parts of the country means that there is less usable space for other activities" (CRMH 1992; 15).

67 The idea of compensating environmental impact is not new. Compensating measures have long been incorporated in nature policy for areas where the natural environment is replaced by other functions. The view of the central government is that area of ecological value may only be replaced by other functions in exceptional cases. Such cases must involve "A substantial collective interest for which no alternative can be found" (LNV 1992; 126). When such exceptions are made, physical as well as financial compensation may be paid. The compensation relates to the planned activity and the

environment in which it will be located (LNV 1995). The Ministry of Agriculture, Nature Management & Fisheries (LNV) uses environmental impact analysis, which indicates that "in addition to finding the most environmentally friendly alternatives, compensatory measures are sometimes proposed" (LNV 1992; 126). This practice is now set out in the Environmental Management Act (§ 7.4: the environmental impact statement, Art. 7.10.4), and may be used by "the competent authority to determine that, if it is not possible to limit all the negative environmental consequences, [...] proposed alternatives should include facilities or measures to compensate elsewhere for the unresolved negative effects" (Environmental Management Act, Wm; 7.10.4). A well-known example is the IJburg residential development on artificial islands in the IJmeer near Amsterdam. IJburg will have approximately 18,000 homes, which will all be located within the National Ecological Network. This will have an impact on local ecosystems. The ROM-IJmeer Plan of Action (Stuurgroep ROM-IJmeer 1996) therefore proposes measures to compensate the loss of ecological features (see also Vereniging Natuurmonumenten and DHV 1996, dRO Amsterdam et al. 1995).

68 The NMP-2 took no new initiatives in relation to NMP-1 and no further action was proposed to substantiate the co-ordination of environmental and spatial-planning policy. This opinion was shared by the Spatial Planning Council (RARO) and others: "Many spatial-planning issues are omitted, the Council has seen no evidence of the third strand of environmental policy, and the plan contains virtually no definite measures relating to the interface between spatial planning and the environment" (1994; 5). The NMP-3 (VROM et al. 1998) – under the motto of self-governance and made-to-measure controls – pointed out that co-ordinating environmental and spatial-planning policy was largely a matter for local authorities. Legislation and regulations would have to be modified accordingly.

69 On 25 May 1989, at the same time as the publication of the NMP-1, a draft bill on environmental policy planning and environmental quality recruitments was presented to the Lower House as a supplement to the WABM. Two sections were thereby added to the WABM: Section 4 (Environmental Policy Plans) and Section 5 (Environmental Quality Requirements'). The WABM was replaced by the Environmental Protection Act, which came into effect on 1 March 1993 (see also Gilhuis 1991).

70 The government followed this development for a brief period during the second half of the 1990s with a view to VROM examining the possibilities for integrating the NMP-4 and the Fifth Policy Document on Physical Planning into a single policy document on the human living environment (Working Group *Visie op Omgevingsbeleid* 1996, VROM 1998). This proposal soon produced prohibitive objections from a number of ministries and other parties. An integrated national policy document for the physical environment is therefore unlikely in the foreseeable future.

71 Surprisingly, traffic and transport were almost completely ignored in the development of physical environment plans and area-specific approaches (Oosterhoff et al. 2001).

72 BUGM, the decree on grants for the implementation of municipal environmental policy, dates from 1989. Its purpose was to provide financial support to encourage municipal authorities to deal with the backlog in enforcement and licensing, and to strengthen these areas. The BUGM scheme was intended as a follow-up to the HUP scheme, with which the government aimed to stimulate programmes for implementing the Nuisance Act. The government also provided grants under the FUN scheme to enable municipal authorities to fulfil responsibilities arising from the NMP framework plan of approach.

The HUP and FUN schemes were replaced in 1993 by the VOGM scheme providing follow-up grants for the development of municipal environmental policy.

73 See, for example, the environmental assessment method for Groningen (*Milieubeoordelingsmethode Groningen*) developed by the Groningen environmental department (Milieudienst Groningen 1993), the manual (*Handboek beoordelingsmethode milieu*) produced by the municipality of Zwolle and the VNG, (Streefkerk 1992) and the VNG method (VNG 1986, Kuijpers 1992).

74 See the 'Administrative Bandwidth' method (*Bestuurlijke Bandbreedte*) developed by the municipality of Utrecht (Gemeente Utrecht 1997) and the Urban Environmental Policy Method (*Methodiek voor Stedelijk Omgevingsbeleid*) developed by the municipality of Tilburg (Milieudienst Tilburg 1994, VNG 1993).

75 See also § 2.6. The United Nations conference on Environment and Development, held in Rio de Janeiro, resulted in Agenda 21, a sustainable development programme for the 21st century. Agenda 21 is primarily geared towards activities at local level. Local authorities were encouraged to draw up their own Local Agenda 21. In the Netherlands, local authorities were given the option of developing a Local Agenda 21 in order to qualify for environmental policy funding from the government. In 1996, approximately 140 local authorities took up this option (VNG 1996).

76 See web site: http://www.vu.nl/english/o_o/instituten/ivm/research/bubble.htm.

77 The Bubble Concept is linked to the idea of tradable emission allowances (Nentjes 1993; 277, Peeters 1993). Total emissions are calculated for an industrial site and divided among the various activities at the site. A company can, depending on its needs, obtain an emission permit for a certain period of time. It may then trade its emission allowance with other companies. However, such a system is only useful if emissions trading does not affect environmentally sensitive areas close to the industrial site, as is usually the case in the Netherlands. If a company close to a residential area obtains an emissions allowance from another company located at the same site but further away from residential areas, total emissions at the site will remain unchanged but the degree of impact on the surrounding area *will* change considerably.

78 The Bubble Concept report focused on only three types of environmental impact: air pollution, noise and external safety, while the goal of the concept was to achieve an integral environmental quality that was not restricted to environmental health and hygiene aspects. It was presumed that litter levels, among other things, would be included as a variable indicator for environmental quality. There were doubts as to whether this was a useful indicator, because litter levels are a variable that can also be used to assess the performance of municipal sanitation departments.

79 Institute of Environmental Studies of the Vrije Universiteit Amsterdam (VU).

80 There are now 24 projects. The Utrecht-Kromhout project failed because the Ministry of Defence was not willing to hand over former barracks to the local authority.

81 The recommendation to take account of environmental health and hygiene aspects as early as possible in the decision-making process appears to contradict the 'strategic choice' approach of Friend and Jessop (1969), which is based on the principle that decisions are not taken until it is strictly necessary. The approach was designed to make the planning process more flexible when there were uncertainties that could only be resolved during that process (Voogd 1995). In most cases it is possible to ascertain the general environmental situation before the decision-making process commences. It is not subject to major fluctuations and is therefore not necessarily an uncertain factor in

the planning process. Furthermore, the environmental situation can also be ascertained through existing legislation, although this legislation does not require this at an early stage (Blanken and De Roo 1998; 281-190).

82 The City & Environment Experiment Act enables municipal councils "to deviate from: (a) environmental quality requirements for soil, noise, air and public safety; (b) procedural provisions and provisions on powers prescribed by or pursuant to the Noise Abatement Act, the Environmental Protection Act, the Soil Protection Act, the Spatial Planning Act, the Housing Act and the Urban and Rural Regeneration Act" (TK 1998; 2). These 'Step-3 decisions' must be fully substantiated (see Art. 8) by the local authority and approved by the Minister of VROM.

83 The method for urban compensation is still far from clear. A number of authors (Kuiper 1995, De Roo 1996c, Walgemoet 1995, Wiersinga et al. 1996) have specified a number of criteria for compensation measures. The guidelines set out by Bartelds (in Bartelds and De Roo 1995; 130-131) were one of the first steps in this direction: (1) individual protection should be guaranteed; (2) priority should be given to source-directed measures and planning-based remediation measures; (3) no negative health effects must result from substitution or compensation and there must be no increase in the level of risk; (4) substitution and compensation must not result in the deflection of environmental impact to other sections of the urban population; (5) compensation must always take place by means of the required active measures; (6) measures designed to compensate pollution must be perceived as such by the local population.

84 The multifunctionality principle should "in theory preserve the potential of the soil to support its different possible functions in the short term and in the long term" (*Leidraad Bodembescherming* 1997; 2). The soil has many functions, including an ecological function, a supporting function, mineral extraction, and cultivation. The soil can also have a natural and cultural-historical value. The concept of multifunctionality was originally the basis of prevention and soil protection. However, for actual remediation "the options were to remove, isolate or control the contamination. However, determining what restoring the multifunctionality of the soil should mean in each particular case is a somewhat complex question" (TCB 1996; 26).

85 See also TCB 1996 and the theme issue *Bodem* no. 4 (1995).

86 This Article related to the provisions that should be included in municipal building regulations.

87 Active soil management can be defined as "the whole range of activities geared towards dealing satisfactorily and efficiently with structural soil decontamination and its consequences in a given area" (*Werkgroep bodemsanering* 1993; 22).

88 The IBC remediation variant was only used when the multifunctional variant was inappropriate, e.g. because there were technical problems or because it was financially impracticable.

89 Soil contamination is classified as serious if, in a volume van 25 m^3 (soil or sediment) or 100 m^3 (groundwater), the average concentration of contaminants exceeds the intervention values.

90 A distinction was made between serious and non-serious soil contamination. The serious cases are then categorised as urgent or non-urgent contamination. The degree of urgency is determined using actual human health risks caused by exposure (expressed as maximum allowable risk), ecology (soil flora and fauna) and spread (more than 100 m^3 per year), especially via groundwater.

91 The latest proposals state that contaminated soil classified as 'serious, non-urgent' no longer has to be cleaned up within a given period of time. The situation is also noted in the land registry. A buyer of the land in question is even deemed to be a 'guilty owner', apparently without consequences... However, buyers and citizens in general will be more profoundly affected than the legislator intended by the fact that their land carries the label 'seriously contaminated'.

92 Hofstra and De Roo's contribution (1997) to the Soil Protection Guidelines (*Leidraad Bodembescherming*) outlined proposals for accelerating the co-ordination of subsurface remediation and above-ground development. Here, 'co-ordination' is taken to mean an "integrated function-oriented and area-specific approach to underground work and above-ground spatial development" (1997; B8-19). That co-ordination is described as 'process-oriented' and 'project-based'. The latter involves the concrete short-term co-ordination of remediation work and planned development. Process-oriented remediation focuses on the longer-term harmonisation of local policy objectives, on communication and on information exchange.

93 This recommendation was in line with the thinking of the time, which was also characterised by concepts such as 'win-win situations', 'policy transparency', 'responsibilities of social actors', 'stimulating the market mechanism', 'chain management' and 'risk reduction'.

94 BEVER = *Beleidsvernieuwing Bodemsanering* (new policy on soil remediation).

95 This was in line with the distinction between plots and cases. A case of soil contamination no longer needed to be dealt with in its entirety, but could be separated into plots (and if necessary into functions), and an appropriate method used for each plot.

96 The approach to cleaning up immovable contamination is somewhat different. In such cases, the principle of cost-effective remediation applies, particularly for groundwater. The aim is to remove as much of the contamination source as possible and implement measures to deal with the spread. The assumption is that each case will be dealt with individually. Standard approaches have not yet been developed. The goal is to create a 'stable end-situation', in other words to halt the spread of the contamination.

97 Recently there has been a worrying increase in the number of indications (in inspection reports and actions by the Public Prosecutions Department) of unsatisfactory clean-ups. These indications relate above all to fraudulent practices. There is, as yet, no indication that environmental quality has been compromised.

98 This procedure applies when environmentally sensitive development is planned for a given location and involves noise levels that exceed the preferred limit value in the Noise Abatement Act. In such cases, the provincial authority must approve the plans. This procedure yields very few environmental gains and is unnecessarily time-consuming.

99 In accordance with Article 8.40 of the Environmental Protection Act, general rules can be drawn up to replace the licensing requirement with a notification requirement. These rules are set out in the Environmental Retail Trade Decree, which applies to an estimated 25,000 businesses (*Milieuvoorschriften* 1971-..; A3-24).

Chapter 6

A Decision-Making Model Based on Complexity

Pilot Projects in Integrated Environmental Zoning as a Source of Inspiration

> What is new about the present activity is not the study of particular complex systems but the study of the phenomenon of complexity in its own right. If, as appears to be the case, complexity is too general a subject to have much content, then particular classes of complex systems possessing strong properties that provide a fulcrum for theorizing and generalizing can serve as the foci of attention (Herbert A. Simon 1996; 181).

6.1 Prelude

In 1994 the Department of Noise and Traffic at the Ministry of VROM commissioned a study into the consequences of integrated environmental zoning for planning (VS-IMZ, see § 5.4). The study was published as a book under the title '*Milieuzones in beweging*' ('*Environmental Zones in Motion*') at the end of 1995 (Borst et al. 1995).

The book pointed among other things to the limitations of the standards system. The conclusion that the consistent strict application of generic environmental standards does not always produce maximum – or even optimum – results could be rationalised by the concept of complexity. The study of the applicability of VS-IMZ showed that the degree of complexity of an environmental/spatial conflict was in reverse proportion to the result obtained from applying the generic standards system. In other words: the more complex the environmental/spatial conflict, the smaller the chance of resolving it by means of stringent environmental standards.

Once this conclusion had been reached, the question was how to translate the 'level of complexity' of a conflict into the most appropriate decision-making strategy. A central, prescriptive method was one of the possibilities, but proved to be less suitable for resolving the more complex environmental/spatial conflicts. The question then arose as to which alternatives were available and – above all – *why* they were suitable. The book '*Environmental Zones in Motion*' distinguished between 'simple', 'complex' and 'very complex' environmental/spatial conflicts. Decision-making strategies then had to be found to match these categories.

It was now a matter of underpinning decision-oriented choices made with a view to developing an effective and efficient approach to environmental/spatial conflicts. This approach would depend on the degree of complexity attributed to the environmental/spatial conflict. The concept of complexity-related decision-making led to a model consisting of three concentric circles (see Fig. 6.1). The inner circle contains those elements that are compatible with a centrally directed and prescriptive strategy for decision-making. The outer ring contains those elements that are compatible with a decentral decision-making strategy adapted to local circumstances. Solutions for conflicts can therefore be found in the circle relating to the relevant degree of complexity, e.g. solutions for straightforward conflicts are located in the inner circle and solutions for the most complex conflicts in the outer circle.

Section 6.2 describes the model in detail. Section 6.3 describes the model as it can, in principle, be applied at a more general level, in terms of the ministerial proposal for a provisional system of integrated environmental zoning (VS-IMZ, see § 5.4). One of the main reasons for using VS-IMZ as an example here is its status as an instrument within the 'framework' for planning-based action (see Fig. 4.8 and 5.12). It is an extreme position, which partly explains why the VS-IMZ has been heavily criticised. From this extreme position it is relatively easy to build on the alternatives in the model. The experiences from the eleven VS-IMZ pilot projects were used as the basis for complexity-related scenarios presented in this chapter (§ 6.4) for dealing with environmental/spatial conflicts. The scenarios provide insight into the practicability of the VS-IMZ instrument and into the applicability of the model.

In general terms, the model is a tool designed to assist decision-oriented planning. It was designed to ascertain the most effective and efficient way of deploying policy instruments in order to solve issues or problems of varying complexity. As such, the model can be used to establish a relationship between (a) the principles of a policy instrument, and (b) the complexity of the issue for which it is intended.

6.2 A Model for Decision-Oriented Action

The model emphasizes the process of matching instruments to issues, and is oriented towards policy development. It is known as the '*IBO*' model. IBO is the Dutch abbreviation for 'instrument analysis for policy development'. The purpose of the IBO model is to make policy choices (and in particular the policy instruments) transparent for which it is not possible to predict all the consequences. This applies especially to choices that have not yet fully evolved and in which no experience has been gained. It also applied to instruments that were still being developed and discussed. The IBO scheme is discussed here against this background.

The IBO model (Fig. 6.1) is represented as three concentric circles reflecting varying degrees of complexity. The circle can be 'cut' like a cake, with each 'wedge' representing a policy theme. A degree of complexity ('simple', 'complex', or 'very complex'), represented by the rings, is assigned to each policy theme.

The IBO scheme can be used to analyse, test and/or evaluate policy instruments by means of the following steps (see also Fig. 6.1):

1. Themes that are under discussion and/or have the most influence on the

functioning of the policy instrument must be identified. The IBO model does not apply to themes that are not a subject of discussion. It is therefore important that these themes are not influenced (or only to a very limited extent) by themes that *are* under discussion.

2. The identified themes are reduced as far as possible to their essential elements.
3. Obviously, these themes are likely to be characterised either as goal-oriented or institution-oriented. Since the instrument to be analysed is still being developed, the basic related questions '*what* has to be achieved?' and 'by *whom*?' still need to be answered.
4. The decision-oriented element (i.e. '*How* will the goal(s) be achieved?') can be identified by dividing up each theme into an equal number of categories (boxes). This means that the theme can be considered from a 'simple', a 'complex', and a 'very complex' perspective, and described in those terms. This usually involves examining the extremes of a theme, which is then categorised on that basis.
5. The relevant themes are entered in the IBO model, reserving a 'wedge' for each theme. The descriptions of 'simple' situations are entered in the inner ring segment, and those for the most complex situation are entered in the outer ring segment.
6. When the IBO scheme has been filled in, scenarios can be developed according to the complexity of the situation. The mechanism can also be applied in reverse, in which case an approach to a specific situation can also be drawn up, based on complexity.
7. The scenarios are analysed, and conclusions drawn as to the feasibility of an instrument. Instruments can then be adapted to produce the most appropriate solution for a specific situation. An instrument may also be found to be inappropriate for the purpose for which it was originally intended. 'Possibilities' must be balanced against 'realities'.
8. When each step has been completed, its consistency in relation to previous steps must be assessed. If the outcome is unsatisfactory, the step must be repeated in order to achieve the desired consistency.

The IBO scheme can also be used to assess the extent to which a new instrument is applicable. This also involves identifying a number of relevant themes. The themes are categorised by complexity, with the middle category being the most characteristic of the new instrument. The scope of the new instrument can be ascertained by considering the situations in the adjoining segments, then estimating their administrative and material implications. In fact, the determining themes for the instrument are set out 'as extremes', allowing different scenarios to be drawn up. As experience is gained with an instrument it can be adapted as necessary or declared impracticable.

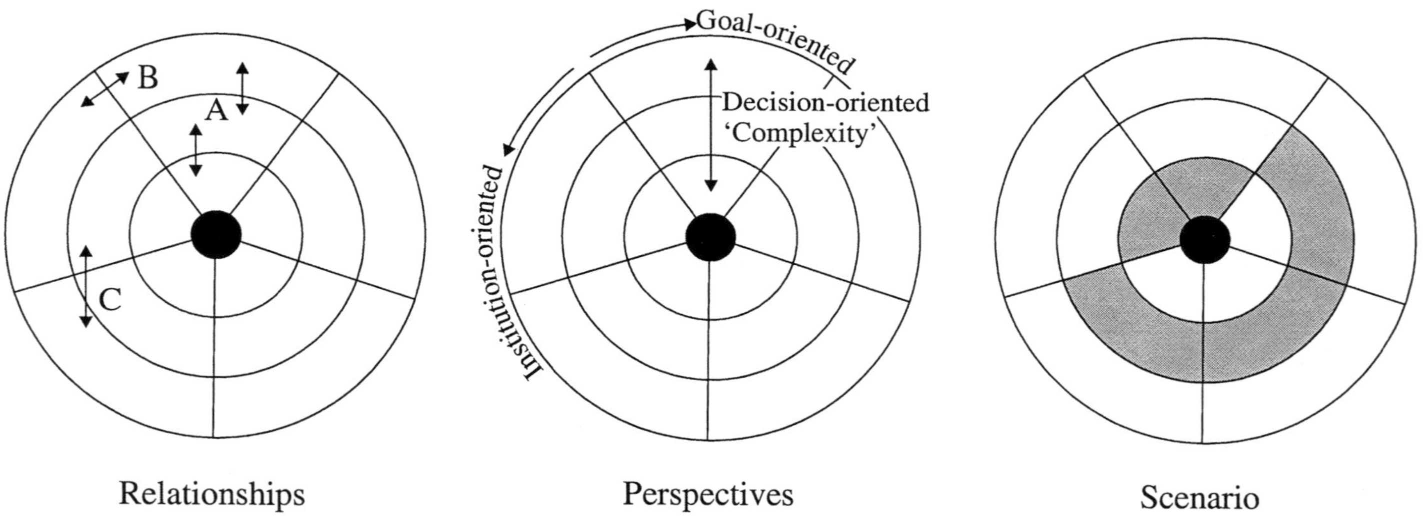

Figure 6.1 Basic structure of the IBO model

Note:
A: Relationship based on content (i.e. relevant policy aspect or theme) and structure (decreasing or increasing complexity)
B: Structural relationship (i.e. comparable level of complexity)
C: No relationship between structure and/or content

Advance assessment will not always be enough to guarantee the optimum performance of a new instrument. Often, an instrument is analysed because it is already in use and is proving to be unsatisfactory (for a discussion on this point see Arts 1998). It was, in any case, this consideration that prompted the development of the IBO model, which was initially designed to analyse the feasibility of integrated environmental zoning (VS-IMZ) in situations of varying complexity (Borst et al. 1995). The IBO model was initially devised in response to discussions of the details of integrated environmental zoning as an instrument for dealing with environmental/spatial conflicts (§ 6.3).

The idea of a circular model containing different themes is not new. We shall discuss two examples here. Nijkamp et al. (1994) developed a 'pentagon model' (see Fig. 6.2), which was intended to promote a policy process whereby "five key factors are considered simultaneously and continuously" (Nijkamp 1996; 139). The pentagon model emphasises the importance of remaining aware of essential aspects of policy development and implementation. This purpose is similar to that of the IBO model.

In contrast to the pentagon model, the IBO model does not presume that the five key success factors are fixed for each specific instrument. The IBO model identifies themes for each individual instrument, depending on the desired goal and the progress of related discourse. The themes are therefore instrument-related as well as issue-related. This does not mean, however, that the issues themselves are not linked. When identifying themes for the IBO model, it is advisable to aim for goal-oriented *and* institutional aspects (see Table 4.1). The decision-oriented aspect is incorporated in the model as a result of the distinction between simple, complex and very complex issues (see Fig. 6.1).

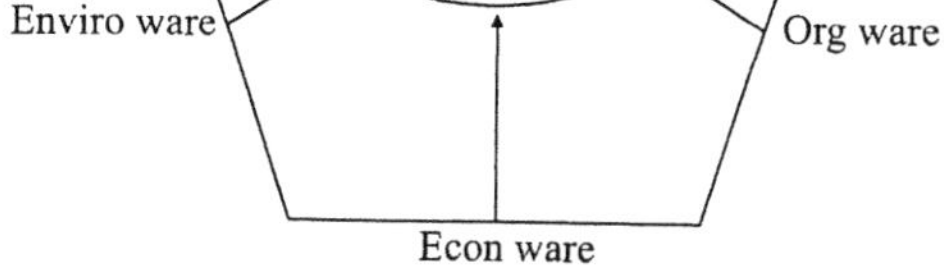

Figure 6.2 'Key success factors' according to the pentagon model of Nijkamp et al. (1994)

Pool and Koopman's model (1990) for strategic decision-making (see Fig. 6.3) is also interesting. As in the IBO model, complexity is seen as a determining factor, although it is not a basic principle of the model. Pool and Koopman also assume that, while decision-making processes are "partly determined by their content and the context in which they take place, there is still a certain freedom of movement for decision-makers" (Pool 1990; 32). However, they see this *a priori* leeway as limited because content and context will be determining factors.

Analysis using the IBO model is not based on policy content and context as determining factors as such, but on their interdependence. The study described in '*Environmental Zones in Motion*' into the implications of integrated environmental zoning for planning showed that, with regard to content and context, standpoints and visions may exist that depend on the *evaluation* and definition of an issue (see also § 5.4). In environmental policy, environmental problems have traditionally taken priority over other issues because environmental health and hygiene were considered to be so important. Environmental problems were defined simplistically according to quantitative standards. When the consequences of a dominant environmental policy became apparent in related policy fields, a shift of emphasis was initiated from a predominant to a shared evaluation. A shared evaluation means, however, that an issue becomes more complex. In short, this evaluation strongly determines the complexity of an issue and the choice of strategy. In the IBO model, therefore, context and content are not only seen as determinants, but also as dependents.

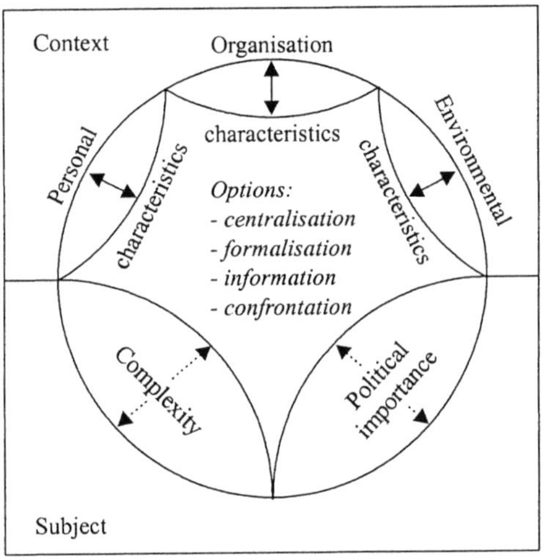

Figure 6.3 Schematic representation of strategic 'options' in decision-making
Source: Pool and Koopman 1990

6.3 The IBO Model Adapted for the VS-IMZ Discussion

In this section, the IBO model is illustrated using the example of integrated environmental zoning (VS-IMZ, see § 5.4). The first step is to select relevant themes or perspectives (i.e. 'cake wedges') that are central to the discussion on integrated environmental zoning, and then to categorise them according to their complexity. In section 5.4, the instrument VS-IMZ is described as a system with a methodical section 5.4, the instrument VS-IMZ is described as a system with a methodical component, a standard-setting component and a procedural component. We concluded that, as a *method*, VS-IMZ has few disadvantages because it provides a reasonable picture of environmental health and hygiene in a given area. VS-IMZ also gives a reasonable picture of the competing demands of environmentally harmful activities/functions and environmentally sensitive functions/areas. The main problems are caused by the choice of standards and the procedural structure of the instrument. The standard-setting component of VS-IMZ was developed mainly according to the principles of the 1979 Noise Abatement Act (§ 5.2). This involved setting generic standards that express the quality of the environment in terms of pollution types such as sound or odour. These are only balanced against other qualities to a very limited extent. The standards also established a framework. The framework-setting and generic nature of the standards system was a constantly recurring theme in the discussion on the practicability of standards in general and of VS-IMZ in particular. The resulting focus on environmental quality meant, for example, that the quality of the spatial environment and the human living environment was in danger of being overlooked. Discussions also centred on the speed with which VS-IMZ would provide results. Neither must we forget the discussion on institution-oriented aspects of VS-IMZ, which focused on the delegation of responsibility between the various parties involved, including local authorities and non-governmental organisations. These aspects and characteristics are incorporated into the IBO model.

Five general aspects can be derived from the above. On the one hand these aspects are subject to criticism, on the other they are determining factors for the decision-making process relating to environmental/spatial conflicts. The aspects are: environmental quality, prescriptive policy (degree of rigidity), generic applicability for spatial functions versus local differentiation, and the temporal and relational aspects. These aspects will be entered in the IBO model.

Quality

In a number of IMZ pilot projects, a high level of quality in terms of environmental health and hygiene was considered too limited a goal (§ 6.4). There are cases in which environmental quality is emphasised at the expense of other qualities, in particular spatial quality. If environmental measures at source are ineffective, this will result in spatial constraints in the area surrounding the source of the pollution, and it will not be possible to site environmentally sensitive functions or activities in that area. It will come as no surprise, then, that there have been calls for local spatial implications to be considered in addition to environmental health and hygiene (see § 5.4), thereby achieving a balance between an increasing number of aspects during decision-making. However, this will not always be an advantage. How can environmental and spatial qualities be considered together? This is even more difficult in cases where environmental issues relate to projects which – in accordance with the ROM designated-areas policy (§ 5.4) – require not only measures for reducing pollution, but also for tackling socio-economic problems, among other things (Kuijpers 1996; 62).

Figure 6.4 Quality as a relevant theme in zoning policy

On the other hand, the scope for developing strategies to deal with such issues is increasing. In the most extreme case, an integral quality for the human living environment could be the desired balanced result.

Rigidity

The zones established around industrial areas have a fixed status under the Noise Abatement Act, and noise zones are prescriptive for spatial-planning policy. Integrated environmental zones resulting from VS-IMZ were also – in accordance with the Noise Abatement Act – designed to be prescriptive for spatial policy. In a number of cases, this placed so many constraints on policy that local governments found it difficult to strike a positive balance between environmental and spatial wishes and requirements (see § 6.4). Local authorities have too little scope to take account of specific local circumstances, which means that the final result may be far removed from the result that could have been achieved under 'optimum' circumstances. Opportunities are missed that could have been utilised if there had been greater administrative flexibility and if standards could have been adapted to local conditions. Instead of defining a fixed environmental zone, variation could have been allowed within a given range, or 'bandwidth'. The Noise Abatement Act allows area-specific deviation from the preferred limit value of 50 dB(A) by means of a 'higher-value procedure' (§ 5.5), whereby the Provincial Executive can allow local authorities a higher limit value of up to 55 dB(A). The most recent proposals on noise policy place greater emphasis on deviation within a permitted range, defined under the responsibility of local authorities for individual local situations (see § 5.5). "A further step towards greater administrative flexibility in the application of integrated environmental zoning can be taken by presenting results in informative but indicative environmental assessment reports" (Borst et al. 1995; 100). In principle, this variant can be compared to environmental impact assessment. Impact assessments provide insight into a situation, but that insight does not necessarily have consequences. A report can be set aside on the basis of well-founded arguments, and clear explanations must be provided of how environmental aspects are to be taken into account. This method allows local

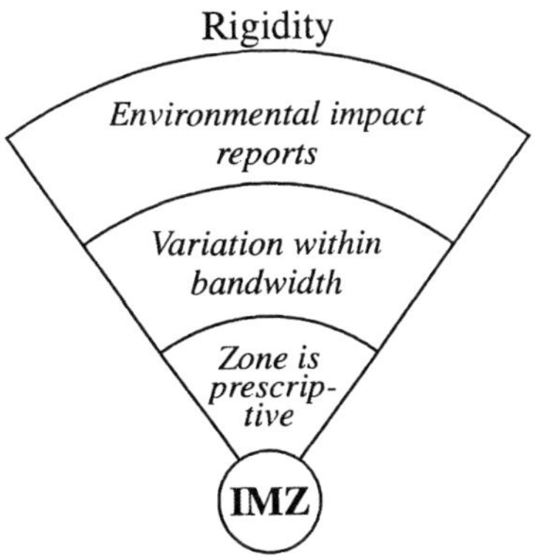

Figure 6.5 Rigidity as a relevant theme in zoning policy

authorities the greatest scope for developing area-specific policy. However, it offers the least guarantee of an acceptable level of environmental health and hygiene in the long term.

Spatial-Functional Perspective

VS-IMZ uses a system of standards that has a generic effect. In principle, the same uniform environmental standards apply throughout the Netherlands. This method is based on a hierarchical concept of governance, whereby central government sets a level of environmental quality for the whole country. In practice there are many exceptions to this rule. One example is the seaport standards, which place port activities under a more flexible regime than other industrial activities (see § 7.4). Furthermore, national standards are not easily attainable everywhere in the Netherlands, partly due to regional differences. In such cases, spatial-functional differentiation of regulations may be appropriate. This can involve defining a general level of environmental quality for given areas in the Netherlands, e.g. by distinguishing between urban, rural and 'other' areas. This is known as 'regional-functional differentiation'. Differentiations in standards according to area and/or function are a realistic option for accommodating the needs of particular areas. This applies to areas with unique environments that require extra protection in order to prevent the loss of that uniqueness and to prevent the quality of those environments 'falling' to the national level (§ 5.3). The Wadden Sea nature reserve is a good example (De Roo 1993). In practice, there are also cases in which 'unique' environmental qualities deviate from the generic standards and are designated in a 'negative' way, although this is not always emphasised as such to the public. One example is the zoning around Schiphol Airport, where environmentally sensitive functions are permitted within the 35 KE, but to a limited extent. Schiphol is rather a controversial example but the idea in itself is not new. However, such regional differentiation may still be insufficient for dealing with environmental/spatial conflicts in a way that takes account of local and specific circumstances. A restaurant terrace on the busy Leidscheplein in the centre of Amsterdam cannot be compared to a back garden in rural Vlagtwedde. It is useful to ask whether the same standards can be applied to busy city centres as to quiet residential areas close to the countryside. Background levels of environmental pollution are higher in busy city centres than anywhere else, and this will influence the perception of nuisance in inner cities. Furthermore, people who live in the cities have often consciously chosen to do so. They value the multifunctionality of the city and accept the reduced quality of the grey environment, or see it as part of a city's charm.[1]

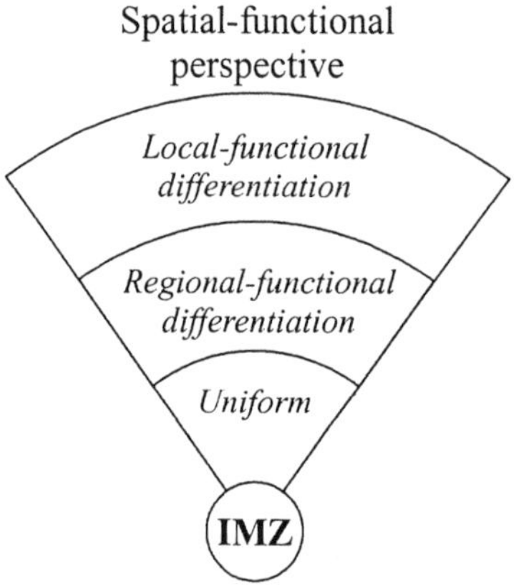

Figure 6.6 The spatial-functional aspect as a relevant theme in zoning policy

Temporal Perspective

In retrospect, the period of time proposed for implementing the IMZ system and achieving targets was a sign of considerable optimism. During the development of VS-IMZ it was assumed that it would be possible to define a provisional environmental zone as soon as the environmental quality of the area surrounding an industrial activity or area had been ascertained (Borst 1994). A conservative policy would then be followed for the site, which would become a definitive integrated environmental zone after a period of one year. The zone would then be incorporated into spatial planning processes in order to influence planning (VROM 1990; 23). Within the one-year period, a remediation method must be chosen and developed. It was the one-year period in particular that proved to be unrealistic, given the scale of the environmental/spatial conflicts in a number of the IMZ pilot projects (Borst 1996 and 1997, Borst et al. 1995, De Roo 1992, Twijnstra Gudde 1992 and 1994). In such cases, phasing was the most obvious option. It is sometimes advisable, particularly where large-scale projects are concerned, to allow phased spatial development in proportion to the rate at which the level of pollution falls. 'Anticipation' is a more rigorous alternative whereby spatial development anticipates expected levels of environmental health and hygiene. This alternative is not directly embraced by current regulation and legislation. It can have undesirable consequences if the expected level of environmental quality is not attained or is attained much later than expected, or when land-use changes in the interim. On the other hand, the option of anticipation can provide opportunities for improving the grey environment through spatial measures.

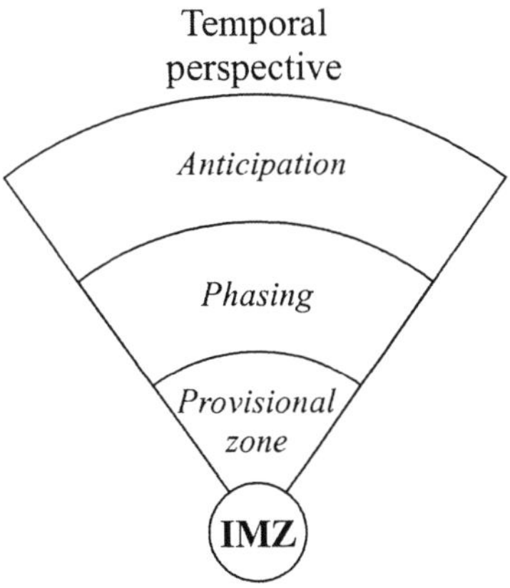

Figure 6.7 The temporal aspect as a relevant theme in zoning policy

Relational Perspective

There was criticism of the relational-communicative approach that derived from the VS-IMZ proposals. VS-IMZ is a programme that can be implemented step-by-step in order to realise the sustainable segregation of environmentally sensitive and environmentally harmful functions. This segregation has been anchored in the spatial-planning process and thus influences land-use development. It is a hierarchical approach with limited flexibility that is geared towards achieving the right balance between source-directed and effect-oriented measures, whereby "pollution must be dealt with in the first place by implementing measures at source, and not until the final stages in the spatial environment" (VROM 1990; 23). Apart from this hierarchical form of governance, a more decentral approach is conceivable in accordance with current developments in environmental policy. This relates to situations in which the local context is such an important factor in determining the strategy for dealing with environmental/spatial conflicts that generic regulations would not be effective enough. A network-based approach may be more successful than a strategy based on central governance if a situation is extremely complex and involves a large number of parties with diverging interests, and if it is difficult for the government to steer the other parties and pursue well-considered single goals (see also § 4.7).

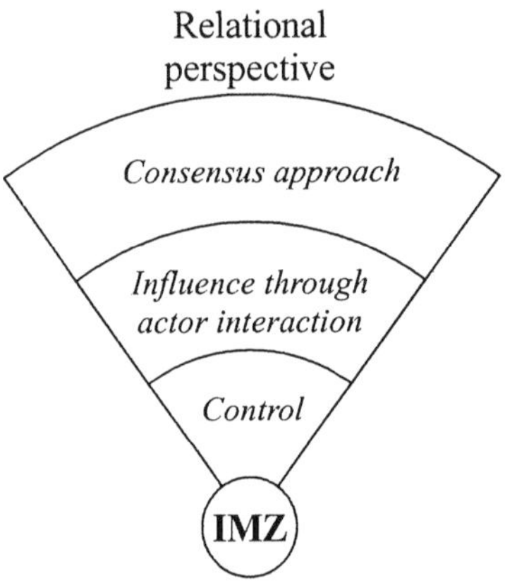

Figure 6.8 The relational aspect as a relevant theme in zoning policy

We have now filled in the IBO model for this particular VS-IMZ discussion. We have also categorised the five themes/perspectives according to their complexity (Fig. 6.9). The inner ring comprises governance elements that are most appropriate for the least complex type of issue. Such issues can be easily separated from their context and – it was thought – could be solved by excluding contextual influences. Given the initial conditions of those issues, it is reasonably likely that the desired goal can be attained, i.e. the level of certainty is relatively high degree. It appears that the policymakers who developed the VS-IMZ instruments have focussed solely on such issues. Unfortunately, not all zoning issues are as simple as this.

The elements in the outer ring of the IBO model can be applied if there is a large degree of interrelation between an issue and its material and administrative context. Such issues are relatively very complex and the initial conditions are obscure. It is also difficult to predict the likelihood of success, and there is a high level of uncertainty regarding efficiency and effectiveness.

This IBO model presents a broad spectrum of options that cover the aspects of the instrument VS-IMZ that are under discussion. The different options are presented with a certain degree of cohesion, which makes it possible to draw up scenarios or make well-founded proposals. The report '*Environmental Zones in Motion*' analyses the experiences from the various IMZ pilot projects on the basis of this IBO model.[2] The feasibility of VS-IMZ is also assessed.[3] The findings are described in the following section.

6.4 Complexity and Decision-Making in Relation to the IMZ Pilot Projects

Between 1989 and 1997, the VROM system of integrated environmental zoning was implemented on a trial basis (VS-IMZ, see § 5.4) through eleven pilot projects involving (1) actual cases of negative environmental impact that (2) negatively influenced the human living environment and (3) had been generated by several environmentally harmful components, and (4) were 'zonable'. For practical reasons the pilot projects also involved (5) pollution caused by large, fixed installations (VROM 1989) and were selected in order to realise (6) an equal geographical distribution throughout the Netherlands. The main aim of the integrated environmental zoning projects was "to attain a high sustainable quality for local and regional environments surrounding one or more large fixed industrial installations" (VROM 1989; 3).

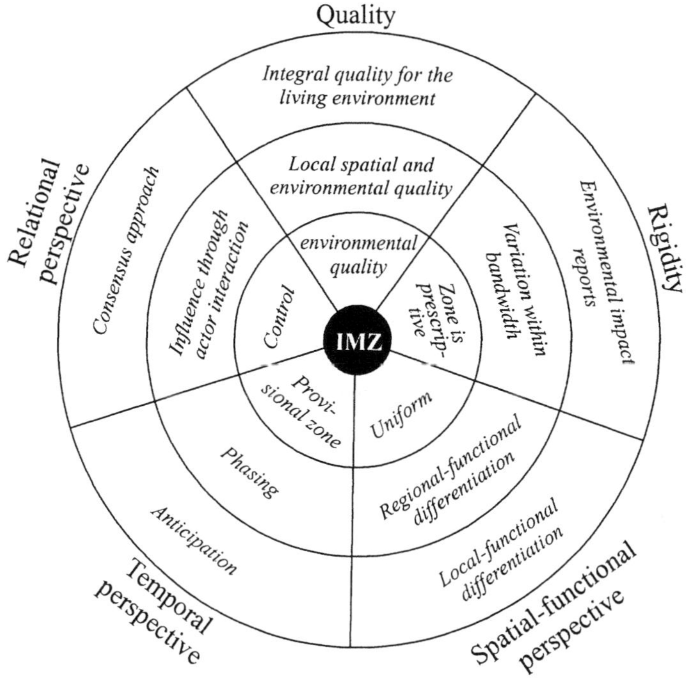

Figure 6.9 The IBO model applied to the Provisional System for Integrated Environmental Zoning

Environmental standards and zones both essential mechanisms on which to base solutions. In addition to meeting the standards set by environmental legislation, the goal was to "find an administrative balance between the development of environmentally sensitive functions (including housing) and the development of environmentally harmful activities (in particular industry)" (VROM 1989; 3). This was a logical position given the administrative climate of the time (see § 5.4) but, in retrospect, it was overoptimistic and was defeated by bureaucratic, spatial and economic barriers.

It was not at all easy to find a solution based on an acceptable balance between residential functions and environmentally harmful activities for all projects. Only the *Hengelo-Twentekanaal* project completed all the proposed phases of VS-IMZ. This project resulted in the amendment of the relevant land-use plans in accordance with an integrated environmental zone and the related planning constraints (Borst et al. 1995, Flohr and Meijvis 1993). The zone provided the framework for the development of the *Twentekanaal* industrial complex and the basis for assessing spatial-planning policy and environmental licensing procedures (S.A.B. 1994). The northern boundary of this industrial area borders on the residential areas of Hengelo, which means that the quality of the environment in the south of the city is not high. This has implications for business location policy at the industrial complex. The complex is located adjacent to valuable countryside, which also explains why it was difficult to realise the expansion that was considered necessary. Insight and clarity regarding the grey environment were lacking, so integrated environmental zoning was embraced as a solution (Stuurgroep IMZ Hengelo Twentekanaal 1991). In addition to the zoning of the industrial complex, the measurement of pollution at the site resulted in a VI[4] integral environmental zone for existing small-scale situations to the east of the complex. Noise, odour and individual risk were the underlying reasons for this. Consequently, it was considered necessary to demolish a number of homes (Gemeente Hengelo 1993). The regulations for new situations extended over a significantly wider area[5] to the north and east of the industrial complex, which meant that building restrictions were placed on large areas of adjacent residential neighbourhoods. Relatively speaking, however, the scale of the sectoral and integral environmental impact was modest and therefore local, so the integral environmental zone and its consequences were accepted as part of the local land-use plan.

The projects in Geertruidenberg, Groningen, and Maastricht were on a similar scale to the Hengelo-Twentekanaal project. The *Groningen-Noordoostflank* (Groningen Northeast Flank) focused on soil decontamination, traffic noise and odour nuisance. The odour pollution was caused mainly by two sugar factories several kilometres away from the project. This pollution was therefore outside the official scope of the VS-IMZ project. The sectoral pollution caused by activities at the site covered by the VS-IMZ could be resolved without too many problems by using sectoral measures. Integrated remediation was not essential, so an integrated environmental zone was not considered appropriate, even for new situations. The step that was supposed to lead to the implementation of a prescriptive integrated environmental zone was therefore considered undesirable. The Groningen municipal authorities preferred to see integrated environmental zoning as indicative, and as a foundation for predicting the consequences of environmental policy and spatial-planning decisions (Gemeente Groningen 1992). In retrospect, the integrated environmental zoning project was important for the Groningen municipal authorities mainly because it allowed them to update their information on local environmental health and hygiene in and around industrial sites, for licensing purposes.

In the *Geertruidenberg* project, industrial pollution – on which VS-IMZ focuses – and other types of pollution caused by traffic and shipping were strongly interrelated. This made it difficult to impose specific measures, which were based on the VS-IMZ inventory, on third parties, while many types of pollution were apparently overlooked. The project extended across the whole municipality. Three industrial sites located around the central residential area of Geertruidenberg were responsible for noise nuisance and air pollution. They also caused environmental pollution that had implications for external safety. Under these circumstances, there was little scope for expanding environmentally sensitive functions. Insight into environmental health and hygiene would have to clarify to what extent the situation would allow expansion of residential areas and leisure facilities on the periphery of the existing residential area. The information gathered through VS-IMZ shows that, for existing and new areas, the 'black' areas were largely restricted to the industrial complexes themselves, with the exception of a number of locations for which spatial development was planned. At-source measures were expected to be sufficient for these locations. Largely because the problems were limited and resolvable within the scope of VS-IMZ – as with the Groningen situation – the emphasis was placed on the inventory phase. Here, too, the situation was considered to be manageable and it was not considered necessary to include rigid frameworks in the local land-use plan.

Figure 6.10 PISA integrated environmental contour showing the 'current situation' for Maastricht

Source: Gemeente Maastricht 1993; Fig. 23

The Maastricht IMZ project for integral remediation (PISA) was the result of the Environmental Policy Integration Project (PIM), the aim of which was to gather data on noise, odour pollution and external safety throughout the city (Fig. 6.10, see also § 5.2 and Gemeente Maastricht 1987). The *PISA-Maastricht* project focused on three areas from the PIM project: Boschpoort, Limmel and Groot Wyck. The study was carried out within the framework of the national VS-IMZ project, but the local authority used its own method for measuring pollution levels. In the three named areas, the environmental pollution levels were high enough to cause a conflict situation. The wish to retain and revitalise the old industrial sites conflicted with the concern for liveability in local residential areas. The problems were therefore not related to existing situations, but to constraints on new construction in terms of noise, odour nuisance and external safety. A large area of the Groot Wyck neighbourhood is classified as a 'grey' area because it is located between the 10^{-6} and 10^{-8} contours for annual individual mortality risk.

Approximately half of the Boschpoort area fell within the odour contour of one odour unit 98 percentile, which meant, according to the VS-IMZ, that new building would be subject to constraints. On the basis of this information, an individual programme of remediation measures was drawn up for each individual problem situation, with particular emphasis on spatial-planning aspects. In other words: an area-specific approach was chosen.

It is interesting to see how, in Maastricht, the emphasis gradually shifted from a broad geographical scale to a local scale as a basis for 'customised' remediation measures and spatial planning. The shift was partly intentional and partly incidental, and took place over a period of approximately ten years. The process was based on strategic planning and the gathering of knowledge and insight in advance. These developments supported and were determining factors in the decision-making process during the various stages of the project. This method was encouraged in the national *Stad & Milieu* (City & Environment) project. However, it was not expected to be the most efficient approach for Maastricht, given the scale of the environmental/spatial conflict and the measures that were considered necessary. The conflict was, after all, a relatively limited one to which prescriptive environmental standards could be applied, and for which the strategic preparations had been carried out by the government and formalised in the VS-IMZ. It is easy to say this in retrospect. Nevertheless, Maastricht has matters firmly under control.

The experiences with the projects discussed above are represented in Fig. 6.11. All these projects involve a category of environmental/spatial conflicts in which pollution can be traced back to one or a small number or sources. It is easy to establish a clear source/effect relationship for the pollution that has occurred. In addition, these are conflicts for which spatial intervention was either unnecessary or kept to a minimum. Instead, emphasis was placed on measures at source. Because there was only a limited number of sources, a clearly delineated spatial structure, and few actors to intervene in the decision-making process, it was relatively easy to define an acceptable environmental zone based on general environmental standards. Prescriptive legal implications can also be applied to the integrated zone, and a link established between environmental licensing and the granting of planning permission. The result is the sustainable segregation of existing industry and future housing.

In section 5.4 we saw that a biased view of the potential scale of environmental/spatial conflict is not always sensible. All too often, greater objective insight into a situation (e.g. by following the Maastricht method) reveals that a situation has been either overestimated or underestimated. In Geertruidenberg and in Groningen, the scale of the environmental/spatial conflict was probably overestimated. We can even ask ourselves whether the Groningen Northeast Flank should have been designated as an IMZ pilot project at all. By contrast, assessments of environmental health and hygiene in projects in Arnhem and the 'Drecht cities' (to be discussed below) revealed that the scale of environmental/spatial conflict had been underestimated.

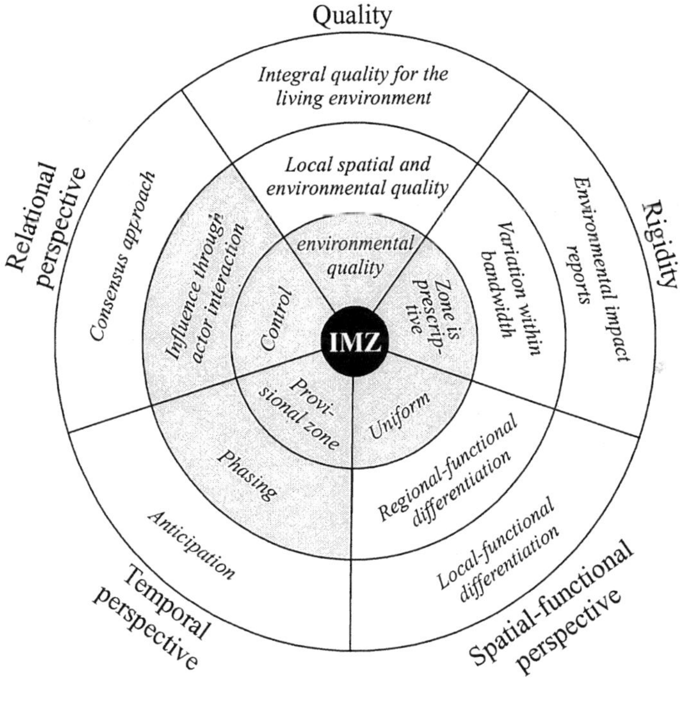

Figure 6.11 The applicability range for simple integrated environmental-zoning projects

Note: The hatched area shows the range: VS-IMZ as an instrument is 'reasonable' to 'good'

The negative findings of these projects inevitably led to heated debates on IMZ and – ultimately – the 'bankruptcy' of the VS-IMZ system. For these major projects it was particularly important to develop a local or regional strategic policy because the strategy proposed by the government and reflected in VS-IMZ has proved inadequate.

This involves more than a clear distinction between projects in which the scale of environmental/spatial conflicts is manageable, and those in which conflicts are too large-scale to be solved by means of strict standards. In a number of projects, the best approach was not simply to apply strict standards. Instead, a balanced implementation of standards and regulation, and greater administrative flexibility would be most likely to produce an acceptable result. The projects in question were the IMZ pilot projects for Burgum/Sumar, Bergen op Zoom, Amersfoort and IJmond.

The IMZ project *Burgum/Sumar Industrial Complex* is located in the Friesian municipality of Tietjerksteradiel, between the residential areas of Burgum and Sumar. The site accommodates various environmentally harmful activities that generate noise, dust and odour, and are a potential threat to external safety. By far the worst form of nuisance is the odour nuisance caused by Rendac (the local rendering plant), which is also seeking to expand its processing capacity. As a result of existing noise and odour pollution, there is an urgent need for remediation measures for homes located on the industrial site itself.

This involved 30 homes in integral Class V and 14 homes under integral Class VI for existing situations. Class VI for new situations covered large areas of Burgum and Sumar (see Fig. 6.12). The extent of the area was determined almost completely by odour emissions from a rendering plant (Witteveen and Bos 1991), which led to constraints on the expansion of the villages of Burgum and Sumar. Hope was fading for the development of Sumar that was felt to be necessary to safeguard liveability in the area. In Burgum, plans for new housing at a number of in-fill locations came under pressure. Because the Burgum/Sumar environmental/spatial conflict centred on emissions from the activities of a single company, it can be characterised as 'relatively simple', yet the scale of the conflict places it in the 'complex' category. Societal values were involved as well as environmental health and hygiene. The solution for this conflict was largely the result of joint action by the parties involved, i.e. a combined effort by local and regional authorities, the national government and industry. The solution was reached through mutual concessions that were based mainly on specific local circumstances. The solution was based on social consensus rather than on a unilaterally imposed 'one-size-fits-all' approach. The rendering plant at the site undertook to reduce odour emissions on condition that it would be allowed to expand. This expansion would lead to increased odour emissions in the immediate surroundings of the installation, but these would decrease further with distance. This stakeholder agreement resulted in the demolition of 36 homes at the industrial site, but the village of Sumar was allowed to expand.

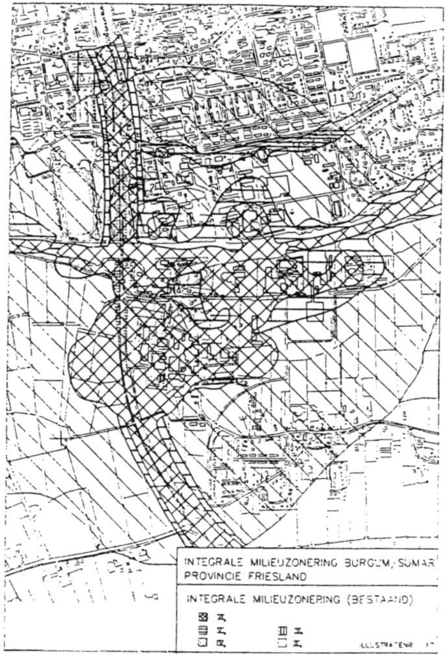

Figure 6.12a Integrated environmental zoning ('existing situations') Burgum/Sumar, according to the VS-IMZ, by Witteveen and Bos

Source: Provincie Friesland 1991

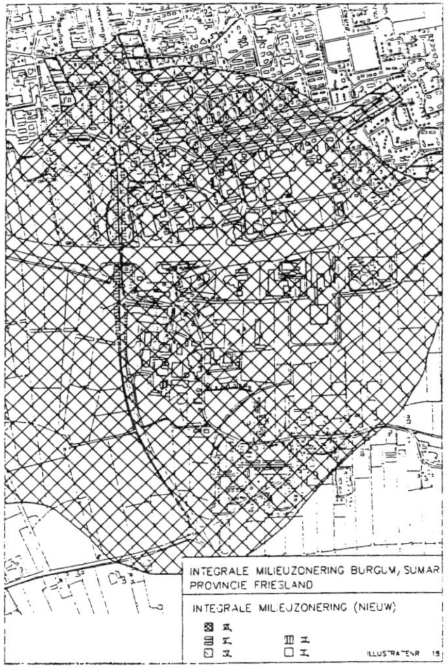

Figure 6.12b Integrated environmental zoning ('new situations') Burgum/Sumar, according to the VS-IMZ, by Witteveen and Bos

Source: Provincie Friesland 1991

At first sight, the environmental/spatial conflict involving the *Theodorushaven* industrial estate in Bergen op Zoom appears relatively straightforward. The conflict centred on a number of connected industrial sites to the west of the main residential area. The pollution with which the IMZ project was concerned was caused by only a small number of companies. The levels of noise and odour nuisance placed the area in the Class VI integrated environmental zone for existing situations. Homes located on the industrial estate, and between the town centre and the industrial estate, were subject to excessive levels of pollution and therefore classified as 'black' areas. The main source of nuisance (i.e. odour nuisance) was an alcohol plant located close to the residential area. This meant that it would not be possible to develop the area in the immediate vicinity – a desirable residential area around the old harbour area of Bergen op Zoom. Class VI for existing situations already has a wide reach and obviously the constraints on future spatial and industrial development extend over an even larger area (Stuurgroep IMZ Theodorushaven 1993). The widespread pollution meant that the *Theodorushaven* pilot project could be classified as a 'complex' environmental/spatial conflict. The local authority blamed the conflict above all on odour standards that supposedly did not correspond to the pattern of complaints.[6] In Bergen op Zoom, therefore, the adaptation of VS-IMZ standards was urgently required and would modify the extensive odour contours according to perceived nuisance levels. The integral zone thus retained its indicative status.

The IMZ pilot for central Amersfoort was inextricably linked to the urban renewal plans relating mainly to the area between Amersfoort central station and the city centre. The urban renewal programme, which was originally the initiative of the Chamber of Commerce and twelve local companies (Hoogland and Kolvoort 1993), consisted of large-scale development including homes and offices. The area in question was therefore a transition area and revitalization area. The proverbial spanners in the works were the contours drawn up for external safety. The contours were largely due to the Koppel and Soesterkwartier industrial estates and the Dutch Railways (NS) depot, and to the presence of industrial noise, air and soil contamination. The accumulation of sectoral pollution also meant that much of the north of the development area came under the integral Classes IV, V and VI for new situations, and pollution had to be reduced before the planned spatial development could proceed. Solutions involved relocating industrial installations and discontinuing the role of the NS shunting and marshalling yard in Amersfoort as a national sorting point. The stakeholders devised the solutions jointly and could therefore recognise the added value of the agreed targets. The 'demolitions' did not lead to amendments in the urban renewal plan, although it came under strong pressure from environmental pollution as a limiting condition. The flexibility, which was essential under the circumstances, was created mainly by realising the programme in phases. Residential development in particular was realised in proportion to the rate at which individual risk was reduced (Gemeente Amersfoort 1992). In addition, anticipatory policy was considered appropriate in order temporarily to allow excessive levels of pollution during transitional phases. The constellation in which this anticipatory policy was followed resulted in only limited uncertainty with regard to the agreed future scenario. In Amersfoort, as in Maastricht, VS-IMZ was not the only instrument used. It was part of a wider planning context that enabled these areas to benefit from the developments and interests that converged within a broad integral planning programme. New opportunities were created because environmental/spatial conflicts were not considered in isolation, but placed in a wider context. Solutions could be partly based on the potential of an area covered by the plan. This meant that the "various parties involved and the authorities responsible recognise the purpose and necessity of reducing environmental pollution, and that the reduction will be achieved with revenues generated by those same large-scale developments" (Borst et al. 1995; 127).

In the *IJmond* IMZ pilot project, a single company – the Hoogovens (Corus) steelworks – was responsible for the environmental/spatial conflict in question. The problem was nevertheless classified as 'complex', as was the case with the DSM Geleen project. The IJmond project involved fairly high levels of noise pollution and odour nuisance over a wide area.[7] Large areas of the residential neighbourhoods in the municipalities of Heemskerk, Beverwijk and Velsen fell within the odour contours of three and ten odour units as a 98th percentile. The presence of fine and coarse dust was also a problem. Fine dust was relevant because of its potential risk to health, and was therefore included in the inventory (Provincie Noord-Holland 1993). A method developed by the province of Noord-Holland was used to measure the accumulated levels of pollution. The results showed that large residential areas were subject to excessive levels of pollution. This was confirmed by a liveability survey among local residents. The survey identified emissions of dust, soot and odour as the most common grounds for complaint (Oliemulders Punter & Partners bv 1993). Because municipalities in the IJmond region wanted to expand their housing stock, it was essential to decontaminate the area concerned. At-source measures against noise, dust and odour were inevitable. Until remediation had been carried out, the provincial authority of Noord-Holland would not, for example, support the large-scale housing development for the municipality of Beverwijk. The provincial authorities were willing to incorporate integrated environmental zoning as a target in their spatial-development plans, which enabled a uniform planning policy to be followed in the various municipalities involved.

The scale of the conflicts in Burgum, Bergen op Zoom, Amersfoort and IJmond is larger due to the extensive environmental pollution in the surrounding areas and the housing density close to the sources of the pollution. The conflicts related to an urban environment in which housing is the dominant function and where, in most cases, further residential developments are planned. These are no longer cases of clear segregation between industry and housing. Even a relatively low level of environmental pollution can be excessive for a large number of homes. It becomes more difficult to decide between source-directed and effected-oriented measures based on a standard-setting approach. At-source measures are inevitable, but their scope limited. In many cases it will not be possible to reduce pollution levels to a level that is compatible with environmentally sensitive functions. Such cases would require spatial-planning measures, which lead to unfavourable local consequences. Here, it is above all the time dimension that made things difficult for the local authorities: it is difficult – if not impossible and therefore undesirable – to define and implement an integrated zone in the short term. There is an almost immediate need for spatial investment, but it is impossible to meet all remediation targets in the short term. A more flexible time frame was the obvious solution to achieving an acceptable level of environmental quality. This would create more opportunities in the short term, particularly in the spatial-planning process, and would inevitably mean that the end result would be less uncertain. It is also becoming increasingly necessary to consider the projects in their local contexts. Notably, the generic odour standard was criticised as being incompatible with perceived odour nuisance. Liveability surveys can be useful supplements for encouraging the flexible and adaptive use of regulations. Examples are shown in Fig. 6.13.

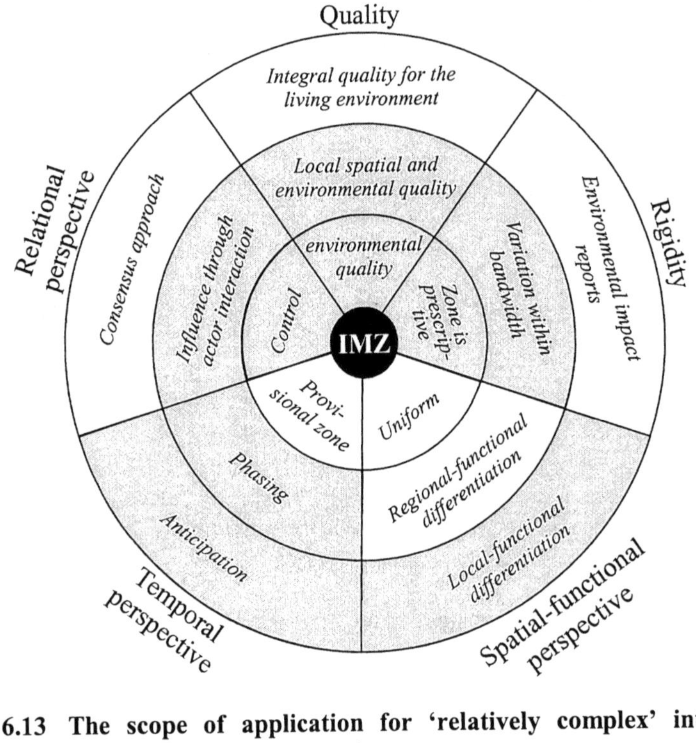

Figure 6.13 The scope of application for 'relatively complex' integrated environmental-zoning projects

Note: The hatched areas show the scope of application: VS-IMZ functions only in part

Because a stringent standard-setting approach does not always produce the best results in more complex environmental/spatial conflicts, local modification and greater flexibility can be more appropriate. Above all, VS-IMZ functions as a guideline for complex environmental/spatial conflicts, and is to be seen as a long-term goal. It is useful to place this guideline in a broader planning context, making it possible to strike an integral balance between the natural environment, the spatial environment, and the economy. Solving complex environmental/spatial conflicts requires strategic commitment from local authorities, which will also have to consider whether it is worth reaching a consensus with the parties involved.

The VS-IMZ programme developed by the Ministry of VROM encountered not only 'relatively simple' and 'relatively complex' environmental/spatial conflicts, but also environmental/spatial conflicts that could be classified as 'relatively very complex'. The pollution inventory in the projects DSM Geleen, Arnhem-Noord and IMZS Drecht Cities revealed large-scale environmental/spatial conflicts. In these projects, extensive environmentally sensitive areas were subject to excessive levels of pollution. This had considerable consequences for existing as well as planned residential areas.

The proximity of a large residential centre meant that the IMZ pilot project *DSM-Geleen* could be classified as a 'complex' or 'very complex' environmental/spatial conflict. This pilot project had a long history and was partly the foundation for VS-IMZ (see § 5.2). Noise, external safety and odour were the main aspects of the project. The aim was to keep current and planned housing outside the 10^{-6} risk contour. The 10^{-8} target-value contour was taken to be 'indicative' (Tesink 1988). A 55 dB(A) contour was also defined. Both contours enclosed existing residential building at certain locations. A remediation programme was drawn up for the problem areas. The two contours were implemented in the *Westelijke Mijnstreek* regional plan (Provincie Limburg 1992). By establishing these zones in advance of remediation, the a desired future scenario was anticipated. Resolving the problem of odour nuisance was a completely different story. DSM successfully appealed against the odour zone on the grounds that it was not well founded. After DSM had undertaken to implement at-source measures, VROM decided to abolish the odour zones around the DSM site. The idea of an integrated environmental contour was rejected following objections by DSM, the provincial authority of Limburg, the municipal authority of Geleen and a large number of local residents. It was feared that the zone would stigmatise the area (Borst et al. 1995). The DSM-Geleen situation was therefore not only a stimulus for a national environmental-zoning system, but also was also an example of a conflict regarding which the environmental department departed from its standpoint regarding a consistent generic policy based on standards and zoning.

The pilot project in Arnhem-Noord became well known because of the pitch-black map that shows the results of applying VS IMZ. The black area depicts the integral Class VI for new situations. The area was placed in this class due to the presence of the odour component H_2S and the toxic substance CS_2 (Hermsen 1991), both of which were causing excessive levels of pollution over a radius of between five and eight kilometres (Boei 1993). The environmental-impact inventory for the *Arnhem-Noord* industrial complex was completed in 1990 and revealed a serious situation to which the solution was not immediately obvious. A conflict had arisen following, on the one hand, years of increasing industrial activity by a number of companies and, on the other, the housing development expanding towards the industrial site. This resulted in an impasse that threatened housing development planned for the future. The standards set out in VS-IMZ did not provide a solution.

Consequently, in the implementation plan (Twijnstra Gudde 1994b) that followed the inventory, the environmental situation was redefined. The situation was eased considerably by abolishing the standards for new situations, and the new spatial development could proceed virtually as planned[8]. The 'black' area with its five-kilometre radius was also abolished. Without any firm agreements being made, the stakeholders accepted a future scenario for the local environment for all later initiatives. The black area that remained (i.e. an area with excessive pollution levels for existing situations) was located for the most part within the boundary of the industrial area. Through this decision-making process, the geographical scale of the environmental/spatial conflict was reduced to some twenty homes on the industrial site itself. The integrated environmental zone for existing situations was used as an indicative assessment framework for spatial development on and around the *Arnhem-Noord* industrial estate.

Even with modification and increased flexibility, none of the constructive solutions within the VS-IMZ framework were considered feasible for the projects Arnhem-Noord and DSM-Geleen. The planning implications of the environmental situation at these locations were largely disregarded. Anticipatory policy was considered, but the result would have been highly uncertain, given that related agreements had no solid foundation, either in the VS-IMZ system or in any legislation. In these situations VROM, which had hitherto been accustomed to thinking in terms of frameworks and guidelines, came up against the political and administrative context of its own standards-based policy. Given the legislative regime of the time, the technical-analytical approach to environmental issues, and the difficulty of and lack of familiarity with participatory policy, an interactive network-based strategy (see § 4.7 and 5.5) was far from appropriate for dealing with issues categorised as 'relatively very complex' despite the fact that, earlier in this study, this was seen as the best approach under such circumstances. Obviously, the relevant aspects of these conflicts are located mainly in the outer circle of the IBO model (see Fig. 6.14).

The *IMZS Drechtsteden* project (*Integrated Environmental Zoning and Remediation in the Drecht Cities*) was unusually large for a zoning project. It covered the municipalities of Alblasserdam, Dordrecht, Hendrik-Ido-Ambacht, Sliedrecht and Zwijndrecht. It was a somewhat atypical project because its regional character meant that it fulfilled an inventory function for the purpose of developing strategic policy. The project also confirmed that VS-IMZ could be a valuable *method* at regional level, and a basis for developing strategic policy (see also Humblet and De Roo 1995), at least in terms of providing information. The strategic nature of the *IMZS-Drechtsteden* makes it very similar to the PIM project in Maastricht that was the forerunner of the actual IMZ project PISA. The objectives of the *IMZS-Drechtsteden* project were therefore general: to provide direction, establish preconditions and help to realise a liveable spatial environment in the Drecht Cities area (Stuurgroep IMZS Drechtsteden 1991; 2). The main difference with the PIM-Maastricht project was that the outcome of the IMZS-Drechtsteden project had far-reaching consequences for spatial development in large areas of the Drecht Cities. The 'black' locations (i.e. those subject to excessive environmental pollution – according to integral calculations) were extensive and covered large areas of the municipalities of Dordrecht and Zwijndrecht (see Fig. 6.15). The locations were classified as 'black' mainly as a result of industrial noise, odour, and external safety risk. Also significant were the toxic and carcinogenic immissions that were present virtually throughout the area. The project focused not only on industrial pollution, but also on traffic/transport pollution and soil contamination, since these forms of pollution also have – from a regional perspective – consequences for homes and therefore also for deciding on sites for new housing developments.

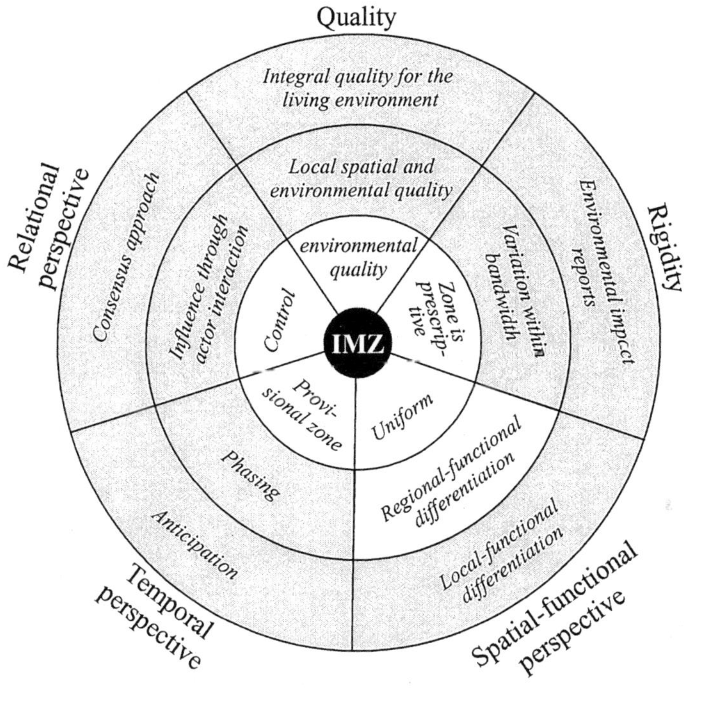

Figure 6.14 The scope of application for 'relatively very complex' integrated environmental-zoning projects

Note: The hatched area represents the scope of application: the added value of VS-IMZ is very limited

If the overview of the environmental situation was taken seriously, it would have inevitable and far-reaching consequences for spatial development. Based on the information in the overview, an outline structure plan for Drechtoevers was drawn up (Projectbureau Drechtoevers 1994). The plan identified three types of location: (1) locations where the environmental situation allowed planned housing development, (2) locations where the environmental situation allowed no urban development whatsoever, and (3) locations which, despite the poor quality of the environment, had high potential in terms of spatial development (Voerknecht 1993). A general remediation survey showed that it should be possible to reduce the environmental burden considerably over a period of five to six years (Stuurgroep IMZS Drechtsteden 1994). Above all, then, there was a need for flexible deadlines in order to anticipate future situations. The provincial authority of Zuid-Holland allowed that flexibility by granting an anticipatory period of five years. Three potentially valuable locations now remained at which it would not be possible to attain the desired level of environmental health and hygiene without relocating pollutant sources. Based on this information, an operational master plan was drawn up (Kerngroep Drechtoevers 1994) for spatial development in the Drecht Cities. Although it would have been the most obvious solution, VS-IMZ was not applied specifically to these three potentially valuable locations, due to their environmental quality.

The complexity and scale of the *IMZS-Drechtsteden* project place it in a different category to the other IMZ projects discussed here. It involves residential areas and environmentally harmful activities and functions, complex intersecting environmental zones, and a diversity of emissions – all on a regional scale. The degree of interrelationship meant that this was considered as a single project. It is therefore logical that the significance of the *IMZS-Drechsteden* project was strategic rather than operational. The strategic policy had to provide a framework for local projects and processes within which – depending on the complexity of the local environmental/spatial conflict – environmental aspects could be considered separately or together with spatial and economic aspects, either with or without consulting local actors.

The strategic nature of the IMZS-Drechtsteden project underlines the message contained in Figs 4.9 and 5.12, which show, directly or indirectly, that the need for strategic policy adapted to local situations increases in proportion to the complexity of environmental/spatial conflicts. However, the pilot projects, categorised by complexity, showed more than this. They also showed that there was a need for procedural flexibility and the modification of environmental-quality regulations. The need for flexibility and modification was predictably difficult to translate into generic solutions because it varies from location to location. Furthermore, the duration of the projects increased in proportion to their scale, and became more qualitative. Deadlines became more flexible, and targets were based on a multi-objective approach (see Fig. 6.16).

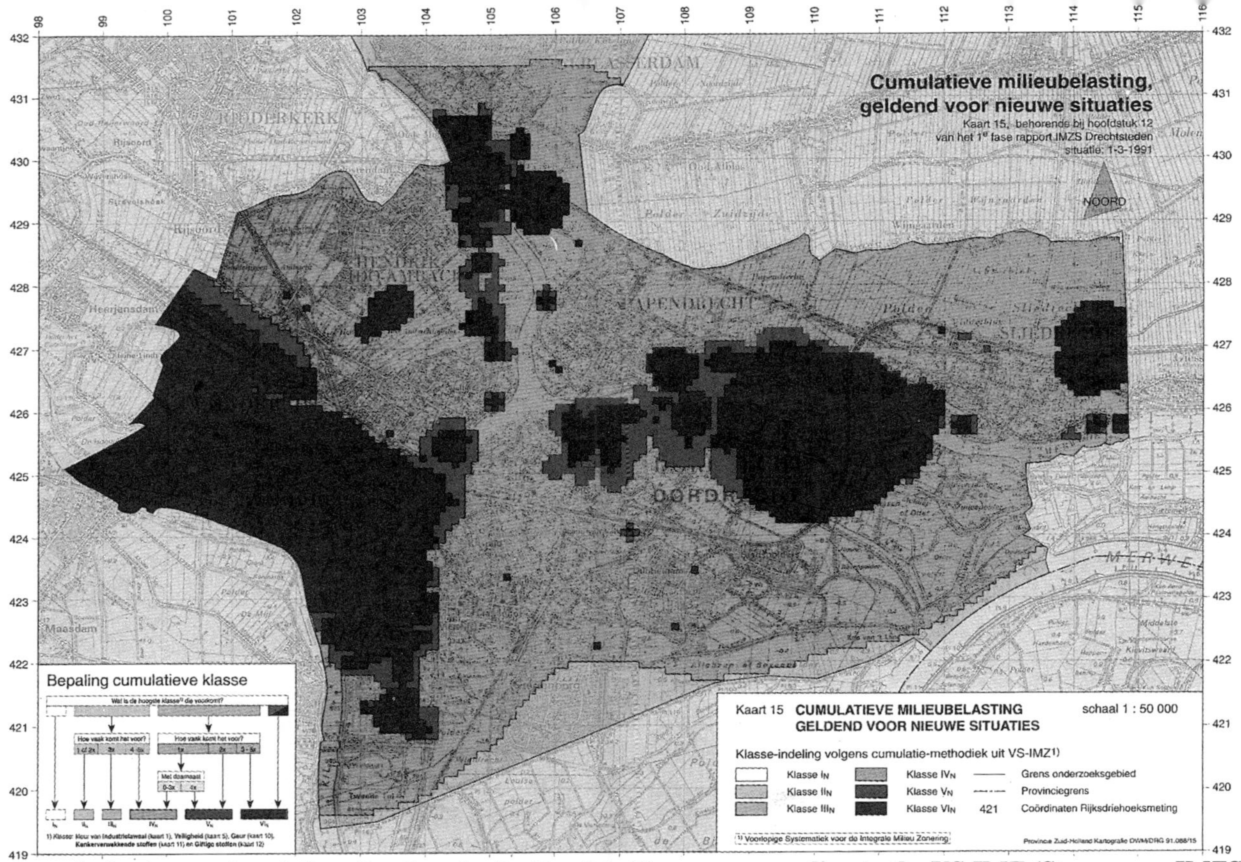

Figure 6.15 Cumulative impact in existing situations in the Drecht-cities area, according to the VS-IMZ (Stuurgroep IMZS-Drechtsteden 1991). *Source:* Stuurgroep IMZS Drechsteden 1991

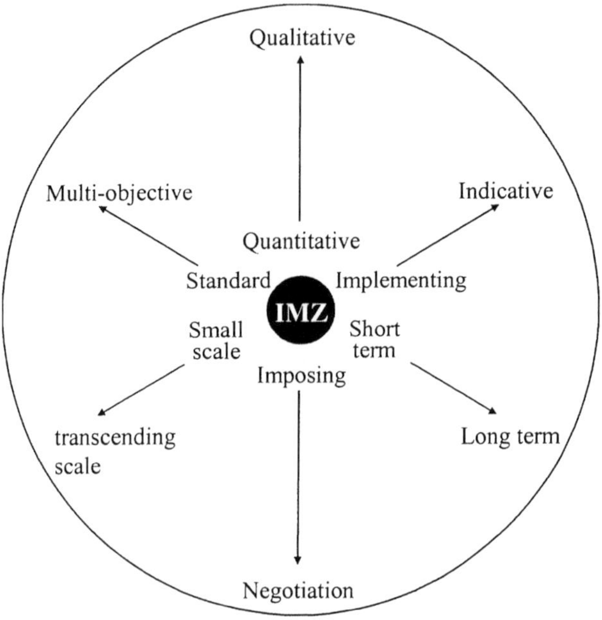

Figure 6.16 Characteristic shifts occurring as the IMZ pilot projects become increasingly complex

6.5 Conclusion

In section 4.8 we set out a number of arguments supporting the use of 'complexity' as a criterion for planning-oriented action. One argument relates to developments in planning-theory discourse. A second argument relates to the discussion of complexity theory. Reference was also made to the present chapter, in which environmental/spatial conflicts are seen in terms of their varying complexity and the consequences this can have for decision-making and planning strategy. The IBO model, which was developed to encompass, 'dissect' and analyse the discussion on integrated environmental-zoning projects, was not only a stimulus and a source of inspiration for examining the relationship between complexity and decision-making. It also provided opportunities to develop specific instruments, – here VS-IMZ is used as an example – for a given group of environmental/spatial conflicts, in this case the IMZ pilot projects, based on the theoretical arguments in Chapter 4.

Analysis of the pilot projects for integrated environmental zoning has shown that the conflicts in question could not always be solved effectively and efficiently using prescriptive and generic environmental regulation. Such regulatory mechanisms are most appropriate for relatively simple environmental/spatial conflicts that have a certain degree of uniformity. This is easy to understand. VS-IMZ is based on a functional-rational approach to defining problems. That approach assumes a direct causal relationship between source emissions, the distance between the source and its surroundings, and the negative unwanted environmental impact caused by immissions. This line of reasoning proved to be too simplistic. VS-IMZ becomes less and less appropriate as the segregation of environmentally harmful activities and environmentally sensitive areas gives rise to spatial, social and/or financial conflicts at local level. Fortunately, this only applies in a relatively small number of cases (see § 5.4). However, in these more complex situations, local spatial and administrative circumstances carry more weight, and generic policy is not sufficiently effective or efficient. This analysis of environmental/spatial conflicts shows that, depending on the complexity of the conflicts, national environmental standards can be *prescriptive*, can function as *flexible guidelines* and/or can provide *insight* into local or regional environmental health and hygiene (see Table 6.1).

Table 6.1 The relationship between the complexity of an environmental/spatial conflict and the feasibility of generic and prescriptive standards

IBO model	Complexity	Norm	VS-IMZ
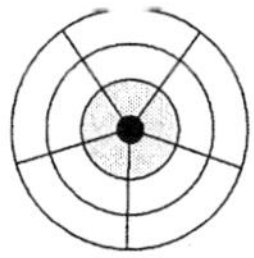	Simple	as framework	feasible
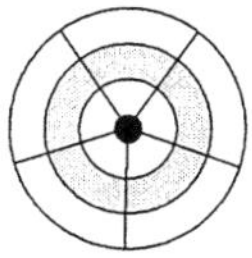	Complex	as flexible guideline	feasible, on condition that...
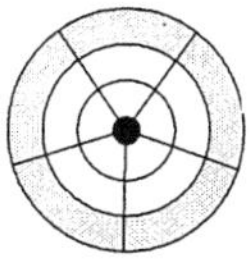	Very complex	to provide insight	not feasible

This is supported by the developments in environmental policy discussed in section 5.5. Under the projects Stad&Milieu (City & Environment), MIG (noise-nuisance policy) and BEVER (soil remediation), greater emphasis was placed on local governance with regard to environmental health and hygiene. The problem-definition process takes place increasingly at the local level. Depending on the definition of the problem, the guideline formulated by the Ministry of VROM is taken as a framework or – if the issue is classified as 'complex' or 'very complex' – the guideline will be replaced by local specific policy and reasons must be given for this. This development can be explained in terms of the varying complexity of environmental/spatial conflicts. It is also a development that answers the needs of local authorities. It was noticeable that local authorities in almost all the IMZ pilot projects – even those where it did not appear to be necessary – were aiming as far as possible for local governance, insofar as legal obligations allowed. From an institutional perspective this is understandable, although disregarding top-down limiting conditions and national frameworks inevitably creates the need for a local strategic vision, formulated as necessary within local limiting conditions. The following chapter will show that it is far from easy to strategically anchor environmental-protection policy in a local policy framework. The chapters will discuss initiatives by the municipal authority of Amsterdam to make environmental policy an integral part of local policy for the physical environment.

Notes

1 There will no doubt be a need to qualify the proposed differentiation based on a distinction between nuisance and risk. The City & Environment Experiments Act (TK 1998) has not made this distinction (§ 5.5).

2 For an analysis of 'pre-emptive soil remediation policy' implemented by the provinces and based on the IBO model, see Roeters 1997. The analysis relates to policy already implemented by provincial authorities on the basis of new soil-policy frameworks to be defined by the government (see § 5.5).

3 The scope of the '*Environmental Zones in Motion*' report extends beyond the eleven IMZ pilot projects. Some 25 potential projects that met the selection criteria are examined in a similar way. (Borst et al. 1995).

4 The integral environmental Class VI category is applied to 'black' areas, i.e. 'inadmissible' areas under the VS-IMZ system. For the individual integral classes, see § 5.4.

5 For an explanation of the distinction between existing and new situations, see § 5.4.

6 Municipal and provincial authorities registered some 1,000 complaints every year relating to odour emissions from the industrial complex (Twijnstra Gudde 1994), so there was clearly an odour-nuisance problem. However, the complaints did not necessarily correspond to the odour-nuisance contour.

7 Pollution caused by toxic substances was not included in the project because of the high background concentration in the area. The companies on which the IMZ project focussed contributed very little to this level of pollution.

8 Environmental health and hygiene at the location was also improved by the closure of most of Biliton, a company located at the site, and the modification of standards for toxic substances.

Chapter 7

Liveability on the Banks of the IJ

Environmental Policy of the City of Amsterdam

> I believe that an exciting, creative environment is the greenhouse in which growth and vitality flourish (Louis Lundburg, in Steiner 1997; 85).

7.1 Introduction

"Amsterdam in 2015. A clean city. A healthy, liveable city on the way to achieving sustainable development [,] a safe and lively place that is still pleasant to be in. [...] Amsterdam in 2015. A 'compact' city in a relatively green area. A city that has enough space for its inhabitants and its visitors. [This] Amsterdam offers its citizens an attractive living environment and is able to control the negative consequences of a concentration of activities. But perhaps the most important thing is that Amsterdam has succeeded in keeping within its boundaries" (Gemeente Amsterdam 1994; 1-2). The reader of the *Milieuvisie* (the vision on the environment for the period 1994-2015) for Amsterdam is thus presented with a target scenario of a compact, sustainable and liveable city in the year 2015. However, a great deal of water will flow under the bridges of the IJ before 2015. In the meantime, there will be increasing demands on space in the city, and attaining the desired scenario will be a formidable task.

The local authority for Amsterdam and environs has heavy commitments under the VINEX (Fourth Policy Document on Physical Planning-Plus, VROM 1996). The municipality of Amsterdam has also developed plans and projects for restructuring former dockland areas and using them as development sites. Here, too, it is a matter of planned development in a dynamic area, whereby a shift occurs from commercial and largely port-related activities to a mix of residential and business functions with a marked urban character. In Amsterdam this process goes hand in hand with conflicting demands between spatial wishes and environmental requirements. Amsterdam has shown considerable creativity, the effects of which are visible beyond its boundaries and have had an impact on national policies for spatial planning and the environment. The Amsterdam approach leads to policy innovation through acting to a certain extent 'in the spirit' of the law, according to a somewhat opportunistic view of policy opportunities and constraints. The 'grey' environment is seen less as a prescriptive limiting condition and more as a quality-determining element of policy on the physical environment, in which spatial and economic aspects are also considered.

The approach of the Amsterdam local authority considers environmental/spatial issues as based on a philosophy that is unique to Amsterdam. This philosophy is linked to the concept of liveability (see also § 2.6), which is the starting point for striking a balance between quantitative policy frameworks on the one hand, and the qualitative benefits of creative and integrated approaches to the physical environment on the other.[1] The aim of this philosophy has been to base policy not on the letter of the law, but on its underlying meaning. This would give the integration of spatial wishes and environmental requirements its own unique dimension, guided not only by general principles, but by considering each location individually. This philosophy is expected to lead to a process in which the different policy interests, including environmental interests, are given equal consideration. This can result in an 'added value' to be found in the unique aspects of the location in question. This chapter will consider whether the Amsterdam approach is a feasible one for dealing with complex environmental/spatial issues.

Although we do not wish to anticipate the result of the analysis described in this chapter, we may conclude that the Amsterdam methodology has resulted in several noteworthy ideas and instruments. Although these have not all been implemented, they have nevertheless contributed to a greater or lesser extent to the discussion on how to tackle environmental/spatial issues. Section 5.5 discussed the Bubble Concept as an attempt to implement an overall approach to environmental health policy. This concept was also an attempt to underpin the notion of compensation, which is also incorporated in the national 'City & Environment' project. Other instruments have attracted national attention, for example the environmental matrix and the environmental performance system (see § 7.5). These instruments ensure that the policy strand 'environment' is given equal consideration within and in addition to the spatial planning strand. They are also designed to provide insight into environmental and spatial interests in spatial planning and development projects.

Amsterdam's liveability strategy dealing with environmental/spatial conflicts has been described and analysed in the context of the IJ Riverbank project (the Amsterdam project for restructuring the docklands for urban development) and the Houthavens redevelopment (part of the IJ project). We will also discuss the role of the standards system in the Amsterdam approach and relate it to the overall approach that Amsterdam wishes to develop based on its liveability philosophy. The spatial planning for the IJ riverbank project and the progress in the Houthavens development will be discussed, the contribution of environmental health to spatial planning will also be analysed. Finally, we shall assess how the concept of 'liveability' is incorporated in spatial planning as the 'carrier' of an integrated development of the local physical environment.

7.2 Liveability as a Policy Philosophy

In Amsterdam the prevailing standpoint is that the quality of the environment on the banks of the river IJ depends on several sub-elements. In addition to the spatial, social-spatial and economic quality of the area, environmental quality has played an important role in the IJ Riverbank project. The extent of the measures taken to achieve the desired quality varies from sub-area to sub-area. Given the nature and location of the IJ riverbank zone, it is likely that extra efforts will be required to raise environmental quality to an acceptable level. In Amsterdam, environmental quality is defined from three angles:

- the physical and chemical quality of the environmental 'compartments' air, soil and water;
- environmental impact resulting from production and consumption;
- people's perceptions and evaluations of the human living environment (Gemeente Amsterdam 1994; 26, Gemeente Amsterdam 1995; 9).

This summary can be related to the concepts 'sustainability' and 'liveability' (§ 2.6). Sustainability depends above all on how and to what extent scarce environmental resources are used. Use of energy and raw materials – and the consequences of this for the future – also play an important role. In Amsterdam, physical space and the way in which it is allocated among various uses is also seen as a factor of sustainability, and as such should be considered in the planning process.

Nevertheless, in the view of Amsterdam's local authority, the current environmental situation is reflected not so much in the concept of sustainability, but in liveability. In striving to create a liveable city, the approach of the Amsterdam local authority has been to consider "environmental pollution (emissions, effects) caused by various functions as well as the environmental quality of the function. The emphasis varies from function to function. For example, for the functions 'work' and 'transport', the emphasis is placed on the effects of work-related and traffic-related emissions on the whole spectrum of environmental quality. For the functions 'living', 'public space' and 'nature', the emphasis is on the [...] function itself" (Gemeente Amsterdam 1995; 9). In the environmental study (*Milieuverkenning*) of Amsterdam, liveability is defined as "the health and well-being of those who live in the city, in their day-to-day environment" (Gemeente Amsterdam 1995; 24). This is the basic expression of the local authority's responsibility with regard to the daily living environment of its citizens.[2] In abstract terms, liveability encompasses the 'here and now' of the local environment (§ 2.6). This is a key concept in Amsterdam and is also used as a criterion for assessing policy on the human living environment.[3]

Elaborating the concept of liveability is looked upon as a challenge because it involves aspects that are not directly related to 'the environment', but reflect the well-being of the population in a broader sense. The environment is therefore considered to be a restrictive element in policy formulation. But working on liveability is also a matter of using opportunities to incorporate in spatial planning as many elements as possible that have an impact on the 'grey' environment. For example, proposals have been put forward to optimise the use of public transport and minimise car use in the city. Other themes to be included in the planning process include ecological housing and the separation and reuse of demolition waste (Gemeente Amsterdam 1991).

The notion of the 'grey' environment not only as something that imposes constraints, but also as something that encompasses potential, was formalised in the *Policy Document on Spatial Planning and the Environment* (BROM)[4] (Gemeente

Amsterdam 1994). The BROM took the first definite steps towards ensuring that spatial planning contributes to liveability as far as possible. The policy document stated that "in spatial planning, too, sustainable development requires a new mindset" (Gemeente Amsterdam 1994; 5). This was essential because "it has become increasingly clear that the spatial and grey environments are strongly interrelated and there is a lack of instruments geared towards a more integrated use" (Gemeente Amsterdam 1994; 5). In aiming to create sustainable and liveable cities, an integrated approach to the spatial and grey environments is seen to be vital.

In Amsterdam liveability is considered to be a multi-faceted overall approach towards the local living environment, whereby a proper balance must be achieved between environmental factors as well as spatial, social-spatial and economic factors. The same overall approach is also considered appropriate for the environmental dimension. The 'grey' environment is no longer seen as synonymous with pollution, since this leads to an approach that is too restrictive. The environmental gains resulting from policy measures must also be considered when describing or ascertaining the level of liveability for an area. The reasoning is that, in this way, the liveability philosophy will lead to a location-specific or area-specific policy that incorporates both a prescriptive and an overall approach, which can be used together or separately as required.

7.3 Liveability and Integrated Area-Specific Policy

In built-up areas, the relationship between spatial planning and the environment is a complicated one. That is why "an environmental policy geared specifically to urban areas is inevitable"[5] (Gemeente Amsterdam 1994; 8). A sectoral policy was not thought to be sufficient to prevent unacceptable levels of environmental pollution due to the development of Amsterdam. Careful consideration of options and an area-specific approach, whereby the diversity of spatial and functional characteristics and the possibilities that are unique to a given area, form the foundation for the highest possible level of environmental quality for the city as a whole (Gemeente Amsterdam 1994).

Integrated area-specific policy would turn Amsterdam into a liveable and sustainable city (Gemeente Amsterdam 1994). An area-specific approach is considered essential in Amsterdam because "it is not possible immediately to improve the quality of the local as well as the regional environment in all areas at the same time" (Gemeente Amsterdam 1994b; 15). Moreover, 'liveability' is seen as a subjective phenomenon, which means that every citizen has his/her own definition of a liveable environment. Policymakers also have to consider the wishes and requirements of those who live in this urban area. An area-specific approach underlines the need for diversity in a city. It is the diversity of forms and functions that determine the appeal of a city. Every area in the city must help to achieve an optimum level of environmental quality for the city as a whole, and local efforts will depend on the specific characteristics of an area. This thinking on the desirability of an area-specific approach *and* the need for a proper balance between all interests have resulted in a unique area-specific and tailor-made strategy formulated by the Amsterdam local authority (Gemeente Amsterdam 1994). This tailor-made overall approach is formalised in instruments such as the Bubble Concept[6] already referred to (see also § 5.5). This concept has never been implemented, but "has made a substantial contribution to thinking" (Interview Cleij, 3 April 1996) on Amsterdam's policy for the physical environment.

The Amsterdam local authority concluded that, for the time being, the Bubble Concept was not a workable measure for the whole city (see § 5.5). Since then, with the publication of the environmental study (*Milieuverkenning,* Gemeente Amsterdam 1995), attempts have been made to arrive at citywide environmental indicators that would produce a picture of urban environmental quality for the city of Amsterdam as a whole.[7] However, increasing emphasis has been placed on districts and neighbourhoods, for which the comprehensive approach can serve on a project basis as a 'coping stone' for urban development.[8] At these levels, area-oriented standards, environmental gains and the concept of compensation have been elaborated.

These new elements of integrated area-specific policy reflect the idea that, in areas that are particularly susceptible to the dilemmas of the 'compact city', environmental policy based on general sectoral standards imposes constraints that can have a negative impact on total environmental quality. These elements also reflect the fact that general sectoral standards do not satisfactorily incorporate environmental interests in the spatial-planning policy strand. The purpose of the proposed new elements for integrated area-specific policy was to formulate an environmental policy that is appropriate to the Amsterdam context, which is characterised by a high level of 'natural' background pollution and limited financial resources. It was proposed that standards should be complied with 'in spirit', by applying them in an area-specific (i.e. location-specific) way and linking them to local liveability, whereby compensation measures play a key role in ensuring that policy remains affordable and local environments achieve a maximum level of liveability.[9] Standards have certainly remained as the starting point in Amsterdam[10] but, when environmental standards lead to unnecessary stagnation, excessive costs or reduced liveability in a given area, they are reviewed in the light of the principles on which they are based.[11] In this sense, Amsterdam's environmental policy is ahead of the general developments that took place in national environmental policy at the end of the 20th century.

This vision on standards and integrated area-specific policy has resulted in the following five general steps:

1. The most appropriate form of spatial and functional development is determined for an area, based on its location, features (design, functional mix) and the expected consequences for the environment.
2. Existing and predicted levels of pollution are determined for new functional and spatial developments.
3. Measures are identified to reduce environmental impact in a given area.
4. The measures that yield the highest environmental gain for the whole city will be selected and implemented.
5. If none of the measures yield an appreciable environmental gain for the area, possibilities for substitution (e.g. tradable emissions rights) and compensation measures will be selected and implemented (De Knegt and Meijburg 1994; 8).

The structure of the Amsterdam strategy for integrated area-specific policy as formulated in 1994 has elements in common with the three-phase approach of the City & Environment project, which was launched at the end of 1995 (see § 5.5, VROM 1995). This approach embraces the principle that "if the thorough implementation of the first two phases does not produce a favourable overall result, [...] it [may be] possible to deviate from the standards under certain conditions unless the negative environmental impact resulting from the deviation is compensated in terms of liveability" (VROM 1995; 11). The national discussion on compensation still continues, and opinions on its workability are still heavily divided. The city of Amsterdam did not wish to wait for the outcome of that discussion, so has acted on its own philosophy.

It is precisely the all-embracing nature of the 'liveability philosophy' that puts it at the centre of Amsterdam's integrated area-specific policy. The instruments this policy is designed to carry must, according to the BROM, be closely attuned to the way in which the people of Amsterdam perceive and value their environment. The IJ Riverbank project is the first attempt to implement this liveability philosophy. In the plans for the Houthavens (the former timber docks that are part of the IJ project) concrete attempts are being made to integrate the philosophy in operational policy.

7.4 The IJ Riverbank Project

The IJ Riverbank project involves the restructuring of the southern embankments of the IJ between the eastern docks and the Houthavens (see Fig. 7.1). The proposed developments were designed to regenerate an area close to the city centre that, although in need of modernisation, still has a great deal of potential. The Amsterdam local authority wants to improve the area's connections to the inner city and "together with the inner city, create an important centre for living, working and shopping [...] for the people of Amsterdam and visitors to the city" (Gemeente Amsterdam 1991; 5). This development will revive the traditional character of Amsterdam as a 'waterside city'.[12]

Along the banks of the IJ, on either side of Amsterdam Central Station, visitors have the feeling that they are standing with their backs to the city. They experience strange feelings of emptiness and desolation although they are only a stone's throw away from the heart of Amsterdam. There could hardly be a greater contrast between the quiet windswept docks and the busy Leidsche Plein. The constant activity of people and cargoes coming and going on the south banks belongs to a bygone age. When the railway line was built some hundred years ago, the centre of Amsterdam was cut off from the IJ and the city turned its back on the river. 'Railway islands' were constructed between the original waterfront, the Prins Hendrikkade and the water in order to provide the city centre with a main station. At the same time, railway yards were built for the docks. The banks of the IJ mainly served the docks and the transport infrastructure. However, over the past few decades these functions have developed elsewhere. Today, the docks, quays, railway yards and industrial sites are largely deserted.

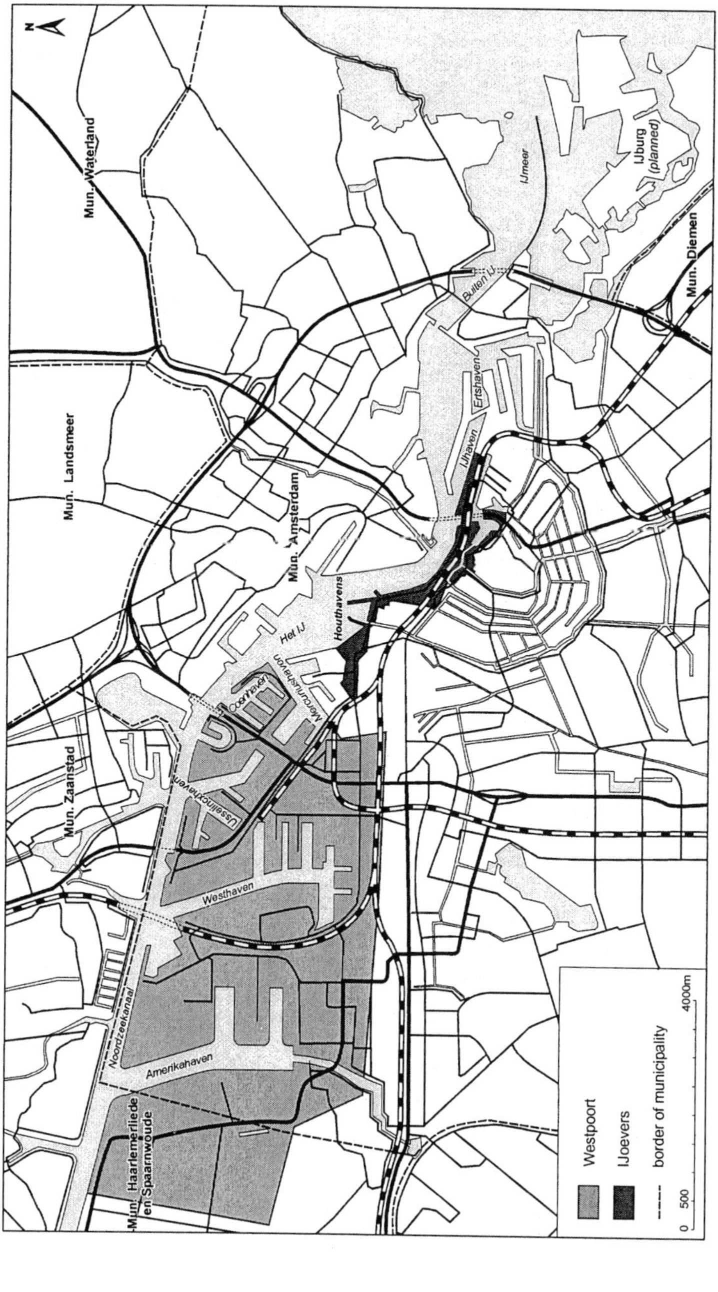

Figure 7.1 The topography of Amsterdam

It is a basic assumption in Amsterdam that "the compact city is the most environmentally friendly variant of urban development and is therefore vital to sustainability" (Cleij 1994). Potential building locations are therefore in great demand. Now that most of the traditional functions of the IJ riverbanks have disappeared and the area requires economic and spatial modernisation, it is logical to use the potential of the area for urban development. Thanks to the proximity of a dynamic city centre, the Central Station and the waterfront, the potential of the IJ riverside area is difficult to ignore.

On 27 June 1991, the municipal council of Amsterdam enacted the revised[13] *Terms of Reference for the IJ Riverbanks*. This document was the foundation for the development of the IJ riverside, including the restructuring of the Houthavens. The Terms of Reference for the IJ Riverbanks were the first step towards the final land-use plan for the banks of the IJ (*bestemmingsplan IJ-oevers*), which was published in 1994. The central objective of the IJ project can be described as 'bringing the city to the IJ'. 'Waterside Amsterdam' denotes a new area of the city that will be joined to its centre, albeit with "its own unique appeal and radically improved access, habitability and ambience" (Gemeente Amsterdam 1991; 14). The banks of the IJ are being transformed into a dynamic area with a wide combination of functions.

"Major urban developments such as the IJ project require an ecological approach" (Gemeente Amsterdam 1991; 95). This introduction to the Terms of Reference anchored the environmental dimension right at the beginning of the spatial planning process for the IJ riverside. Although the terminology suggested a deeper approach than was feasible on closer examination, this was nevertheless an expression at an early stage that the spatial development of the area would be partly determined by environmental aspects. The Terms of Reference (Gemeente Amsterdam 1991) pointed to the constraints imposed by environmental pollution:

- that was (still) present in the area;
- that would be caused by future activity in the area;
- that originated in other areas.

These constraints imposed by pollution have affected the plans for housing as part of the IJ project. "Further away from Central Station, the 'city-centre' character diminishes and relatively more homes will be built. The largest residential development will be the Houthavens. [...] The character of this area, in terms of the existing selection and mix of functions, will be more like the Spaarndam district than the city centre" (Gemeente Amsterdam 1991; 16). The residential areas were to be realised at the two outer edges of the plan area. However, despite the pressure to build for the housing market, the residential function will occupy "of necessity a modest place" (Gemeente Amsterdam 1991) within the plan area.[14] Other functions, above all culture and recreation, must contribute to improved liveability in the area and to the "desired diverse cityscape" (Gemeente Amsterdam 1991; 14). Between both outer edges of the plan area, a large area has been allocated to commercial services, especially expanding companies for which a city-centre location is no longer appropriate because of accessibility problems, businesses for which a waterside location is appropriate, shops and distribution centres.[15]

The *Terms of Reference for the IJ Riverbanks* consist mainly of urban planning proposals that give priority to spatial planning (i.e. form) and the quality of the spatial environment (i.e. content). The plan also refers to environmental restrictions, constraints and limiting conditions that needed to be taken into account in the subsequent planning process. Most of these focal points were subject to statutory regulations. In addition to noise nuisance, soil contamination, routes for transporting hazardous substances and measures for dealing with local industrial nuisance, wind nuisance is also mentioned as a factor possibly requiring attention outside the existing legal framework. Intentions were also set out with regard to improving the environmental situation, liveability and sustainability. The plan proposed "a conservative policy for motorised traffic and parking" (Gemeente Amsterdam 1991; 95), waste separation and sustainable, ecological building. The basic principles relating to the interface between the natural and built environments, however, were still modest:[16]

- Large-scale soil remediation will be carried out at a number of locations in order to improve the environment. Projects will then be implemented to remove the contaminated soil and clean it up or re-use it where possible (Gemeente Amsterdam 1991; 98).
- The IJ Riverbank project will take account of existing port operators and the employment they provide, whilst complying with environmental requirements (Gemeente Amsterdam 1991; 98).

In 1991, however, local authorities had not yet reached the stage of accepting responsibility for policy dealing with issues at the interface between the natural and built environments. Problems were resolved mainly by reference to the relevant environmental regulations, but hardly any attention was paid to excessive environmental impact and its possible consequences. This was especially true for pollution originating outside the plan area, including the Westpoort docklands (see Fig. 7.1).

The draft structure plan '*Amsterdam open stad*', published in 1994 (Gemeente Amsterdam 1994d), radically changed the place of the environment in spatial planning policy. This change took place mainly under the influence of the recently published policy document on spatial planning and the environment: *Beleidsnota Ruimtelijke Ordening en Milieu* (BROM) (Gemeente Amsterdam 1994) (see § 7.2). The underlying philosophy of the policy document on integrated area-specific policy is clearly reflected in the draft structure plan.

The structure plan stated that the contribution made to urban environmental quality by an integrated area-specific environmental policy would differ from level to level and sub-area to sub-area. It would also depend on area-specific spatial and functional characteristics and on the gains from environmental measures.[17] The plan also introduced the environmental performance system (see § 7.5) as an integrating and location-specific instrument for the local level (i.e. the level of the land-use plan).

With the publication of the *IJ Riverbanks Land-use Plan* in 1994, the 'contours' of Amsterdam's liveability strategy for integrating the natural and built environments became even clearer. Although the Bubble Concept described in the BROM lacked a concrete form and the workability of the environment matrix[18] had not yet been tested at the structure-plan level, integrated area-specific policy began to take on a definitive form in the land-use plan, largely thanks to a new tool called the Environmental Performance System (EPS, see § 7.5 and Appendix 7.1). Environmental considerations were clearly represented in the outline of the land-use plan, the explanatory notes and the appendices. At the same time, the environmental component is incorporated in the plan's general objectives as well as the objectives for the sub-areas. The requirements and limiting conditions derive largely from the EPS (see § 7.5) and its reference to the environmental priorities included in the land-use plan.[19] The explanatory notes (*Toelichting*) to the land-use plan state that "broadly speaking [...] the environmental aspects described in the Terms of Reference for the IJ Riverbanks are still relevant" (Gemeente Amsterdam 1994e; Toelichting 21). The main problems are soil contamination, noise nuisance, traffic-related air pollution, industrial noise from the western docklands, moderate to poor water/soil quality and a lack of urban green space"[20] (Gemeente Amsterdam 1994e; Toelichting; 21).

7.5 The Environmental Performance System

The Amsterdam *environmental performance system* (see Appendix 7.1) was introduced in the IJ Riverbanks land-use plan (Gemeente Amsterdam 1994e). The local authority devised the system in order to guarantee a given level of environmental quality for new spatial development, based on the principles of flexibility, freedom of choice and compensation. Furthermore, the system and related environmental priorities[21] can be seen as environmental requirements supplementary to the land-use plan. According to the local authority, this supplement is neither permitted nor expressly forbidden in the Spatial Planning Act (Humblet and De Roo 1995, Timár 1994). However, this is open to argument since the performance system does not refer to land-use in itself.[22]

The EPS has a straightforward structure. Based on seven themes, performance indicators are set for aspects relevant to the project. If the themes 'compact city', 'mobility', 'noise nuisance', 'sustainable building', 'energy', 'water', 'green space' and 'waste' are looked at in detail, the indicators relate to sustainability, aspects that improve liveability, and environmental limiting conditions (Humblet and De Roo 1995).

The IJ Riverbanks land-use plan states that, depending on the situation in the area in question, measures and indicators could be chosen by theme. In order to "ensure that the spatial and functional development can take place within the area in accordance with the limiting conditions set down by the principle of sustainability, a minimum environmental performance level (EPL) has been set for the entire area" (Gemeente Amsterdam 1994e; 4). For the IJ riverbank area, this level was set at 20 points. A location-specific EPL could be defined for sub-areas if required. The EPL of 20 points was arbitrary and composed of several scores that were equally arbitrary and hardly comparable. To clarify this, the notes to the IJ Riverbanks land-use plan stated that "the environmental performance system is still just a policy instrument and not a scientific method" (Gemeente Amsterdam 1994e; 22). Such a system allows in advance for location-specific circumstances and wishes with regard to sustainability and liveability to be included in the planning process, while at the same time allowing a certain freedom of choice, whereby "flexibility is guaranteed for the development of the area, no matter which of the several planning scenarios is selected" (Gemeente Amsterdam 1994e; 4).

The EPS incorporated in the land-use plan for the IJ riverbank area is interesting in terms of how it relates to existing regulations and legislation. The method partly aims to improve local liveability without resorting to central regulation. Extra measures can be specified for a particular area (e.g. secure bicycle facilities) or certain choices can be 'enforced' (sustainable woods instead of tropical hardwoods). The extra efforts will contribute above all to improved liveability and promoting certain options can have a positive effect on sustainability.

The environmental theme 'disturbance' (§ 2.5) is included in the package of measures for the EPS. However, this element has not been addressed in a balanced way because of the focus on noise nuisance in the area. In the EPS, the theme 'noise' is divided into measures for 'road-traffic noise', 'railway noise' and 'industrial noise'. If the noise level is equal to the maximum permissible level for noise,[23] a 3-point penalty will be imposed.[24] Noise levels that do not reach the permitted preferred limit value[25] earned 2 bonus points. The theme 'noise' in the EPS makes it clear that the system contains not only flexible measures (i.e. measures with positive scores), but also prescriptive measures (i.e. measures with negative scores). The IJ riverbank area suffers not only from noise pollution, but also dust and odour nuisance, but these forms of environmental pollution are ignored in the EPS used for the IJ Riverbank land-use plan. This is somewhat surprising, given that there are legal compliance requirements for noise nuisance, while these hardly existed for dust and odour or were not sufficiently prescriptive, which could justify the clarification and tightening-up of the EPS.

The Amsterdam local authority attaches great value to the principle of environmental compensation for spatial development[26] and the EPS is one way of introducing it. The above example relating to noise is also relevant. When construction takes place in an area where the noise level has reached its permissible maximum, the three penalty points must be compensated with three bonus points earned through other measures, in order to obtain a positive final score of 20 points. These are pioneering methods introduced by the Amsterdam local authority. They are an implicit indication that existing legislation and regulations are not satisfactory in all cases, or do not address local circumstances in terms of guaranteeing a certain level of environmental quality. The Amsterdam local authority wishes to – and believes it should – take responsibility for this if local circumstances require. It is therefore anticipating the opportunities that a new approach to environmental policy will bring (§ 5.5).

If compensation is adequate, the EPS allows a fairly high maximum level of pollution. Of course, not all pollution originates from the area in which it is present. This is also true for homes that have not yet been built. However, this does not mean that no effective measures can be taken. A number of the EPS measures geared towards compensating pollution are therefore geared towards the method of constructing new homes. Noise pollution can thus be partly compensated through measures under the theme 'sustainable building'.[27] However, private-sector developers pass on the costs of such measures in house prices. This means that the costs of improving liveability are not borne by the parties that cause the pollution (i.e. the industry in question and the local authority that wishes to develop the area), but by the homeowners.

The EPS is the concrete expression of Amsterdam's philosophy of liveability, based on three starting points mentioned above (see § 7.2): the physical and chemical characteristics of the area, the perception and evaluation of the human living environment and the demands placed on the environment by production and consumption. The EPS makes liveability and environmental quality not only tangible/measurable and direct, but also flexible and connected to spatial planning with a certain level of self-regulation. This can be seen as a breakthrough in the integration of environmental and spatial planning.

7.6 Houthavens: Environment and Spatial Structure

The Houthavens (timber docks, see Figs. 7.1 and 7.2) are located between the western port area of Westpoort and the historic centre of Amsterdam. They are included in the IJ Riverbank land-use plan. The Houthavens area is therefore a transitional area between the city (residential) and the port (work). The area has virtually lost its port function and, with its proximity to the old city centre, its accessibility and waterfront location, it has all the potential required for a first-class city area. However, the development potential and liveability are clouded above all by pollution from the adjacent Westpoort industrial area. According to the Amsterdam philosophy, this loss of liveability must be compensated by enhancing the quality of other aspects. The following questions need to be addressed: what is the nature and scale of the pollution in the Houthavens, how can this pollution and the related loss in environmental quality be analysed and compensated using a standards-based and multi-objective policy, and what are the consequences for spatial development.[28]

The Houthavens docks were built more than a century ago for the handling and storage of roundwood and sawn timber. Over time, road and rail transport gradually took over the transport function for this sector. As a result, the Houthavens fell into disuse and the harbour basins were filled in (Gemeente Amsterdam 1995b). There are now plans to use the area to extend the development of the city towards the River IJ. The intention is to join the development to existing residential areas and the Spaarndam and Zeehelden districts. These areas 'behind the railway' must become 'neighbourhoods on the IJ'. The Houthavens area must become multifunctional, so that the functions 'living', 'work' and 'recreation' are all fully catered for. Some 1,500 homes are planned for the area. There is also a plan with up to 8.5 hectares of land reserved for functions other than residential. Most of this land will be for small businesses (see Fig. 7.2). The *Programme of Urban Planning Requirements for the*

Houthavens area (Gemeente Amsterdam 1995b) proposes re-excavating the filled-in harbour basins in order to make maximum use of the riverside location. The *Programme of Requirements* is the last in a series of plans for the Houthavens (see § 7.4) that contains concrete proposals for development and function of the former timber docks, and which contains elements that integrate spatial and environmental aspects.

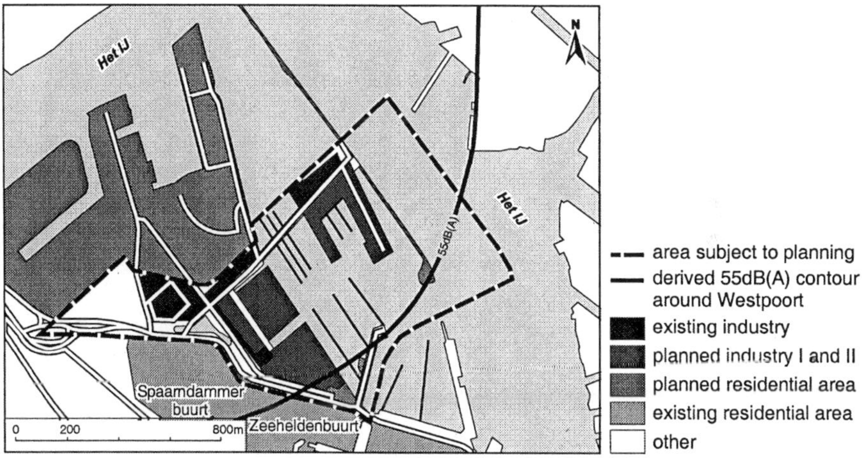

Figure 7.2 **Proposal for the redevelopment of the Houthavens timber docks**
Source: Gemeente Amsterdam 1995b
Note: The 55 dB(A) contour around the Westpoort industrial area is shown

Despite the need to restrict residential development on the banks of the IJ because of pollution levels, the Houthavens area is destined to become a residential neighbourhood. Located between the city and the Westpoort docks – where a great deal of commercial activity is and will remain located – Houthavens fulfils the function of a 'bridge' between the two areas. In 1991, at the beginning of the planning procedure, the most important problems were considered to be soil contamination and noise from industry, trains and trams, and these would have to be taken into consideration during the planning process. The location adjacent to a large body of open water also meant that wind nuisance was a relevant planning theme.[29] The pressure on the environment in the Houthavens area is resulting in conflicts with plans for residential development. The spatial/environmental dilemma is that of how to create a highly desirable residential development and environment without imposing constraints on nearby industry.

In addition to compensation as an instrument for guaranteeing liveability for local residents, the programme of requirements also defined a zone for industry that does not cause nuisance (see Fig. 7.2). The zone is located between the Westpoort area and the Houthavens residential locations, and will accommodate businesses that "from the point of view of nuisance, cannot be located *in* residential areas, but need to remain close to them. [...] Increasing the concentration of commercial activity between the new Houthavens residential area and the industrial area in the port area serves a double purpose" (Gemeente Amsterdam 1995b; 11) according to the Amsterdam local authority. "A strip of commercial development 2-3 storeys high for smaller businesses and a cluster for medium-sized businesses will form the 'bridge' between the residential and working districts of the Houthaven area" (Gemeente Amsterdam 1995b; 25). This strip of commercial development lies between Westpoort and the homes in the plan section 'land'[30] (see Fig. 7.2). "The situation regarding noise nuisance means that buildings in this development strip must be no less than 12 meters high" (Gemeente Amsterdam 1995b; 27), thereby allowing the area to function as a sound wall between the residential area and the Westpoort industrial area.

The Houthavens project certainly integrated environmental interests in the spatial planning process at an early stage,[31] particularly from 1994 onwards, when the Amsterdam local authority included the aspects liveability and sustainability in its spatial plans, and placed them in a workable context within the EPS. Yet this approach, which is based on this foundation and has a framework of clear limiting conditions, did not prevent environmental/spatial conflicts. On the contrary; the development plans were temporarily halted by a conflict relating to the interface between the 'grey' and built environments, namely between existing industry at Westpoort and the development plans. The main cause of the conflict was the way in which the sound strategic preparation is elaborated and consequently clashes with the recalcitrance of national environmental regulations. Measures for improving the physical and chemical quality of the compartments 'air' and 'soil' are thus restricted to the *desired* situation instead of the *actual* situation.

7.7 Houthavens versus Westpoort

Despite the advanced planning and guarantees of liveability and sustainability, building projects still could not commence in the Houthavens. The main obstacles were the physical and chemical quality of the soil and air in the area. Furthermore, existing businesses in the Westpoort district saw the spatial development of Houthavens as a competitive threat. The local authority adhered to the view that "residential development is acceptable for this location [...] and desirable from the point of view of sustainability" (VROM 1995, Appendix 3; 2), albeit on condition that "the (incidental) nuisance from the western port area must be compensated as fully as possible in order to guarantee liveability for future residents" (VROM 1995, Appendix 3; 2).[32] On the other hand "housing construction at Houthavens [...] must not obstruct the continued existence of businesses in the western port area" (Gemeente Amsterdam 1995b; 11). Herein lies the Houthavens dilemma. Despite all its good intentions, the Amsterdam local authority was confronted with an environmental/spatial conflict that caused the Houthavens plans to stagnate.

The Houthavens project ground to a halt because the discussion between the local authority and the business community – in particular Cargill – in the nearby Westpoort port and industrial area reached a deadlock (Gemeente Amsterdam 1995c) and the provincial authority of Noord-Holland did not assess the Houthavens plans until after the new environmental licences were issued to Cargill (Provincie Noord-Holland 1995b; 4).[33] It was not yet sufficiently clear how the excessive pollution, in particular odour and noise nuisance, caused by the activities at Westpoort could be dealt with in order to allow housing development to go ahead. Nor was it clear to what extent the plans would hamper the businesses located at Westpoort (Gemeente Amsterdam 1995c). It was not until the issue of environmental permits to Cargill was reviewed that it became clear precisely how large an area was covered by the permits and what proportion of the planned homes would be located outside that area. In Amsterdam the assumption was that, for this reason, the Council of State would not approve the land-use plan for the IJ riverbanks. The Municipal Executive of Amsterdam therefore decided to consider the environmental licensing procedure for Cargill and the decision-making process for the housing development at Houthavens as separate issues and to make them consecutive. A decision would not be taken on the residential development until agreement had been reached on the interpretation of Cargill's environmental permit (Commissie voor Volkshuisvesting, Stadsvernieuwing, Ruimtelijke Ordening en Grondzaken 1995).

A number of businesses in the Westpoort port area feel threatened by possible problems in the future and constraints on expansion. The businesses are expecting an increasing number of complaints because residents moving into the new homes will be less used to living close to such an area than the people living in the existing residential neighbourhoods. The businesses are afraid that the complaints will mean that the local authority places even more requirements on them, but they are also concerned about the consequences of any negative publicity. The local authority claims that "This can be avoided to a large extent by providing balanced information to prospective homebuyers and tenants" (Gemeente Amsterdam 1995c; 2) under the motto "You are moving to a dynamic portside environment, with all the advantages and disadvantages that brings" (Gemeente Amsterdam 1995c; 2). The business community wants the local authority to guarantee that "no further environmental requirements will be imposed on businesses for 10 years, even if standards are tightened-up elsewhere".[34] Such a decision by the local authority requires the approval of the provincial government (the body that issues environmental permits) and the national government (the body that sets environmental standards). The local authority is also in favour of a 'boundary agreement'. Once the definitive environmental permits have been issued, the fixed environmental contours will determine the boundaries for residential development. "Further residential development to the west is prohibited for the next 50 years. The boundary will naturally be included in the land-use plan" (Gemeente Amsterdam 1995c; 4). It is clear, without exploring the legal consequences of this strategy relating to commercial activity and the development of the Houthavens area, that industry required a large degree of certainty regarding agreements made. The local authority was prepared to give this, but met with a certain ambivalence towards its proposals.

In the meantime a number of reports were published describing the physical and chemical situation in the Houthavens and the relationship between the source and the area receiving the immissions (Zeedijk 1995, Provincie Noord-Holland 1994,

Sandig en Vossen 1994). However, these reports do not discuss the spatial confrontation between environmentally sensitive requirements and environmentally harmful activities. Nor do the reports draw conclusions with regard to policymaking, procedures or the consequences for spatial development. Nevertheless, the results of these environmental studies were such that pollution originating in the Westpoort area was a threat to housing development plans in the Houthavens area (§ 7.8 and 7.9). The programme of requirements for Houthavens already refers to these reports in order to emphasize the possible need to protect against and compensate noise nuisance. The programme also cautiously anticipated the possible implementation of the 'seaport standards'[35] (see Gemeente Amsterdam 1995b).

7.8 Noise Zones surrounding the Westpoort Area

The IJ riverbank land-use plan provides for large areas of "mixed residential and employment zones with a city-centre character" (Gemeente Amsterdam 1994e; 23). The Amsterdam local authority drew up regulations for industry with regard to land-use plans. Companies in nuisance categories 1 and 2 (Fig. 7.2) are permitted under these regulations.[36] Plans for companies in nuisance category 3 (zone III) adjoining a residential area can only be drawn up in exceptional cases. These regulations were elaborated for the Houthavens area in the Programme of Requirements (Gemeente Amsterdam 1995b). 'A-locations' under the Noise Abatement Act[37] are excluded from the IJ riverbank area. However, Houthavens is still covered by a noise zone for the ADM-NSM-Tomassen[38] industrial site, where activities have "almost completely disappeared" (Gemeente Amsterdam 1994e; 24).[39] The land-use plan states that "an application will be submitted for a maximum permissible level of 55 dB(A) for construction in the area, in order not to restrict future business expansion"[40] (Gemeente Amsterdam 1994e; Appendix 4A 22). The maximum permitted level for noise generated by traffic in the Houthavens area close to the IJ boulevard is 65 dB(A). On the basis of this standard, the land-use plan specifies that residential development is prohibited within a radius of 44 metres of the IJ boulevard.

The environmental performance system (EPS) in the IJ riverbank land-use plan makes provision for the compensation of local noise nuisance, namely by "minimising other forms of noise" (Gemeente Amsterdam 1995b; 30 and Appendix 7.1). Partly for this reason, the Houthavens area will be pedestrianized. Other compensatory measures include sound insulation to reduce nuisance caused by neighbourhood noise (see also § 7.5).

The new homes in the Houthavens area are almost all located within the 55 dB(A) noise zone[41] (see Figs 7.2 and 7.3) for Westpoort. This means that, not on the basis of the seaport exemption but on the basis of the current exemption facilities for industrial noise, the maximum permitted level of 55 dB(A) for residential building is exceeded.[42] The 55 dB(A) contour is derived from the 50 dB(A) contour around the Westpoort port and industrial area[43] (see Fig. 7.3). The 50 dB(A) contour is a statutory zone around an industrial location, "outside which the noise level with respect to that location may not exceed 50 dB(A)" (Art. 53, Noise Abatement Act). The provincial government of Noord-Holland expected the noise-reduction measures for existing homes up to the 55 dB(A) contour derived from the 50 dB(A) contour to be successful.[44] However, the municipal authority had plans to build 1,500 homes inside the 55 dB(A) contour.

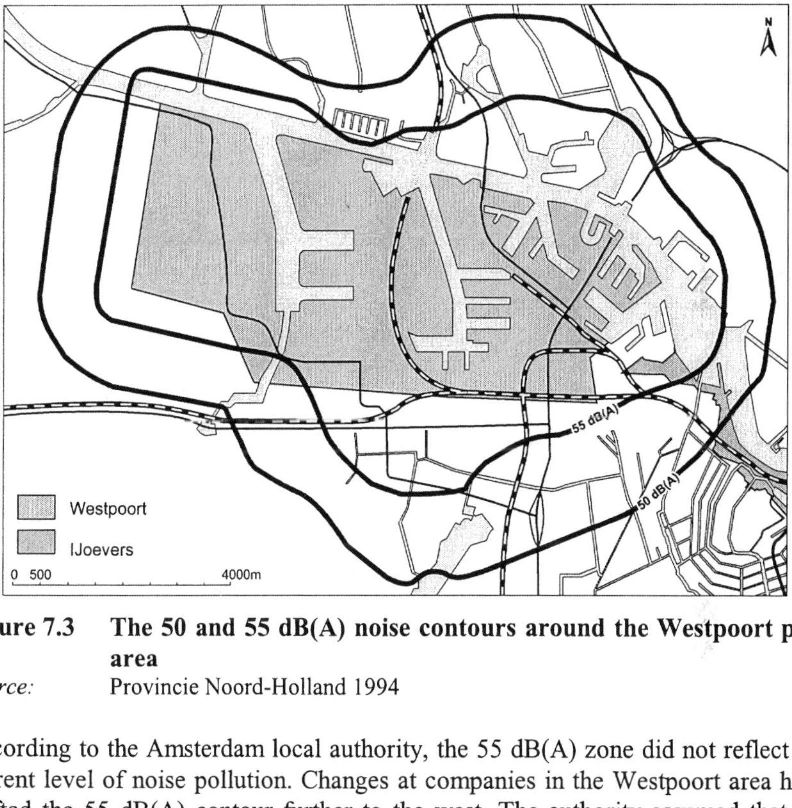

Figure 7.3 The 50 and 55 dB(A) noise contours around the Westpoort port area

Source: Provincie Noord-Holland 1994

According to the Amsterdam local authority, the 55 dB(A) zone did not reflect the current level of noise pollution. Changes at companies in the Westpoort area have shifted the 55 dB(A) contour further to the west. The authority assumed that the contour would shift even further following noise-abatement measures. The provincial government of Noord-Holland held a different view. The Cargill company, for example, makes use of the full permit range, which means that the slightest change can lead to a noise problem (Bakker interview, 1996). The Amsterdam local authority had hoped that sufficient noise screening could be realised. On the basis of this expectation, the municipal authority applied for exemption from the provincial government, which granted it[45] up to 53 dB(A) (Gemeente Amsterdam 1995b; 15). The municipal authority assumed that the 55 dB(A) contour would shift to the west of the Houthavens area, whereby screening provided by buildings[46] would deliver a 'profit' of 2 dB(A), hence the decision to opt for 53 dB(A).

In the IJ riverbank land-use plan, the west of the Houthavens area is designated for commercial premises (therefore not for homes) and can also be used to screen noise nuisance originating in the west. Buildings that provide screening, namely industrial buildings in the land area and a hotel or offices on the waterfront, between the residential development and Westpoort, could shield some of the noise pollution.

The Amsterdam theory that noise pollution in the Houthavens area would fall by a satisfactory degree, and that it was possible to erect adequate screening buildings, explains the application for exemption up to 'only' 53 dB(A). However, the screen provided by non-sound sensitive buildings provides insufficient protection for the homes on the south-western periphery of the 'waterside' plan area (Rangelrooij and Spaans 1995).

Moreover, the provincial authority of Noord-Holland questioned the need for offices or a hotel at such a location (Arents interview, 1996), thereby casting doubt on the promised sound insulation. According to the provincial authority, the Amsterdam line of reasoning was unworkable and a maximum exemption application (55dB(A)) for the municipality of Amsterdam as well as businesses in the Westpoort area would have yielded a more favourable result. The Amsterdam local authority now had to instigate a new 'higher noise-level procedure' because the level applied for – 53 dB(A) – was too low for building purposes. However, a new application meant delays.

The fact that the higher value was set at 53 instead of 55 dB(A) imposes restrictions on expansion for businesses at Westpoort. The exemption up to 53 dB(A) became the new criterion for the provincial authority for Westpoort companies wishing to expand in the future. A maximum exemption of 55 dB(A) would also restrict commercial expansion at Westpoort.[47] From any point of view, therefore, residential development in the Houthavens area has had and will have an effect on noise allowance relating to commercial expansion. It was specified that, if no homes were built at Houthavens, "the noise level, after remediation, will be allowed to increase to 57 dB(A). The noise allowance is [...] 53 dB(A)"[48] (Provincie Noord-Holland 1995). Now that homes *are* being built in the area, the maximum permitted noise level is 55 dB(A), giving a noise allowance of 45 dB(A)[49] for future commercial expansion in the vicinity.

The stalemate in the Houthavens planning procedure could possibly be solved if the Amsterdam local authority applied for exemption up to 55 dB(A) and bore the cost of additional measures and remediation costs incurred through every company expansion in Westpoort which, without extra measures, results in noise levels higher than the 45 dB(A), up to a maximum of 53 dB(A) (interview Bakker 1996). In the event that "a higher noise allowance than [...] 45 dB(A) is necessary at Houthavens, this must be compensated. [...] It is obvious that the businesses in the area cannot be required to bear the cost of these provisions"[50] (Provincie Noord-Holland 1995). This effectively defeated Cargill's 'constraint argument' (Bakker interview 1996), so 'noise' was no longer an obstacle to local-authority plans for developing Houthavens, although the need remained for a sizeable screen of buildings to keep the noise level below 55 dB(A).

In the programme of requirements for Houthavens (Gemeente Amsterdam 1995b; 81), the municipal authority of Amsterdam had also reached the conclusion that its policy would not achieve the desired results, now that "the calculations show that there are only limited possibilities for remaining below the 53 dB(A) level"

(Gemeente Amsterdam 1995b; 81). The programme therefore proposes a smaller reduction in the noise level, based on the principle of "a reduction to 60 dB(A) outside the home to a level of 35 dB(A) inside the home" (Gemeente Amsterdam 1995b; 30).[51] This shows that the Amsterdam local authority is prepared to retain the seaport standard as the maximum noise level (60 dB(A)) for Houthavens.

Following problems with the restructuring of former dockland areas, the Rotterdam local authority proposed raising the permitted level of noise caused by dockland activities relating to housing construction. Rotterdam believes this to be essential with regard to the development of urban residential areas (Gemeente Rotterdam 1994). The Minister of VROM was sympathetic to the problems with the restructuring of the docks and introduced the 'seaport standard' as an amendment to the existing legislation. The seaport standard raised the permitted noise level at housing construction sites to 60 dB(A) on condition that the new homes are part of a restructuring project or planned infill in existing residential areas. Despite the fact that parliament felt this to be a piece of 'convenience legislation', the amendment was nevertheless approved and came into effect on 1 March 1993.

The possibility of resorting to the 'seaport standard' is, however, a limited one. According to this exemption, residential development must not spread towards industrial areas. Strictly speaking, the residential development at Houthavens can be considered as 'spreading', which means that the seaport standard cannot apply, according to the letter of the law. The Amsterdam local authority, however, sees the seaport standard as a safety net, because "we want to build and we want to do it properly, based on our own philosophy" (Cleij interview, 1996) by making liveability a priority. If the philosophy is unworkable without the extra noise allowance provided by the seaport standard, this will be implemented[52] (Cleij interview 1996). Given that the municipal and provincial authorities both wish to build homes at Houthavens, the authorities responsible will interpret the regulations relating to seaport standard as broadly as possible (Cleij interview 1996; Arents interview, 1996).

In January 1995, the provincial authority cited residential development spreading towards the industrial area as one of the grounds for refusing to set the higher value requested by the municipal authority of Amsterdam (Provincie Noord-Holland 1995b; 5, Provincie Noord-Holland 1995c; 2). Several months later, in the defence submitted to the Council of State in response to Cargill's claim that the subsequent decision to set the higher value was contrary to the provincial policy for preventing residential development spreading towards the industrial area, the provincial authority wrote that "we would point out, perhaps unnecessarily, that the issue here is not the fact that residential development is spreading towards the industrial area, but that the homes are planned equidistant from and next to existing homes" (Provincie Noord-Holland 1995d; 7).

7.9 Other Forms of Pollution Originating at Westpoort

In addition to noise nuisance, soil contamination (which will not be discussed here) and odour nuisance were also present in the Houthavens area. Studies have also been made of possible dust nuisance and health risks from local industry and the transportation of hazardous substances.

Risk

The Houthavens area lies outside the individual risk contour of 10^{-6} deaths per year relating to the transport of hazardous substances by road and water, and relating to the high-risk industry in the surrounding area (VROM 1995). However, part of the Houthavens area lies inside the indicative zone for group risk. This zone "can be used as an optimisation instrument when weighing up the options for residential development, in order to minimise the group risk" (Gemeente Amsterdam 1995b; 82).

Dust

The transhipment activities at Westpoort mean that there are a number of sources of coarse and fine dust which can affect the air quality in the Houthavens area.[53] It is necessary to distinguish between coarse and fine dust because coarse dust[54] (largely generated by the activities of the companies I.G.M.A. and Cargill) is above all a cause of visual nuisance. Fine dust particles smaller than four micrometers,[55] on the other hand, also constitute a health risk. Recommended levels only exist for fine dust, for example in the Netherlands Emission Guidelines[56] (Stafbureau NER 1992). These advisory targets are attained virtually throughout Amsterdam (Zeedijk 1995). The fine dust generated by I.G.M.A., Cargill and OBA[57] is taken up in the ambient level, which means that the provincial authority of Noord-Holland does not need to place restrictions on residential development in the Houthavens area (Schoonebeek interview, 1996).

The Netherlands does not have standards for coarse dust. Germany uses standards based on annual dust-deposit averages: 0.35 $g.m^{-2}$ per day, 50-percentile and 0.65 $g.m^{-2}$ per day, 98-percentile. These German levels are not exceeded in the Houthavens area (Zeedijk 1995) but, given the range of these standards, the provincial authority questions their usefulness (Schoonebeek interview, 1996). The VNG publication on industry and environmental zoning ('*Bedrijven en milieuzonering*') specifies a distance of 700 meters for transhipment companies emitting dust (Kuijpers 1992), but this distance is only indicative (see § 5.2). In his report on dust ('*Stof in de Houthavens*', 1995) Zeedijk concluded that no problems were to be expected in relation to the VNG indication because the distance between Houthavens and the nearest major dust-emitting source was 1,200 meters.

The province of Noord-Holland had not been familiar with the issue of dust, and had no experiences of making decisions on whether, on the basis of measured levels of dust, housing development should be allowed. A study was therefore carried out of nuisance perception in the neighbourhoods close to the Houthavens area (Schoonebeek interview, 1996). Zeedijk subsequently concluded that "there presently appears to be no dust nuisance in the residential areas adjacent to the port area. The calculated dust concentration in the Spaarndammerbuurt area is similar to what can be expected for the Houthavens area" (Zeedijk 1995; 27). The provincial authority therefore concluded that dust nuisance would not impose constraints on residential development at Houthavens.

Odour

Noise and odour are the main determining factors in the environmental relationship between Westpoort and Houthavens. The provincial authority of Noord-Holland, as the licensing authority, initially decided on an integral approach that considered all environmental factors, in line with the Bubble Concept (interview Arents 1996). The application was initially rejected because the total integrated environmental quality appeared unacceptably high (Pijning letter, 1996). It appeared during the ensuing objections procedure that the situation at Houthavens was not as bad as at first thought. The decision was subsequently reviewed (Arents letter, 1996).

Figure 7.4 shows the odour contours for immission concentrations around Westpoort for the year 1993. The Houthavens area lies between 8 ou/m^3 and 5 ou/m^3 as a 98th percentile.[58] Because the proposed national legislation on noxious odours was rejected in the Memorandum on Noxious Odours (VROM 1992) and the Revised Memorandum on Noxious Odours (VROM 1994), there were no prescriptive consequences for assessing development plans (see also § 5.5). However, the proposed standards provided insight into the level of odour nuisance that could be expected. "Odour emissions are such that serious nuisance can be prevented, but this is not certain"[59] (Sandig and Vossen 1994; 28). The expectation was that, in the year 2000, the Houthavens area would be between the 6 ou/m^3 and 3 ou/m^3 contours, but "we cannot be certain whether there will be a serious odour-nuisance problem" (Sandig and Vossen 1994; 28). Measuring odour levels has proved difficult in practice. "It is difficult to quantify odour problems because of the many types of odour, the scale, the level of emissions and the subjective character of odour perception" (Borst et al. 1995; 51). In any case, such figures do not tell us a great deal.

Early on, the Amsterdam local authority shunned national odour standards and, on the basis that liveability in the Houthavens was a priority, formulated its policy on the assumption of 'serious nuisance'. The local authority also wished to incorporate the predicted developments with regard to odour-emitting industrial activity (Pijning interview, 1996). A survey into living conditions was subsequently carried out by telephone to provide the local authority with information on odour perception and related nuisance (Sandig and Vossen 1994). The results of this survey into living conditions should provide an accurate picture of the level of odour nuisance in the Houthavens area.

The telephone survey included questions on the frequency and extent of odour nuisance in residential neighbourhoods situated within the same odour contours as Westpoort, and in neighbourhoods situated farther away from odour-emitting sources[60] (see Fig. 7.4). On the basis of this comparison, Sandig and Vossen (1995) concluded that "serious nuisance can be prevented if the immission concentrations exceeded as a 98th percentile are below 3 ou/m^3. [...] Given the results of the survey, it is unlikely that serious nuisance will occur in the Houthavens area in the year 2000," (Sandig and Vossen 1995; 35). At the "target odour level (3 ou/m^3 as a 98th percentile) the percentage of the population experiencing nuisance is approximately 12 per cent" (Sandig and Vossen 1995; 36). If an odour level of 3 ou/m^3 can be attained, this would meet the two requirements (mentioned above) set by the national government. However, given odour levels in 1993 and the predicted levels for 2000, it was expected that the level of 3 ou/m^3 as a 98th percentile would be exceeded in the Houthavens area, thereby causing a nuisance problem.

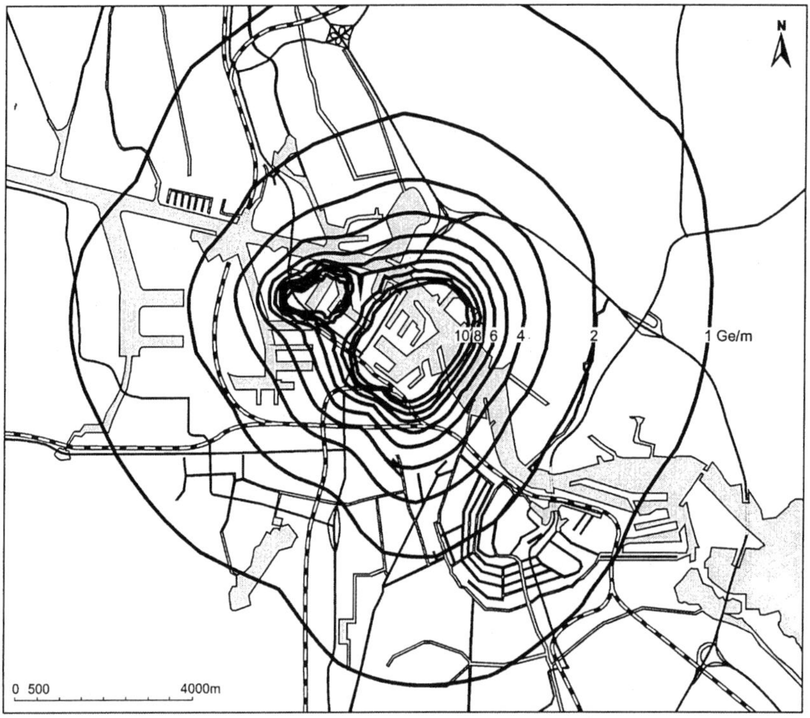

Figure 7.4 Odour emissions originating from the Westpoort port area
Source: Sandig and Vossen 1994

Decision-making on residential development at Houthavens entered a 'grey' phase due to excessive odour levels, and the question was which authority would take responsibility for the building plans. The provincial authority of Noord-Holland, as the licensing body for businesses at Westpoort and the body that gave official approval for the IJ riverbank land-use plan, ultimately took the decision to delegate responsibility to the municipal authority of Amsterdam. The provincial authority did not wish to block the construction plans set out in the IJ riverbank land-use plan.

7.10 Conclusion

In Amsterdam, liveability represents a comprehensive tailor-made approach for the local environment. Liveability is a broad concept. In the decision-making process, therefore, the environment is not treated as a separate theme, but is balanced against other factors affecting the human living environment. Neither is the environment limited to restrictive standards as a framework for traditional environmental protection policy. By contrast, it is represented through supplementary policy on liveability and sustainability. The policy in Amsterdam has been progressive and geared towards the development of a policy that is above all flexible and creative. This is most clearly evident in the Amsterdam environmental performance system, which is incorporated in the land-use plan for the IJ riverbanks.

The Amsterdam strategy has evolved beyond the traditional technical and analytical approach. Local strategic policy is developed on the basis of self-governance, not according to the letter of the law, but according to the spirit of the law. The liveability philosophy is well-founded and elaborated in the spatial planning process for the IJ riverbanks. Although the strategic vision on the environment is covered by the necessary guarantees and incorporated in the various strands of policy, it does not dovetail satisfactorily with more detailed national technical and procedural legislation. At the beginning of the IJ riverbank project, environmental aspects were discussed mainly in the context of general intentions. As the project progressed the consequences of the actual environmental situation in the area, and the related legislation, were underestimated to a certain extent. This was one of the factors that caused the spatial development of the Houthavens to stagnate.

Despite the carefully formulated, strategic, comprehensive and tailor-made approach based on liveability, it has not been easy at Houthavens, given the environmental situation, to achieve an acceptable level of liveability based on national principles that allows for environmentally sensitive spatial development. This is largely due to the fact that it has not been possible to comply with the standards-based framework of the Noise Abatement Act, which resulted in a clash between overall policy and standards-based policy. Moreover, noise-nuisance problems have had a negative impact on liveability in the area. Extra measures to improve liveability have had to be taken in the Houthavens area itself because businesses in the neighbouring Westpoort port area could not be expected to make extra efforts to reduce emissions of pollutants.

The environmental performance system (EPS) is perhaps the most acceptable method for improving the liveability of the local environment. The EPS is intended to be a prescriptive but flexible instrument for realising liveable residential neighbourhoods. This is done partly through the principle of compensation Factors causing negative impact must be compensated for by measures that improve local environmental quality in other ways. The system aims not only for compensation, but also for a level of quality that exceeds minimum statutory requirements. This makes the instrument an innovative one in Dutch environmental policy, despite everything that can be said about it. It remains to be seen to what extent compensatory measures are actually perceived as such by local residents, who will also foot part of the bill through house prices and rents. The EPS is above all a verifiable instrument that puts into practice the local authority's good intentions with regard to creating a high-quality human living environment.

The Amsterdam local authority has shown in an innovative and – for the Netherlands – unique way that liveability can be a valuable principle in spatial planning to give full consideration to environmental factors in spatial development. At the same time, it has shown that it is willing to take responsibility for developing a strategic environmental policy at the local level. Despite the problems caused by the legal compliance requirements of standards-based policy, the various stakeholder authorities have not aimed to oppose residential development at Houthavens. The issue is no longer the standards themselves but the political decision as to where responsibility for the area should lie. There was enough mutual confidence/trust to place the responsibility for approving or rejecting residential development with the municipal authority of Amsterdam.

Chapter 4 discussed three perspectives on planning action: goal-oriented action, decision-led action and institution-oriented action. Chapter 5 showed that, depending on the complexity of the issue towards which policy is geared, effective and efficient policy can be achieved if the three perspectives show a certain degree of coherence. This coherence was discussed further in Chapter 6. The Amsterdam approach shows a balance between decision-led action – i.e. adopting liveability as a philosophy – and goal-oriented action, i.e. by retaining liveability as an overall objective. With regard to institution-oriented action, there is a degree of decentral governance, but this can only be described as 'shared governance' to a limited extent. This also means that decision-making is not collective. At best there is a degree of consultation in the various planning processes. Amsterdam has anticipated national developments in environmental policy relating to the goal-oriented, decision-led and institution-oriented aspects of policy-making. It is an interesting 'investment', but it has not yet borne fruit in the decision-making on the development of the Houthavens area because, at the time of the decision-making process, national developments were still very much a future scenario as opposed to a reality. In this chapter, which discusses the Amsterdam liveability philosophy and the resulting comprehensive strategy, it is shown that, although a balance has been achieved between the three planning perspectives, conflicts can nevertheless arise. Such conflicts are due to an imbalance between centralised standards-based policy and the principles and objectives at the local level, namely housebuilding in a liveable environment.

What does this teach us about shared governance, the focus on local individual situations and the benefits of an overall and tailor-made strategy? The Amsterdam approach shows that such a policy strategy is actually possible. It has also shown that this objective can be attained with the help of a situation-specific policy that creates the right conditions. The Amsterdam approach shows, too, that local authorities may be required to make commitments that are not always feasible. It is a vastly different approach from the conventional approach with regard to the environment. Without successful examples from a large authority such as Amsterdam, it will be far from easy to fully incorporate an environmental policy based on shared governance into a local strategic and operational policy geared towards the physical environment. The example of Amsterdam shows that creativity can play an important role in innovative policymaking. The important thing is to give that creativity a chance.

Notes

1 Cleij interview, 3 April 1996.

2 The Amsterdam local authority defines a 'liveable' society as "a society in which every person is able, within reason, to meet his/her own needs without preventing others from doing the same" (Gemeente Amsterdam 1995; 24).

3 Cleij interview, 3 April 1996.

4 The BROM was produced jointly by the Environmental Health & Hygiene Department and the Spatial Planning Department of the Amsterdam municipal authority.

5 "Area-specific urban policy can be described as environmental policy that is formulated for specific urban areas and is geared towards the preservation, recovery or development of the functions or features of those areas" (Gemeente Amsterdam 1994b; 11).

6 A method designed for elaborating the bespoke approach for individual areas and for developing compensation measures.

7 "It is no longer considered feasible to calculate 1 indicator that expresses environmental quality for the whole of Amsterdam. [...] A total figure [...] ultimately provides no information at all. It does not tell us which parts of the city have a high environmental quality, neither does it tell us where the problems lie" (Gemeente Amsterdam 1995; 19).

8 The comprehensive concept that is tailored to local situations has been more or less formalised from a spatial-planning perspective, but is particularly relevant when considered in terms of underpinning the compensation principle (see § 5.5).

9 After the 'norm storm' on standards had subsided (see Chapter 5) and subjects for discussion had assumed more normal proportions, the Amsterdam local authority imposed a number of restrictions on the proposals for the city. For example, the comprehensive approach must not be realised by sacrificing the individual protection afforded to a certain extent by the standards system. In this context, a restriction was also imposed with regard to non-compliance with health and hygiene standards (Gemeente Amsterdam 1994d; 26). It was also emphasised that area-specific standards do not imply that Amsterdam is no longer required to comply with national environmental standards.

10 Deviation from the national statutory standards was not permitted.

11 Interview with Cleij and Meijburg, 3 April 1996.

12 The IJ project is not an isolated initiative in Amsterdam. There are similar plans for adjacent areas. Most of the KNSM (Royal Dutch Steamship Co.) Island is now a residential area, the development of Java Island is well under way, and discussions have begun on how precisely the IJburg island – for which 18,000 homes are planned – is to be developed.

13 In January 1990, a draft of the Terms of Reference was presented to the Amsterdam municipal council (Gemeente Amsterdam 1990). The motions submitted at the presentation were included in the revised document of 1991.

14 The standpoint of the municipal council in 1991 was that "The constraints imposed by the Noise Abatement Act are such that homes can only be built at a limited number of

locations. This inevitably means that these will be considered as luxury homes. The emphasis is [...] on the top end of the housing market" (Gemeente Amsterdam 1991; 14). Nevertheless, in the Houthavens area, much of the space for housing has been allocated to less expensive homes, regardless of the pollution level in the area.

15 In constructing the new IJ infrastructure, the local authority has aimed to utilise the possibilities for rail transport. The proposed 'IJ-rail' would join the Nieuw Oost (New East) area with Sloterdijk (a neighbourhood in the west of Amsterdam) and a north-south rail link was also planned. The intention was to construct as many as possible of the transport axes underground (Gemeente Amsterdam 1991), but this later proved unworkable due to the costs involved.

16 The Terms of Reference set out seven principles for environmental aspects. The two terms of reference referred to here reflect environmental intentions.

17 Similar ideas are discussed in the BROM in the context of the Bubble Concept (see § 7.2).

18 The *environmental matrix structure plan* (Gemeente Amsterdam 1994c) was a tool designed to play a strategic role in the spatial-planning policy of the Amsterdam local authority and was intended as a general assessment preceding a definitive structure plan (Hoogstraten and De Laat 1994). With the matrix, the 'grey' environment can be incorporated in the spatial planning process at an early stage. The ultimate aim is to incorporate this instrument "as a structural part of the (spatial) planning process" (Gemeente Amsterdam 1994d; 26). The matrix consists of a list of environmental aspects that are considered relevant and a list of spatial aspects. Because the environmental effects of land-use forms vary, distinctions have to be made between the different functions in the plan area. The essence of the matrix is therefore "to illustrate the connection [...] between environmental aspects, spatial aspects and types of land use" (Humblet and De Roo 1995; 48).

19 According to the Amsterdam local authority, the environmental requirements and limiting conditions would preclude the need for an environmental impact report (EIR) as a "statutory consideration of environmental interests" (Gemeente Amsterdam 1994e; 21), i.e. the 'guarantee' aspect of an EIR. Moreover, "according to current environmental legislation [...] the scale of the plans for the IJ riverbanks does not require an EIR" (Gemeente Amsterdam 1994e; 21).

20 A high-rise impact report (HIR) was also required in order to assess effects on the immediate surroundings in terms of light, wind nuisance and visual nuisance. The IJ Riverbanks land-use plan states that "an HIR is required for buildings higher than approx. 30 meters that extend 50 per cent or more above the prevailing building height [...]" (Gemeente Amsterdam 1994e; 25).

21 For certain measures, the environmental performance system refers to a list of environmental priorities, which is included in the IJ Riverbanks land-use plan together with that system (see Gemeente Amsterdam 1994e; appendix) and the Programme of Urban Planning Requirements for the Houthavens area (see Gemeente Amsterdam 1995b). The environmental priorities set out the criteria for construction projects in the area.

22 There were no objections to the inclusion of the ESP in the land-use plan for the IJ riverbanks. A confrontation with the Council of State, the highest court in the Netherlands for planning and environmental issues in the Netherlands, was therefore avoided.

23 The maximum permissible level for industrial noise nuisance is 55 dB(A).

24 According to the EPS as set out in the land-use plan for the IJ riverbanks and on the Programme of Urban Planning Requirements for the Houthavens, deviation from the maximum permissible level is possible if sufficiently compensated. However, the law does not permit this. The municipal authority of Amsterdam responded with the view that the formulation was incorrect and that '>'should be replaced by '=' (Interview Meijburg 25 April 1996). The amended system is shown in Appendix 7.1.

25 The preferred value for industrial noise is 50 dB(A).

26 Cleij interview, 3 April 1996.

27 The current structure of the system does not allow a noise situation categorised as relatively unsatisfactory to be compensated with measures that take effect within the broad spectrum of problems under the theme 'disturbance', such as odour, dust, etc. However, the system permits compensatory scores within the theme 'noise' for road and rail traffic and industry.

28 Because of the complexity of environmental/spatial interrelations, the Houthavens project was included as a pilot project in the 'City & Environment' report (VROM 1995, Appendix 3). Although it is still largely a theoretical exercise, the project expresses the Amsterdam local authority's vision of how to deal with complex environmental/spatial issues.

29 "The climate on the boulevard, particularly the wind effect at ground level caused by high-rise buildings, will be studied in more detail" (Gemeente Amsterdam 1991; 95).

30 The Houthavens project comprises developments on the IJ Riverbanks (the land part) and in the IJ (the water part).

31 In accordance with Step 1 of the 'City & Environment' approach.

32 The programme of planning requirements for the Houthavens district states that "the advantages of a location adjacent to a port, [...] [should] be compensated by extra efforts in terms of urban planning and environmental policy. [...] Housing environments should therefore have 'extra' qualities that are just as essential to liveability as legal compliance requirements for maximum environmental nuisance" (Gemeente Amsterdam 1995b; 15). This extra quality must be realised through the EPS that is incorporated in the land-use plan for the IJ riverbanks.

33 The review of environmental permits is completely separate from the local authority's housing development plans.

34 The Amsterdam local authority presents such an agreement as "the fourth logical step to follow on from the three steps of City & Environment. It is not only residents who have a right to an environment in which nuisance is restricted to an acceptable level and a scheme for maximum compensation. The business community is also entitled to operate in an area that is allocated to them and preserved by law and is affected only periodically and by small steps to be taken each time" (Stadig 1995; 6).

35 Under specific conditions relating to open-air harbour activities (see§ 7.8), an additional noise load is acceptable by law up to a maximum of 60 dB(A) as a condition for housing construction.

36 See Appendix 3 (list of categories) to Article 9 of the regulations for the IJ Riverbank land-use plan (Gemeente Amsterdam 1994e).

37 Category A has now been incorporated in the Establishments and Licenses Decree of the Environmental Protection Act (Article 2.4).

38 ADM-NSM-Tomassen is located on the northern bank of the IJ, opposite Houthavens.

39 The noise zone around the site of the former Minerva shipyard covering the Houthavens still exists. Minerva was located in the area of the IJ riverbank plans. However, the current land-use plan excludes (former) A-locations, which means that, in a legal sense, there is no longer a noise zone at the site.

40 The Amsterdam municipal council was the body responsible for granting exemptions. The port area is the only area for which the provincial government has not given the municipal authority a mandate.

41 The western port area contains premises within the meaning of Article 2.4 of the Establishments and Licenses Decree of the Environmental Protection Act. Noise zones have been established around these premises on the basis of the 50 dB(A) contour. However, dispensation is possible up to 55 dB(A) in the event that noise reduction measures are not effective enough. The 55 dB(A) contour is derived from the 50 dB(A) contour.

42 "The maximum permitted noise level, originating at the industrial site, measured at the front of the homes that are within a zone at the time of enactment of a zone, or are under construction or planned, is 55 dB(A)" (Art. 65, Noise Abatement Act).

43 The 50 dB(A) contour around the Westpoort area was set by the Minister of VROM on 23 June 1993 in compliance with Articles 53 in conjunction with Arts. 64 and 59 of the Noise Abatement Act.

44 After abatement measures have been taken, the procedure for establishing a higher value can begin. The Minister of VROM will have to grant exemption for approximately 2,000 homes around the western port area (Bakker interview 1996).

45 On 18 March 1994, in the context of the IJ riverbank land-use plan, the Amsterdam local authority applied for higher values under Art. 67 paragraph 3 in conjunction with Art. 47 paragraph 1 of the Noise Abatement Act, in conjunction with Art. 10 of the Limit Values for Industrial Sites Decree. The provincial authority initially rejected the application, but subsequently reviewed the decision and set a limit for up to 130 homes in the Houthavens area. In addition, the provincial authority of Noord-Holland incorporated as a condition the shield of buildings proposed by the municipal authority (Provincie Noord-Holland 1995b).

46 Variant 6 as calculated by the DGMR, "An extra shield in the form of an office building", (Rangelrooij and Spaans 1995; 8) is 26 meters high and 400 meters long!

47 We should note here that there is scope for this level of noise at Houthavens itself but not at adjacent locations, where homes are located at a comparable distance from commercial premises.

48 53 dB(A) is equal to the difference between 57 dB(A) - 55 dB(A) (Bakker interview 1996).

49 Making full use of the 55 dB(A) contour up to 55.49 dB(A) allows an increase in the total level of noise originating at Westport – 55 dB(A) – of 0.49 dB(A). The noise allowance is therefore 55.49 - 55 dB(A), which is 45 dB(A) (Bakker interview, 1996).

50 Up to a maximum noise allowance of 53 dB(A) (Bakker letter, 1996).

51 "Closed-off recessed balconies are a possible solution that combines the benefits of sound insulation and healthy outdoor space. In such cases, the façade (within the meaning of the Noise Abatement Act) is not the exterior façade, but the wall between the living quarters and closed-off external spaces. An alternative is to construct a false front with an opening between the roof and the façade, which forms an acoustic barrier for the internal walls of the homes" (Gemeente Amsterdam 1995b; 30).

52 The Amsterdam local authority prefers existing permitted noise levels to be exceeded, rather than increasing the permitted level (namely the seaport standards), because the latter option, in contrast to exceeding an existing level, is not an incentive to reduce noise (Cleij interview 1996).

53 The Amsterdam local authority commissioned the Eindhoven University of Technology to conduct a study into the spread of dust in the port area.

54 With measurements above 10 micrometers (Zeedijk 1995).

55 "Fine dust is a pollutant, particularly black dust that is insoluble in water. The bond can be so strong that pollution can be permanent or clean-up costs high" (Zeedijk 1995; 8).

56 The Health Council has to give its opinion on recommended values before they can be considered as limit values.

57 "The further reduction of dust emissions from existing sources appears to be hardly achievable" (Provincie Noord-Holland 1995e).

58 The Cargill company is one of the main sources of odour emissions in the vicinity of Houthavens. In 1993 its emissions exceeded the permit limit (Sandig and Vossen 1994), but the company has since taken measures based on the ALARA principle (Pijning interview 1996).

59 In general it can be said that "research into the relationship between exposure to odour and the nuisance experienced as a result has shown that exposure below 1 ou/m^3 as a 98th percentile causes serious nuisance. At this level of exposure, between 2 per cent and 12 per cent of residents perceive the odour as nuisance. At a level of odour concentration below 1 ou/m^3 as a 99.5th percentile, no nuisance is perceived. If the concentration is above 10 ou/m^3 as a 98th percentile, there is a serious odour-nuisance problem. Odour can also be a problem for residents at levels below 10 ou/m^3 as a 98th percentile. Immission concentrations between 1 and 10 ou/m^3 as a 98th percentile constitute a 'grey' area. Spatial planning decisions cannot be taken until the specific relationship between odour levels and perceived nuisance has been established" (Sandig and Vossen 1994; 6).

60 Cargill assumed that residents in existing neighbourhoods were, to a certain extent, accustomed to industrial odour emissions, but that this would not apply to those moving into the new homes. It therefore expected a possible increase in the number of residents experiencing odour nuisance in the Houthavens area (Pijning interview, 1996).

Appendix 7.1 Environmental Performance System for Amsterdam

Environmental performance per planning zone in points		
	Performance	Points
THEME: COMPACT CITY		
Floor Space Index in gfa/plot:	> 2	1
	> 3	2
	> 4	3
THEME: MOBILITY		
Walking distance (in meters) to nearest public-transport stop:	< 300	1
	< 200	2
	< 100	3
Parking standard for residential areas (no. of spaces/home):	≤ 1.0	1
	≤ 0.8	2
	≤ 0.6	3
Parking standard for work locations (no. of spaces per /250 m^2):	2 (B location)	1
	1 (A location)	2
		3
Parking spaces in built car parks, exchangeable for Call-a-car amenities (no. of cars):	> 5 cars	1
	> 10 cars	2
	> 20 cars	3
Secure bicycle storage within 100 meters (no. of places/home):	1	1
	2	2
	3	3

THEME: NOISE NUISANCE		
Road traffic, noise level	max. exemption value	-3
	for each 3 dB(A) reduction	+1
	< preference value	+2
Rail traffic, noise level	max. exemption value	-3
	for every 3 dB(A) reduction	+1
	< preference value	+2
Industry, noise level	max. exemption value	-3
	for every 3 dB(A) reduction	+1
	< preference value	+2
THEME: SUSTAINABLE CONSTRUCTION		
one or more elements realised with an alternative product from the BWA environmental preference list[a] (column):	column 5	-4
	column 4	-2
	column 3	0
50% of elements realised with an alternative product from the BWA environmental preference list[a] (column):	columns 1 and 2	+4
90% of elements realised with an alternative product from the BWA environmental preference list[a] (column):	columns 1 and 2	+7
THEME: ENERGY		
Heating and/or cooling by means of cogeneration:	yes	3
	no	-1

Energy performance standard for utility buildings, according to NEN 2916 (in m^3 aeq/m^2 gfa):	< 30	1
Energy performance standard for homes (in m^3 aeq):	800	1
	750	2
	700	3
THEME: WATER / GREEN SPACES		
Discharge of rainwater (from roof) to:	100% to sewer	-1
	<100% to sewer	+1
Compliance with recommendations on saving water[b]:	no	-2
	yes	+2
Planted roof or roof garden (not terrace):	yes	+1
THEME: WASTE		
Access to waste-collection site (milieuabri) (max. distance from homes, in meters):	<100	1
MAXIMUM POSSIBLE SCORE:		47

a For the BWA (soil / water / atmosphere) environmental preference lists, see Appendix D of the IJ riverbank land-use plan (Gemeente Amsterdam 1994e).

b See parapraph 4.3 of the explanatory notes to the land-use plan (*Toelichting bestemmingsplan IJ-oevers*) (Gemeente Amsterdam 1994e).

Source: IJ Riverbank land-use plan (Gemeente Amsterdam 1994e)

Chapter 8

From 'Command-and-Control' Planning to Shared Governance

Final Observations on the Link between Complexity and Decision-Making in Environmental Planning

> Effective planning begins by confronting the problem at hand and assessing conditions of uncertainty, rather than misapplying theories and methods without regard to particular problem conditions. By matching planning processes to problem characteristics, planning offers a chance to overcome, or at least reduce uncertainty (Karen S. Christensen 1985; 63).

8.1 The Heart of the Matter: Straightforward and Complex Issues

This study has considered the *degree of complexity* of environmental/spatial conflicts as a possible criterion for decision-making in environmental policy. These envirornmental/spatial conflicts are clashes between the 'grey' environment (environmental health and hygiene) and spatial developments. As part of the study we looked at the transition in the Netherlands from a strongly goal-oriented environmental policy to one increasingly based on shared governance. This turning point centred on a choice between hierarchical central governance and decentralised decision-making. This development was a consequence of the growing dissatisfaction in the Netherlands with the use and implications of generic and restrictive environmental standards: 'command-and-control planning'. This turned out be a form of planning that was '*too good to be true*'. In response to this development, the Dutch government sought a new governance philosophy for environmental policy. By opting for 'self-regulation within frameworks' and a policy based on local issue-related 'public consensus', the door was opened to a more flexible policy that would allow for local circumstances and place more responsibility for the environment with local authorities. A new governance philosophy should also result in an approach that 'acceptable solutions within the environmental policy sector can produce the most favourable result for a comprehensive policy that is also area and situation-specific, and relates to the local context of the issue to be addressed: 'tailor-made comprehensive planning'.

This problem definition based on actual situations led us to ask the following question in section 1.2: *can environmental quality be defined as a strictly delineated theme, or should it be considered in the light of overlapping issues?* Chapter 5 of this study shows that for a number of decades new environmental problems arose one after another in the Netherlands, forcing the Dutch government to consider environmental quality separately from other issues relating to the physical environment. This view is becoming less and less prevalent, however, because of the changing way in which environmental quality is evaluated – at least within the government apparatus – in conjunction with other qualities of the physical world in which we live. This study is based on the assumption that, when environmental quality as a policy issue is considered outside its context, it is an issue that will remain 'the heart of the matter'. We refer to such cases as a 'relatively straightforward'. Now that environmental quality is increasingly balanced against other issues relating to the physical environment, the context within which environmental quality has to be ascertained. This study therefore focuses on *the consequences of planning-oriented action* when environmental quality is either defined as a separate theme or dealt with in conjunction with other issues.

8.2 Complexity and the Environmental/Spatial Conflict

A large part of this study is about environmental policy in transition in the Netherlands, approaches to environmental/spatial conflict, and the ways in which they influence each other. We have considered the degree of complexity of environmental/spatial conflicts from a somewhat abstract perspective and how decision-making and solution strategies dovetail with them. It is, then, a question of finding ways in which to deal with environmental/spatial conflicts through policy, in addition to the current standards-based approach. These developments were a response to the criticism of conventional prescriptive and generic environmental legislation. Not all stakeholders were in agreement with this form of central governance. This is also true of the way in which environmental quality is evaluated in relation to other qualities attributable to the physical world. Local authorities in particular have pointed out that generic and prescriptive environmental legislation has a contextual effect that imposes unnecessary constraints on other sub-areas of policy on the physical environment. Policymakers were apparently overlooking opportunities for a more balanced policy at local level. Such a policy could benefit environmental and other qualities.

This position requires some explanation. Chapter 6 points out that prescriptive legislation proposed by the government – i.e. command-and-control planning – can solve the majority of environmental/spatial conflicts in a relatively satisfactory and straightforward way. These are conflicts in which the scale of environmental pollution is limited or in which there are few or no environmentally sensitive functions in the vicinity of environmentally harmful activities. In such cases it is sufficient to maintain a given distance between the source of the pollution and environmentally sensitive areas. Generic standards allow a routine approach to these conflicts, which can be categorised as 'relatively straightforward'. The emphasis in this approach therefore lies on how it is implemented.

There are nevertheless a large number of conflicts that cannot be satisfactorily resolved using prescriptive legislation. These are conflicts in which it is difficult to separate environmentally sensitive activities from environmentally harmful activities, with the result that creating a safe distance between the two would have a negative impact on local interrelationships or interests. It is particularly difficult – and possibly even undesirable – to find solutions in the short term through prescriptive legislation. These characteristics come into play during periods of transition or restructuring in areas with environmentally harmful activities to make way – at least in part – for environmentally sensitive functions. Investment in spatial development is the obvious solution for such areas. This is even a desirable solution, given the current preference for development in line with the 'compact city' concept. However, necessary environmental remediation causes a stalemate in the meantime. In these circumstances, adhering to generic environmental-policy frameworks can have major spatial consequences locally. If, in such cases, environmental standards are not used as prescriptive frameworks but as guidelines, opportunities are created for adapting the generic standards to the local situation, thereby creating greater flexibility for procedures. This in turn creates more possibilities for a tailor-made comprehensive approach to the conflict.

A limited number of environmental/spatial conflicts can be categorised as 'relatively very complex'. When environmentally sensitive and environmentally harmful activities are closely interwoven, when environmental pollution is difficult to trace to specific sources and when the pollution has dispersed over a wide environmentally sensitive area, maintaining a given distance between the source and the affected areas is simply impracticable because of the lack of space. In such cases, the system of environmental standards may provide insight into the problem but it serves no purpose to derive administrative and physical spatial measures from this system. Generic policy is ineffective and inefficient in such conflicts, even when it is implemented with a degree of flexibility and nuance. A more logical solution is to take the local spatial and administrative situation as the starting point, and then to devise a solution strategy whereby environmental quality is no longer considered in isolation, but in conjunction with other local interests: tailor-made planning and shared governance.

In this study, therefore, we have pointed out that environmental/spatial conflicts differ mainly in terms of *complexity* due to the varying degrees of *contextual influence*, i.e. the extent to which the conflict is influenced by local administrative and material spheres. Environmental legislation designed to support environmental objectives clearly affects other sub-areas of policy geared towards the physical environment. Indeed, in a number of cases it can have far-reaching spatial consequences. It comes as no surprise, therefore, that this has resulted in heavy criticism of traditional hierarchical environmental policy. This criticism has made it clear that the answer to the question as to whether the 'grey' environment should be dealt with as an entirely separate theme depends on how it is *appraised*. Furthermore, given that there is contextual influence, that appraisal is also important for other themes relating to the physical world. This observation places the value judgement – particularly that of decision-makers – at the centre of the decision-making process.

8.3 The Relationship between the Conflict, Complexity and the Decision-Maker

Traditionally, environmental policy in the Netherlands has been not only hierarchical, but also technical/functional in character. It has always been strongly oriented towards direct cause-and-effect relationships (*causa proxima*) between the source of pollution and its resulting impact. As a consequence of this, aspects outside the cause/effect relationship were given only partial consideration, for example contextual influences. The analysis of environmental policy in this study shows that, by excluding external influences through the use of prescriptive policy, and adhering to a more or less technical/functional (linear) planning process a large degree of certainty can be ensured, so that the goals which have been set can be fully realised – so long as this does not have a noticeable effect on the contextual environment. If the consequences are more far-reaching than expected, or if a re-evaluation of the contextual environment takes place in society or the government, contextual influences will come to be seen as part of the issue, and possible indirect influences on the issue will be given greater consideration. In such cases, the complexity of the issue will increase. It can be argued that *there is a positive relationship between the degree of complexity of an environmental/spatial conflict and the influence of the material and administrative context of the problem.* Nevertheless, the determining factor is the decision-maker's appraisal of the complexity of an issue.

This analysis of environmental policy shows that there are two general reasons that explain why environmental/spatial conflicts have been defined as 'relatively straightforward' for such a long time in the Netherlands. The first reason is that this was a conscious choice made with a view to gaining insight into environmental issues separate from the context in which they arose, in order to produce effective short-term results. The second reason is the lack of knowledge about the complex background to environmental pollution, which therefore required a 'simple' problem definition. The most logical approach was to take the relationship between source and effect as a starting point for policy. While this approach continued to product satisfactory solutions, it was not necessary to seek a more complex problem definition. More generally, every decision-maker will try to see the issue with which he/she is confronted in as straightforward a way as possible by identifying the direct cause-effect relationships between the aspects involved in the issue. If this approach proves unsatisfactory, alternatives must be sought. This development also occurred in environmental policy, among other things after it was found that compartmentalised policy could result in 'passing the buck' between the compartments ('air', 'soil' and 'water'). In environmental policy, the compartmentalised approach was eventually replaced by a more integrated approach based on environmental themes. This involves dealing with environmental issues together using a comprehensive approach, and elaborating them according to the source-oriented and effect-oriented strands of policy. If a relatively straightforward problem definition is no longer workable, and more (f)actors need to be incorporated, an increase in complexity is not the only consequence. The problem will also be redefined, and the decision-making process and approach will be modified. As the decision-makers' problem definition increases in complexity, policy strategy must be amended accordingly. Furthermore, as issues become more complex, the end result becomes increasingly uncertain. At the same time, however, there will be more possibilities for devising solution strategies. The conventional single-track can then be replaced by many alternatives for achieving goals or desired situations.

Developments in environmental policy over time are not only the result of evaluations of policy performance, which conclude that issues should indeed be seen as more complex, considering contextual and other influences (an *a posteriori* choice that reflects the complexity of the physical or policy environment). It is also possible that a conscious decision is taken to define an issue as 'relatively straightforward', even though there may be reasons for considering it as 'complex'. This study points out that an urgent issue can be categorised as 'simple' and goal-maximisation given priority by deliberately stating that a straightforward approach will be followed and that the issue will be considered separately from its context. This is an *a priori* choice that reflects the desire to see an issue as straightforward, despite an intervening context. In such cases, maximum results are the most important consideration. This is particularly true in the event of emergencies, i.e. crisis planning. The contextual consequences of following this approach are then seen as irrelevant and less important than achieving the goals that have been set. Partly as a result of the wide public concern about the quality of the physical environment, the policy trend for many decades has been strongly goal-oriented. Far-reaching prescriptive standards were developed as a deliberate policy choice. These standards were largely based on environmental health and hygiene criteria, without a great deal of consideration for other aspects of the physical environment. One of the consequences of this approach was that, for a long time, environmental/spatial conflicts were reduced to their most basic and urgent aspects, and were often 'lifted' from their spatial-functional context.

Whereas it is stated above that the material and administrative contexts are given greater consideration as environmental/spatial conflicts become more complex, this study makes it clear that the reverse is also true: as the *material and administrative context of an environmental/spatial conflict increases, the complexity of that conflict also increases*. There is a two-way relationship between the appraisal of the material and administrative contexts and the complexity of an environmental issue. This appraisal and, consequently, the degree of complexity in fact a matter of choice for the decision-makers(s). This is an essential feature of what is referred to in this study as 'decision-led action' (see also § 8.5).

8.4 The Significance of Complexity from a Systems-Theory Perspective

In this study, the degree of complexity attributable to an issue is expressed in terms of systems theory. That degree of complexity depends on the extent to which the problem definition and solution strategy for an issue are considered from the point of view of the various aspects of the issue, the interrelationship between those aspects in relation to the whole, or from the point of view of how the aspects interact with the issue as a whole and its context. As we have pointed out above, an issue can be categorised as 'straightforward' if it can be seen in terms of a direct causal or linear relationship between its various aspects. In such cases, a *functional-rational* approach is satisfactory. Such an approach assumes that the conditions on which the final result will depend are already present in the initial situation. The relationship between the various aspects of the issue and its context is assumed to be stable. Under these circumstances, the end result can be predicted to a large extent. However, such circumstances are exceptional when societal issues are involved, and the analysis of developments in Dutch environmental policy confirms this.

The less evident the direct cause-and-effect relationship between the elements of an issue, the more complex it will be (internal complexity). This is also true of the extent to which issues are influenced by the context in which they have arisen (external complexity). In such cases, planners must take into account the fact that the situation is a dynamic one characterised by 'disequilibrium', in which it is not clear how the contextual environment influences the aspects that make up the whole. This is illustrated by the environmental/spatial conflicts categorised earlier in this study as 'very complex'. The same is also true of the urban context – described in Chapter 3 – in which environmental/spatial conflicts can arise. In such circumstances, where urban space is under increasing pressure from development based on the 'compact city' concept, it is more difficult to deal with environmental/spatial relationships by using a straightforward cause/effect approach. These conditions require a degree of creativity on the part of decision-makers, who will have to think increasingly in terms of opportunities instead of clear predefined 'ends'. This is particularly true of issues for which the problem definition and solution strategy are arrived at by means of a participative decision-making process involving a large number of stakeholders with diverse interests. It is not strange, therefore, that the theoretical framework that is needed to gain knowledge and insight into such complex situations is described as *communicative-rational*. This does not mean that objectivity and controllability become less important, rather that the approach is a more interpretative one applied to intersubjective phenomena that can nevertheless be considered (at a certain level) from an objective perspective and understood, predicted, and expressed in models.

In this study, the comparison between government proposals for an integrated environmental-zoning system and the ROM designated-areas policy – as instruments from the empirical problem field – has served to illustrate the concept of complexity, supported by arguments from the discipline of systems theory. Both instruments help to illustrate how environmental policy has evolved. The Ministry of VROM proposed the instruments more or less simultaneously, and their feasibility was assessed. The instruments – which are diametrically opposed on almost all fronts – are discussed in detail in section 5.4. The differences between them – it is argued here – can be understood for the most part by considering the complexity of the issues for which they are intended.

The ministry's proposed integrated environmental-zoning system could have been the jewel in the crown of traditional standards-based environmental policy. The system was based on the principles of the environmental standards system and functional-rational theory. The environmental standards already embraced the conditions that determined which measures were required and what the end result would be. The integrated environmental-zoning system was based on a direct cause/effect relationship, whereby the system provided accurate insight into the conflict between environmental pollution at a given location or in a given area, and the existing spatial constellation. In a number of cases, however, this system can only produce the predicted result if contextual influences are excluded through the use of prescriptive environmental standards. In other words, if the desired level of environmental quality has to be attained regardless of the spatial-planning consequences, then so be it. In relatively complex environmental-spatial conflicts those consequences may be far-reaching and may include, for example, the closure of industrial plants and the demolition of residential areas.

It is a completely different story for the ROM designated-areas policy. The policy is geared towards more than an acceptable level of environmental quality. Environmental health and hygiene problems in a given area are addressed *together with* a number of other issues by means of an integrated tailor-made and comprehensive approach. This approach is largely free of the clearly defined rules that characterise the integrated environmental-zoning system. Top-down regulation is replaced by 'self-governance', apparently giving the ROM designated-areas policy a great deal of leeway. Problem definitions are not provided in advance, but jointly agreed upon by stakeholders. The flexibility of the initial situation means that the scope for devising solutions is not fixed either, and there is only a limited amount of certainty with regard to the ultimate result. In contrast to the zoning system, the ROM policy seeks opportunities for developing a collective integrated approach that takes account of the unique, specific and contextual circumstances of a location. The ROM policy is not based on a linear process, but has the characteristics of a network-based approach. Although it does not produce clearly defined rules that determine the subsequent planning process, the network approach has a number of implicit rules that cannot simply be ignored. These relate mainly to the way in which participation and interaction are to be incorporated and allowed to function within the decision-making and planning processes.

This comparative analysis of integrated environmental zoning and the ROM designated-areas policy shows that the conceptual framework of systems theory ('constituent parts, whole and context') provides insight into the scope of application for policy instruments. The analysis illustrated that policy issues can be distinguished according to their complexity, which in turn determines how the issues are dealt with. Consequently, if a policy instrument is to be efficient and effective, it must relate to the degree of complexity attributed to an issue as explained, among other things, by systems theory.

'Complexity' was not, of course, a main consideration when integrated environmental policy and the ROM designated-areas policy were developed. This would not have been logical, given that the concept of complexity is never – or hardly ever – used as a policy criterion. Nevertheless, this study considers complexity as a criterion in decision-making. We have taken this view not only because it was possible to distinguish environmental/spatial conflicts in terms of complexity in the empirical problem field of this study and compare the scope of application of two 'rival' policy instruments on the basis of complexity. It has also been possible to operationalise complexity-related decisions based on planning-theory arguments for the purpose of planning-oriented action.

8.5 'Complexity' from a Planning-Theory Perspective

In order to operationalise the concept of complexity for the purpose of planning-oriented action, this study has looked for possibilities for anchoring 'complexity' – expressed in systems-theory terms – in a planning theory context. Again, the relationship between integrated environmental zoning and the ROM designated-areas approach was a source of inspiration. Integrated environmental zoning is an instrument whereby the initial conditions and frameworks are such that the end result is predictable within reason. In systems-theory terms we can speak of an instrument that is strongly geared towards the 'constituent parts of the whole'. Contextual influences are excluded or ignored as far as possible. The result: a growing aversion to zoning policy. At the same time, the ROM policy found increasing support. From the point of view of systems theory, the ROM policy was not so much geared towards the constituent parts of the whole but towards the whole in relation to its context. The initial conditions no longer conferred certainty. However, the increasing uncertainties brought a greater number of opportunities for shaping the planning process. Issues dealt with using ROM policy were categorised at a higher level of complexity than the issues that integrated environmental zoning was designed to solve. This was an indication that, as issues become more complex, a shift occurs from goal-setting to participative policy. This indication resulted in greater insight into the relationship between complexity and planning-oriented action.

When we refer to goal-*setting* policy, we are implicitly referring to a hierarchical institutional structure. Just as goal-setting policy involves more than the goal-oriented aspects of planning, interactive decision-making involves more than the institutional structure. When decision-making is interactive, there *is* an orientation towards a goal, albeit a less obvious one. In an interactive decision-making process, the goals are not predefined or prescriptive but are agreed upon collectively. In this study, this argument has led to the conclusion that, regardless of the degree of complexity attributed to an issue, planning-based action will always be goal-oriented and institution-oriented to a certain extent. However, the degree of complexity determines the characteristics of both the goal-oriented and institution-oriented aspects.

This observation has enabled us at least to distinguish between two perspectives on planning: a goal-oriented and an institution-led perspective. The *goal-oriented perspective* refers to the object-based dimension of planning and the various steps in the planning 'cycle'. It refers to the stages at which decisions are taken and the ultimate effects of those decisions, whereby the aim is to bring the desired scenario and actual end result as close together as possible. This means that effectiveness is a priority in planning-based action. This study summarises goal-oriented action in an abstract way as a continuum between two extremes (see also

Fig. 8.1). One extreme is a single, fixed goal, to be achieved by the prescriptive standards that have traditionally been part of environmental policy. At the other extreme we have multiple composite and dependent goals. In terms of environmental policy, this extreme is reflected in the proposals of the Bubble Concept (§ 5.5) and the ROM designated-areas policy (§ 5.4).

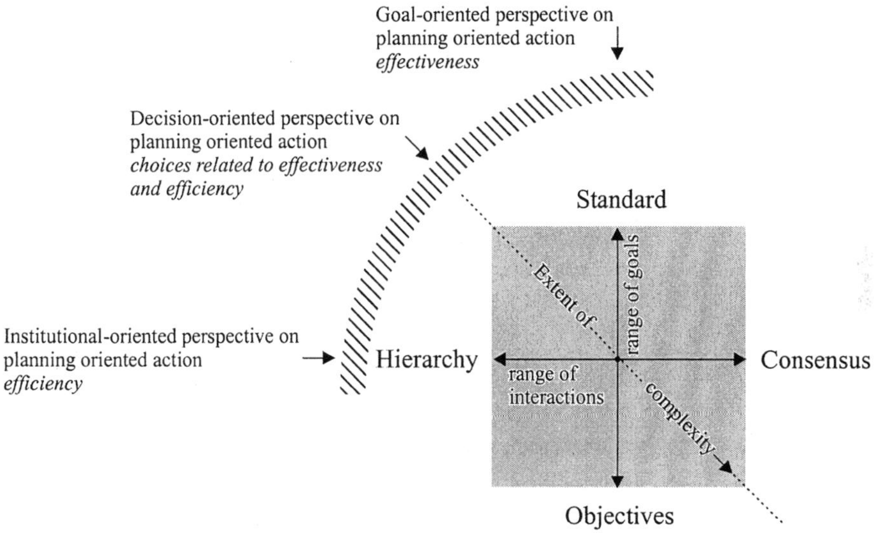

Figure 8.1 The three perspectives on planning action, shown together in a model for planning-oriented action

The *institution-oriented aspect of planning* relates to the intersubjective dimension of planning, namely the organisation of, communication about and participation in decision-making and policy. The key to this is the way in which stakeholders participate and interact, and the *efficiency* of planning-oriented action. This study discusses the two extremes within which stakeholders participate and interact. At one extreme we have a hierarchical structure and at the other we have a consensus-based consultative structure (see Fig. 8.1). In environmental policy the hierarchical structure is evident in the way in which the Dutch government – in this case the national Department of the Environment – organises its approach to environmental health and hygiene issues. For many years, the department has used environmental standards to steer policy in the desired direction. By contrast, the ROM designated-areas policy has been based on a network-like form of interaction between stakeholders, whereby a more or less equal degree of interdependence exists between stakeholders representing different interests. In this context, stakeholders (including non-government stakeholders) are expected to reach a level of consensus with regard to the problem they collectively experience and their joint approach to it.

In this study the goal-oriented and institution-oriented facets of planning are regarded as having a structuring function. Both must ensure that decision-makers' intentions are actually implemented. The substance of the argument, however – as discussed here – lies in the *decision-led action* taken by decision-makers. This relates above all to the rationalisation of decisions regarding the efficiency and effectiveness of planning-based action. As discussed above, the choices are based on the decision-makers' appraisal of the planning issue, where necessary in conjunction with related issues. We have seen that these decision-led choices determine the goal-oriented and institution-oriented facets of planning. The interrelationship between the goal-oriented, institution-oriented and decision-led aspects of planning is shown in Table 4.1 in the context of discussions on planning theory. The table shows that the degree of complexity attributed to an issue determines how decision-led choices are rationalised. It also determines the relationship between the decision-led, institution-oriented and goal-oriented aspects of planning. This argument is summarised in Fig. 8.1.

A model for planning-based action has been drawn up in this study to express the relationship between the three facets of planning (Fig. 8.1). In the model, the goal-oriented and institution-oriented facets are linked by complexity, which serves as a decision-led criterion for planning-based action. The level of complexity can be read from a notional diagonal axis that extends from the upper-left quadrant to the lower-right quadrant. This axis also represents the transition from a functional-rational approach to a more communicative-rational approach. The upper-left quadrant therefore represents issues in the 'relatively straightforward' category, while the lower-right quadrant represents issues in the 'relatively very complex' category. This proposed model provides a foundation in planning theory for the concept of complexity and the substance of the argument it represents (Table 4.1) follows on logically from past and present planning-theory discourse.

8.6 Complexity, Coherence and Consensus in Environmental Policy

The proposed model for planning-based action offers a planning-theoretical perspective that assumes general applicability. The model for planning-based action can also be seen as comprehensive. It can encompass the different visions on planning-based action, the various forms of decision-making and related policy strategies, based on the criterion 'complexity'. Each element in the model can then be 'tuned' to the issue in question. In this study, we have used the model to analyse developments in environmental policy, based on the vision of complexity as a criterion for planning-based action. This is illustrated by the instruments of integrated environmental zoning and the ROM designated-areas policy. The environmental zoning system can be placed in the upper-left quadrant of the model, i.e. in the quadrant that represents the extreme of goal-oriented and hierarchical policy. Zoning is, after all, an instrument that derives from traditional centralised and prescriptive environmental policy. Following the line of reasoning in this study, zoning is an instrument that can be applied to environmental/spatial conflicts categorised as 'relatively straightforward'. By contrast, the ROM designated-areas policy represents an approach based on individual local circumstances, which determine the approach to be taken. That approach is a participative one that takes account of circumstances considered relevant by stakeholders. The aim is to achieve a positive multifaceted and tailor-made result, whereby a desired level of environmental health and hygiene is no longer seen as a goal in itself but as a constituent part of a larger desired scenario. The ROM designated-areas policy can be placed in the lower-right quadrant of the model, and should therefore be applied to environmental/spatial conflicts categorised as 'complex' in varying degrees. We have thus seen that both instruments occupy somewhat extreme positions and have their own basic principles. They cannot therefore be combined in a single environmental policy, at least not when one is unaware of the principles of both instruments, which – it is argued here – derive from their varying degree of complexity.

This study has focussed on shifts in environmental policy that occurred during the 1990s in the Netherlands. The 'old' policy, which was based on technical-functional principles, has been transformed into a decentralised policy based on the principle of shared governance. However, participative decision-making is not yet fully accepted. Neither has policy evolved to the stage where only multifaceted goals are set (C in Fig. 8.2). The developments are summarised in Fig. 8.2 as a movement from A to B in the model for planning-based action. According to the arguments put forward in this study, this means that environmental policy is no longer geared towards issues in the 'relatively straightforward' category. Neither is it directed towards issues in the 'relatively very complex' category, but rather towards issues in the category 'relatively complex' (B in Fig. 8.2).

In this study we have pointed out that, as early as the end of the 1970s, policy-makers' approach to environmental issues had already come to be regarded as too simplistic and a new and more integral approach was required to deal with such issues more efficiently and effectively. As a result, the compartmentalized approach was phased out at the end of the 1980s and became part of a theme-based environmental policy. The policy themes were the foundation for source-directed and effect-oriented policies. These proposals were elaborated in the first National Environmental Policy Plan (NMP-1), which appeared to complete the process of internal integration for environmental policy. However, as this study points out, this was only true for the decision-led facet of planning action (see Fig. 8.2). The decision to replace technical-functional policy with an interactive and cohesive policy for the various policy elements did not have a direct effect on the goal-oriented and institution-oriented facets of planning. The goal-oriented aspect of planning was based on generic and prescriptive standards until well into the 1990s. This did not change until, in the first half of the 1990s, a number of examples convinced policy-makers that prescriptive standards could have far-reaching consequences for spatial planning. At the same time, local authorities were successfully fighting against the diminishing flexibility with regard to taking account of local circumstances. At a national level, the call for decentralization was increasing and the support base for the 'grey' environment that had seemed so strong in the 1980s began to shrink in favour of spatial development. The institution-oriented aspect of planning, which was based on a hierarchical approach, did not begin to change until the beginning of the 1990s. The 1990s were the years of participative decision-making and this naturally affected environmental policy in the Netherlands. Cautiously at first, and later with more confidence, the support base for participative decision-making (e.g. in the form of the ROM designated-areas policy) increased as stakeholders saw that it was flexible and allowed greater leeway for taking account of local situations. This study argues (see Fig. 8.2) that this development resulted in experimental network-based strategies for dealing with relatively very complex issues (i.e. an institution-oriented approach), while environmental legislation remained largely technical-functional and goal-oriented and, in terms of effectiveness and efficiency, remained geared towards relatively straightforward issues (i.e. a goal-oriented approach). At the same time, the decision-led aspect of planning was becoming increasingly integral in response to sectoral and technical-functional conventions. The result was a shift away from policy geared to relatively straightforward issues towards a policy focusing on relatively complex issues. This argument points to a disequilibrium in environmental policy in the 1980s and 1990s.

Continuing this line of argument, we can attribute a degree of cohesion to the proposals made in the context of the City & Environment project, which was implemented in the second half of the 1990s (see B in Fig. 8.2). This project proposed two radical changes in traditional standard-setting policy. In the first place it became possible to deviate from prescriptive environmental regulations, provided that there were sound reasons for doing so. In the second place, responsibility for taking such a decision was placed at the local level. In short, environmental standards remained in place but were no longer seen as prescriptive in all cases. It is now possible to deviate from standards and adopt an area-specific approach. This trend has been even more pronounced in the policy areas of noise abatement and soil protection. National noise-abatement standards have become indicative rather than prescriptive, and are fixed at the local level. National standards prevail, however, if a particular local authority does not wish to develop its own standards-based policy. Multifunctionality is no longer an aim of soil-protection policy, on the contrary: spatial-planning needs are predominant and soil remediation measures are determined according to the function allocated to the site in question.

The government summarised these changes in environmental policy as a shift 'away from a policy that is sound in the technical sense towards a policy based on public consensus'. This study adds a number of qualifications to that description. One could easily argue that even a policy described as 'technically sound' was based on a degree of consensus, i.e. the support base that evolved out of concern for the environment – a concern shared by many people in the Netherlands. This development was regarded as the cornerstone of hierarchical and prescriptive policy – that is until the first National Environmental Policy Plan (NMP-1) in 1989. Therefore, the consensus-based policy that was subsequently promoted as the counterpart to a technically optimum policy relates primarily to consensus among stakeholders in the planning and decision-making processes. In this context, the 'consensus' or 'support base' refers to the need for more participative decision-making in environmental policy. The aim is therefore to achieve greater consensus among stakeholders with regard to problem definition, solution strategies, resource deployment and the degree of involvement in implementation. The planning process would thus become more efficient. However, if we consider the proposals and efforts made in environmental policy, it is evident that developments have been predominantly vertical, i.e. there has been a shift away from central governance towards decentral governance in the Netherlands. Few developments have taken place in the horizontal context. It remains to be seen whether non-governmental stakeholders will become directly and actively involved in decision-making on local environmental issues and environmental/spatial conflicts.

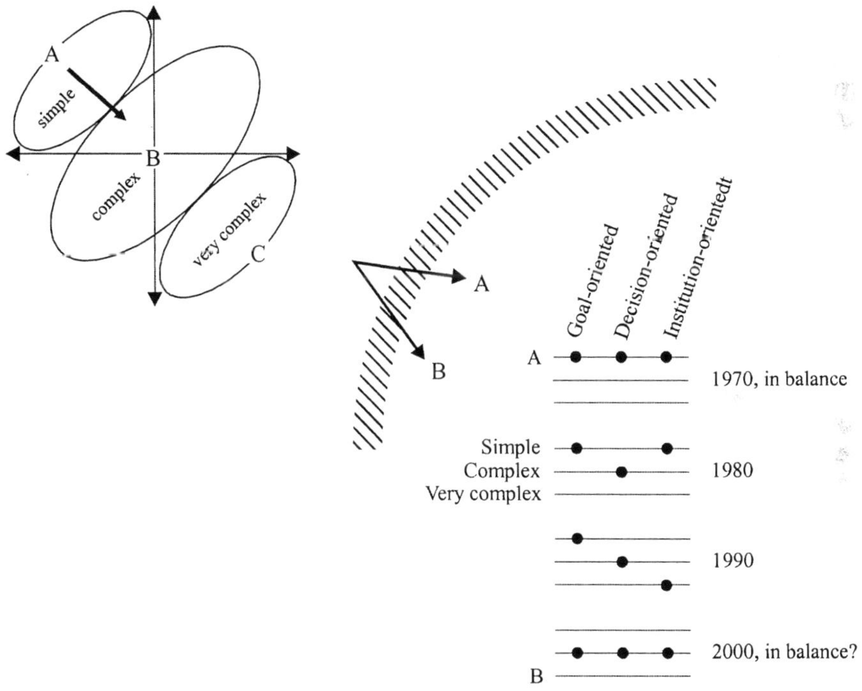

Figure 8.2 Shifts in the emphasis of environmental policy over time
Note: They are seen from the perspective of the three aspects of planning-based action, and set against the relative levels of complexity that can be attributed to environmental issues (for A, B and C, see text)

Nevertheless, from the perspective of the three facets of planning, the desired changes appear to have resulted in a reasonably cohesive policy that contrasts with the policy of the 1980s and the first half of the 1990s. The decision-led aspect of planning was already based on the realisation that environmental issues could not all be categorised as 'relatively straightforward'. Not only was there a need to reform compartmentalized policy, the realisation was also growing that a generic approach did not take sufficient account of local circumstances in every case. Goal-oriented planning made greater allowance for local aspects because standards have shifted somewhat towards being area-specific and indicative in nature rather than generic and prescriptive. In institution-oriented planning, this shift can be seen in the move towards local governance and, as an extension of this, greater consultation at the local level.

The developments described above have produced a policy context that can be expected to result in reasonably effective and efficient measures for dealing with issues categorised in this study as 'relatively complex'. Although changes in environmental policy are driven by issues that are more complex than this, their complexity will nevertheless vary. This means that we must qualify the standpoint underlying environmental policy reform, i.e. the principle of subsidiarity, which means that issues arising at a local level should also be resolved at that level. This applies above all to the more complex issues. However, as this study points out, generic policy may be more appropriate for issues categorised as 'relatively straightforward', particularly when these issues occur nationwide in a uniform context. An added advantage of this shift of focus in environmental policy towards relatively complex issues is that it seeks a 'middle way', whereby measures for dealing with issues on the scale 'relatively straightforward' to 'relatively very complex' need not differ too widely. This 'middle way' is interesting from a strategic point of view and renders environmental policy more flexible. From this middle-of-the-road position stakeholders can choose between a generic approach, an area-specific approach or a more participative approach, depending on the nature of the issue.

This argument leads us to the conclusion that knowledge, insight and appraisal relating to the complexity of environmental/spatial issues can contribute to the development process of environmental policy. A policy geared to the complexity of environmental issues will be more effective and efficient. This must be reflected in planning-oriented measures. We should note that, now environmental targets are no longer enforced within rigid frameworks in the Netherlands, it is increasingly necessary to strike a balance between environmental aspects and other aspects relating to the world we live in. Compared to the traditional approach this can yield more environmental gains on a local scale, but we should not deceive ourselves that such cases will automatically prove to be the rule rather than the exception.

8.7 Environmental Policy at the Local Level: towards Shared Governance?

This study points out that, as the context in which environmental/spatial issues arise becomes more important, there will be a greater need for a policy that is geared to local situations, rather than generic prescriptive policy. This has a number of consequences. In complex situations, areas of policy geared towards the various aspects of the physical environment will overlap to a certain extent. If environmental standards are prescriptively enforced in such cases, we can expect this to have worrying consequences for spatial functions. If an approach is chosen that takes account of local circumstances, the local authority must formulate a strategy based on location-specific characteristics and local consensus. However, local authorities will not have a 'store' of strategic principles to draw on because, under a traditionally prescriptive and hierarchic policy, their primary responsibility was to enforce top-down legislation. Local authorities are therefore relatively inexperienced when it comes to fulfilling their new responsibility for shared governance.

This study looks at Amsterdam's experience with developing overall policy at the local level with the aim – among other things – of actively preserving, and where possible improving, environmental quality in the city. The aim of the Houthavens development, part of Amsterdam's IJ Riverbank project, is to build housing for the top end of the market in the former dockland area. The policy formulated in this case is not based completely on national environmental principles. By contrast, the local authority of Amsterdam has developed a local strategic environmental policy based on internal limiting conditions and principles. Links have been established with other forms of policy for the physical environment. Visions such as the Bubble Concept and the environmental matrix were elaborated in a series of strategy plans. These visions were designed to ensure that environmental interests are not ignored in spatial planning and projects. This also required a redefinition of the concept 'environmental quality'. A new and more progressive interpretation has taken its place next to the traditional, restricted interpretation. The new definition is related to extra policy-based efforts designed to improve the liveability of a location or area and incorporates the concept of sustainability. This distinction in environmental quality is elaborated in concrete proposals that influence and have implications for spatial planning. Problems were encountered with Amsterdam's progressive approach, however, when it clashed with national legislation, which is changing at a much slower pace. Consequently, the Houthavens project is still experiencing procedural conflicts and conflicts arising from the belief that national environmental standards are too stringent.

We can conclude from this study that, although it is possible to implement decision-led, goal-oriented and institution-oriented planning in a coherent way at a local level, this approach may nevertheless fail if it is incompatible with planning measures implemented by 'higher' levels of government. This study also shows that in Amsterdam, however innovative and complex the issues may have been, the chosen approach was not a participative one. The relational aspect of the case has not been discussed in detail and the goal-oriented facet of planning is still predominant. The Amsterdam local authority can benefit from the scale of its organisation, which should be capable of implementing a consistent and innovative policy that is area-specific and multifaceted. Few local authorities in the Netherlands are in a position to maintain such an administrative structure, a fact which adds a further qualification to the principle that issues should be resolved at the level at which they arise.

The above arguments do not lead automatically to the conclusion that local policy for dealing with environmental/spatial conflicts will make a greater contribution if it is incorporated in a larger whole. The total 'added value' attributed to policy geared towards separate issues and sub-issues will have to be measured against the value of an integrated approach. The result of this process must in turn be measured against the effort required to implement either a sectoral or an integral approach. Obviously, it is no easy task to assess in advance an approach that, by definition, involves a large number of uncertainties.

8.8 Conclusion

This study is an abstract discussion of the question of whether policy issues can be categorised according to their complexity and of the significance of this categorisation for decision-making. We have observed that the complexity of environmental/spatial conflicts varies to such an extent that generic prescriptive policy developed to solve them has been unsatisfactory. Environmental policy in the Netherlands has proved to be primarily oriented towards issues categorised in this study as 'relatively straightforward/simple'. This policy was developed within a functional-rational theoretical framework based on the direct cause-and-effect relationship between environmental pollution, its sources and its spatial consequences. Such a policy leads to a uniform fixed (dogmatic) approach, which is referred to here as command-and-control planning. In this study, relatively straightforward issues are distinguished from more complex issues, for example by establishing the extent to which an issue relates to its contextual environment. The greater the interrelationship, the more important the context and the greater the degree of complexity attributed to the issue. We have developed this argument into a workable form, namely the model for planning-based action presented in this study. The model is a comprehensive one, based on the idea that planning always involves decision-led, goal-oriented and institution-oriented aspects. The nature of those aspects depends on the complexity of an issue. This does not mean that more complex issues require more complex decision-making or planning methods. However, a change in decision-making is required. This vision of planning was the basis for this study of environmental policy. We have shown that the three facets of planning in environmental policy have not been cohesive, but have nonetheless resulted in a shift away from command-and-control planning towards tailor-made area-specific overall planning, i.e. the form of planning in which a specific aspect – in this case environmental quality – is regarded as part of a greater whole, and in which the results of the planning process are largely unpredictable due to the level of uncertainty. In such cases, therefore, planning should be based on network configurations rather than direct cause-and-effect relationships. Consequently, the dynamic interaction between the elements of the issue and its context result in a considerable degree of uncertainty with regard to procedures and outcomes. The interaction is not only between the objects of planning, but also primarily between actors (stakeholders), which means that the theoretical framework for this type of planning is largely based on communicative rationality. Using complexity as a criterion for decision-making does not mean that we regard communicative and functional rationality as two theoretical extremes for decision-making and planning, rather that they should be regarded as overlapping extensions of each other. The extent to which they overlap will depend on the complexity of the issue in question. In such cases, 'complexity' is the bridge between functional and communicative rationality.

References

Chapter 1

Borst, H., G. de Roo, H. Voogd, H. van der Werf (1995) *Milieuzones in Beweging; eisen, wensen, consequenties en alternatieven*, Samsom H.D. Tjeenk Willink, Alphen aan den Rijn.

Gemeente Groningen (1996) Concept Milieubeleidsplan In Natura, Milieudienst, Groningen.

Gleick, J. (1988) *Chaos: Making a New Science*, Penguin, New York.

Kaufmann F.X., G. Majone, V. Ostrom (eds) (1986) *Guidance, Control and Evaluation in the Public Sector: The Bielefeld interdisciplinary project*, De Gruyter, Berlijn.

Kooiman, J. (1996) Stapsgewijs omgaan met politiek-maatschappelijke problemen, in P. Nijkamp, W. Begeer en J. Berting (eds), *Denken over complexe besluitvorming: een panorama*, Sdu Uitgevers, Den Haag.

Kramer, N.J.T.A., J. de Smit (1991) *Systeemdenken*, Stefert Kroese, Leiden.

Lugt, F. de (2000) *De ramp van Enschede; zaterdag 13 mei 2000*, De Twentse Courant, Tubantia/Oostelijke weekbladpers BV, Enschede.

Meadows, D., D. Meadows, J. Randers, W. Behrens (1972) *Rapport van de Club van Rome, De grenzen aan de groei*, Uitgeverij Het Spectrum, Utrecht.

O'Riordan, T. (1976) *Environmentalism*, Pion, London.

Pigou, A.C. (1920) *The Economics of Welfare*, Macmillan, London.

Prigogine, I., I. Stengers (1990) *Orde uit chaos; De nieuwe dialoog tussen de mens en de natuur*, Uitgeverij Bert Bakker, Amsterdam.

RARO (Raad van advies voor de ruimtelijke ordening) (1994) Advies over het Tweede Nationale Milieubeleidsplan, nr. 166, SDU Uitgevers, Den Haag.

TK (Tweede Kamer) (1993) Nationaal Milieubeleidsplan 2; Milieu als maatstaf, Vergaderjaar 1993-1994, 23560, nrs. 1-2, Den Haag.

TK (Tweede Kamer) (1996) Decentralisatie, brief van de minister en staatssecretaris van Volksgezondheid, Ruimtelijke Ordening en Milieubeheer, Vergaderjaar 1995-1996, 22236, nr. 36, Den Haag.

VROM (Ministerie van Volkshuisvesting, Ruimtelijke Ordening en Milieubeheer) (1984) Meer dan de Som der Delen; Eerste nota over de planning van het milieubeleid, Tweede Kamer, 1984-1985, 18602, nr. 2, Den Haag.

VROM (Ministerie van Volkshuisvesting, Ruimtelijke Ordening en Milieubeheer) (1993) Vierde Nota over de ruimtelijke ordening extra (Vinex); deel 4: Planologische Kernbeslissing Nationaal Ruimtelijk beleid, Den Haag.

VROM (Ministerie van Volkshuisvesting, Ruimtelijke Ordening en Milieubeheer) (1995) Waar vele willen zijn, is ook een weg; Stad & Milieu Rapportage, Directoraat Generaal Milieubeheer, Den Haag.

VROM (Ministerie van Volkshuisvesting, Ruimtelijke Ordening en Milieubeheer) (2000) Concept Vijfde Nota Ruimtelijke Ordening, RPD, VROM, Den Haag.

World Commission on Environment and Development (1987) *Our Common Future*, Oxford University Press, Oxford.

Chapter 2

Adriaanse, A., R. Jeltes, R. Reiling (1989) Information Requirements of Integrated Environmental Policy Experiences in The Netherlands, *Environmental Management* 13/3, pp. 309-315.

Ashworth, G.J., E. Ennen (1995) Het Groninger Museum: een ruimtelijke inpassing van een ongewenste activiteit? in B. van der Moolen en H. Voogd (eds) *Niet in mijn achtertuin, maar waar dan? Het Nimby-verschijnsel in ruimtelijke planning*, Samsom H.D. Tjeenk Willink, Alphen aan den Rijn, pp. 78-90.

Ast, J.A. van, H. Geerlings (1993) *Milieukunde en milieubeleid*, Samsom H.D. Tjeenk Willink, Alphen aan den Rijn.

Bakker, H. (1995) Nimby - reden voor planologisch zelfonderzoek, in B. van der Moolen en H. Voogd (eds) *Niet in mijn achtertuin, maar waar dan? Het Nimby-verschijnsel in ruimtelijke planning*, Samsom H.D. Tjeenk Willink, Alphen aan den Rijn, pp. 182-192.

Barrow, C.J. (1995) *Developing the Environment; Problems and Management*, Longman Scientific & Technical, Harlow, UK.

Bartelds, H.J., G. de Roo (1995) *Dilemma's van de compacte stad; uitdagingen voor het beleid*, VUGA, Den Haag.

Beatley, T. (1995) Planning and Sustainability: The Elements of a New (Improved?) Paradigm, *Journal of Planning Literature*, Vol. 9, No. 4., pp. 383-395.

Beatley, T. (1995b) The Many Meanings of Sustainability: Introduction to a Special Issue of JPL, *Journal of Planning Literature*, Vol. 9, No. 4., pp. 339-342.

Berg, G.P. van den (1994) *Hoogspanningslijnen gevaarlijk? De resultaten van bevolkingsonderzoeken*, Natuurkundewinkel, Rijksuniversiteit Groningen, Groningen.

Bergh, J. van den, B. Doedens, S. Frijns, M. Groot, M. Hoogbergen, F. Krimp en M. Opmeer (1994) *Leefbaarheid in de compacte stad; Onderzoek naar het indiceren van leefbaarheid*, Universitaire Beroepsopleiding Milieukunde, Instituut voor Milieuvraagstukken, Vrije Universiteit Amsterdam, Amsterdam.

Blowers, A. (1990) Narrowing the options: political conflicts and locational decision making, in ASVS-congrescommissie (eds) *Milieu en ruimte: verslag van het 13e ASVS-lustrumcongres*, ASVS-Publikatiereeks, nr. 11, Amsterdamse Studievereniging voor Sociaal-geografen, Amsterdam, pp. 94-102.

Borst, H., G. de Roo, H. Voogd, H. van der Werf (1995a) *Milieuzones in Beweging; eisen, wensen, consequenties en alternatieven*, Samsom H.D. Tjeenk Willink, Alphen aan den Rijn.

Borst, H., G. de Roo, H. Voogd, H. van der Werf (1995b) *De planologische konsekwenties van Integrale Milieuzonering; Analyse van knelpunten en oplossingsrichtingen op basis van de IMZ-proefprojecten*, Deelonderzoek A, Faculteit der Ruimtelijke Wetenschappen, Rijksuniversiteit Groningen, Groningen.

Bouwer, K., P. Leroy (eds) (1995) *Milieu en ruimte; Analyse en beleid*, Boom, Meppel.

Bouwer, K., J.C.M. Klaver (1987) *Milieuproblemen in geografische perspectief; Een geografisch overzicht van de milieuproblematiek veroorzaakt door maatschappelijke activiteiten in Nederland*, Van Gorcum, Assen/Maastricht.

Breheny, M. (1992) Towards Sustainable Urban Development, in A.M. Mannion, S. Bowlby (eds), *Environmental Issues in the 1990s*, John Wiley & Sons, Chichester, UK, pp. 277-290.

Brimblecome, P., F. Nicholas (1995) Urban air pollution and its consequences, in T. O'Riordan (ed.) *Environmental Science for Environmental Management*, Longman Scientific & Technical, Harlow, UK, pp. 283-295.

Cammen, H. van der, L.A. de Klerk (1986) *Ruimtelijke ordening; Van plannen komen plannen*, Aula, Uitgeverij Het Spectrum, Utrecht/Amsterdam.

Camstra, R., J. van der Craats, W. Reedijk, B. Timmermans (1996) *Verder dan de voordeur; Woningcorporaties en de leefbaarheid van wijken in Nederland*, Afdeling Onderzoek en Ontwikkeling, Nationale Woningraad, Almere.

Carson, R. (1962) *Silent Spring*, Houghton Mifflin, Boston.

Cavalini, P.M. (1992) *It's an Ill Wind that Brings no Good: Studies on odour annoyance and the dispersion of odorant concentrations from industries*, University Press Groningen, Groningen.

Copius Peereboom, J.W., L. Rijenders (1989) *Hoe gevaarlijk zijn milieugevaarlijke stoffen*? Boom, Meppel.

Cutter, S.L. (1993) *Living with Risk; The geography of technological hazards*, Edward Arnold, London.

Dankelman, I., P. Nijhoff, J. Westermann (1981) *Bewaar de aarde: in het perspectief van de World Conservation Strategy*, Meulenhoff Informatief, Amsterdam.

Ehrlich, P., A. Ehrlich (1969) *The Population Bomb*, Ballantine, New York.

Ellis, D. (1989) *Environments at Risk; Case Histories of Impact Assessment*, Springer-Verlag, Berlijn, Heidelberg.

Ettema, J.H. (1992) Uitgangspunten voor normstelling, in W. Passchier-Vermeer (eds), *Geluidoverlast en gezondheid, deel B van Lawaaibeheersing*, Handboek voor Milieubeheer, Samsom H.D. Tjeenk Willink, Alphen aan den Rijn, pp. B6000-1-B6000-9

Gemeente Amsterdam (1995) Milieuverkenning Amsterdam, Milieudienst, Amsterdam.

Gezondheidsraad (1994) Geluid en gezondheid, publikatienummer 1994/15, Commissie Geluid en gezondheid, Den Haag.

Haggett, P. (1979) *Geography: A modern synthesis*, Harper & Row, New York, London.

Hardin, G. (1968) The Tragedy of the Commons, in *Science* 162, p. 1243-1248.

Healey, P., T. Shaw (1993) Planners, Plans and Sustainable Development, *Regional Studies*, Vol. 27.8, pp. 769-776.

Hoeflaak, H., H.A.P. Zinger (1992) Externe integratie in de ruimtelijke ordening, *Stedebouw en Volkshuisvesting*, nr. 6, pp. 10-15.

Hollander, A.E.M. de (1993) Algemene inleiding: fysieke omgevingsfactoren en gezondheid, in D. Ruwaard en P.G.N. Kramers (eds), *Volksgezondheid Toekomst Verkenning; De gezondheidstoestand van de Nederlandse bevolking in de periode 1950-2010*, RIVM, Bilthoven, pp. 600-608.

Hough, M. (1989) *City form and Natural Process; Towards a new urban vernacular*, Routledge, London.

Humblet, A.G.M., G. de Roo (eds) (1995) *Afstemming door inzicht; een analyse van gebiedsgerichte milieubeoordelingsmethoden ten behoeve van planologische keuzes*, Geo Pers, Groningen.

Ike, P. (1998) The spatial impact of building in concrete versus building in wood, in B. van der Moolen, A.F. Richardson and H. Voogd (eds) *Mineral Planning in a European Conext: Demand and Suply, Environment and Sustainability*, Geo Press, Groningen, pp. 269-278.

IUCN (International Union for Conservation of Nature and Natural Resources) (1980) *World Conservation Strategy: Living resource conservation for sustainable development*, IUCN, Gland.

Johnston, R.J. (1989) *Environmental Problems; Nature, Economy and State*, Belhaven Press, London/New York.

Kaufmann F.X., G. Majone, V. Ostrom (eds) (1985) *Guidance, Control and Evaluation in the Public Sector*, De Gruyter, Berlin/New York.

Koning, M.E.L. de (1994) *In dienst van het milieu; Enkele memoires van oud-directeur-generaal Milieubeheer prof. ir. W.C. Reij*, Samsom H.D. Tjeenk Willink, Alphen aan den Rijn.
Kooiman, J. (1996) Stapsgewijs omgaan met politiek-maatschappelijke problemen, in P. Nijkamp, W. Begeer en J. Berting (eds), *Denken over complexe besluitvorming: een panorama*, Sdu Uitgevers, Den Haag, pp. 31-48.
Kuijper, C.J. (1993) *Bedrijven en milieuzonering*, Vereniging van Nederlandse Gemeenten, Den Haag.
Linders, B.E.M. (1995) Weerstand rond de uitbreiding van Schiphol, in B. van der Moolen en H. Voogd (eds), *Niet in mijn achtertuin, maar waar dan? Het Nimby-verschijnsel in ruimtelijke planning*, Samsom H.D. Tjeenk Willink, Alphen aan den Rijn, pp. 122-128.
LNV (Ministerie van Landbouw, Natuurbeheer en Visserij) (1990) Natuurbeleidsplan, TK 1989-1990, 21149, nrs. 2-3, Den Haag.
Lugt, F. de (2000) *De ramp van Enschede; zaterdag 13 mei 2000*, De Twentse Courant, Tubantia/Oostelijke weekbladpers BV, Enschede.
Malthus, T.R. (1817) *An Essay on the Principles of Population*, Murray, London.
Marshall, A. (1924) *Principles of Economics*, Macmillan, London.
McDonald, G.T. (1996) Planning as Sustainable Development, *Journal of Planning Education and Research*, 15, pp. 225-236.
Meadows, D., D. Meadows, J. Randers, W. Behrens (1972) *Rapport van de Club van Rome, De grenzen aan de groei*, Uitgeverij Het Spectrum, Utrecht.
Midden, C.J.H. (1993) *De perceptie van risico's*, Intreerede, Technische Universiteit Eindhoven, Eindhoven.
Miller, D., G. de Roo (eds) (1997) *Urban Environmental Planning; Policies, instruments and methods in an international perspective*, Avebury, Aldershot, UK.
Mishan, E.J. (1967) *The Cost of Economic Growth*, Staples Press, London.
Mishan, E.J. (1972) *Cost-benefit Analysis, Unwin University Books*, George Allen & Unwin Ltd, London.
Miura, M. (1997) *The Housing Pattern in the Residential Area in Japanese Cities and Wind Flow on Prevailing Wind Direction*, Shibaura Institute of Technology, Department of Architecture and Environmental System, Fukasaku, Omiya-Shi, Japan.
Moolen, B. van der (1995) *Ontgrondingen als een maatschappelijk vraagstuk*, Dissertatie Faculteit der Ruimtelijke Wetenschappen, Rijksuniversiteit Groningen, Groningen.
Moolen, B. van der, A. Richardson, H. Voogd (eds) (1998) *Mineral Planning in a European Context*, Geo Press, Groningen.
Nelissen, N., J. Van der Straaten, L. Klinkers (eds) (1997) *Classics in Environmental Studies*, International Books, Utrecht.
Netherlands National Committee for IUCN/Steering Group World Conservation Strategy (1988) *The Netherlands and the World Ecology: Towards a national conservation strategy in and by the Netherlands, 1988-1990*, Amsterdam.
Niekerk, F. (1995) De ruimtelijke inpassing van lokaal ongewenste activiteiten, in G. de Roo (eds) *Milieuplanning in vierstromenland*, Samsom H.D. Tjeenk Willink, Alphen aan den Rijn, pp. 165-177.
Nijkamp, P. (1996) De enge marges van het beleid en de brede missie van de beleidsanalyse, in P. Nijkamp, W. Begeer en J. Berting (eds), *Denken over complexe besluitvorming: een panorama*, Sdu Uitgevers, Den Haag, pp. 129-146.
Opschoor, J.B., S.W.F. van der Ploeg (1990) Duurzaamheid en kwaliteit: hoofddoelstellingen van milieubeleid, in Commissie Lange Termijn Beleid, *Het Milieu: denkbeelden voor de 21ste eeuw*, Kerckebosch BV, Zeist, pp. 81-127.

Page, G.W. (1997) *Contaminated Sites and Environmental Cleanups; International Approaches to Prevention, Remediation, and Reuse*, Academic Press, San Diego.

Pearce, D., E. Barbier, A. Markandya (1990) *Sustainable Development: Economics and environment in the Third World*, Elgar, Aldershot, UK.

Pigou, A.C. (1920) *The Economics of Welfare*, Macmillan, London.

Pinch, S. (1985) *Cities and Services; The geography of collective consumption*, Routledge & Kegan Paul, London.

Provincie Groningen (1990) Bij nader inzien, Milieubeleidsplan 1991-1994, Groningen.

Pruppers, M.J.M., G.J. Eggink, H. Slaper, L.H. Vaas, H.P. Leenhouts (1993) Straling, in D. Ruwaard en P.G.N. Kramers (eds), *Volksgezondheid Toekomst Verkenning; De gezondheidstoestand van de Nederlandse bevolking in de periode 1950-2010*, RIVM, SDU Uitgeverij, Den Haag.

Ragas, A.M.J., R.S.E.W. Leuven, D.J.W. Schoof (1994) *Milieukwaliteit en normstelling*, Handboeken milieukunde 1, Boom, Meppel.

Reade, E. (1987) *British Town and Country Planning*, Open University Press, Milton Keynes.

RIVM (Rijksinstituut voor Volksgezondheid en Milieuhygiëne) (1988) Zorgen voor Morgen, RIVM en Samsom H.D. Tjeenk Willink, Bilthoven en Alphen aan den Rijn.

RIVM (Rijksinstituut voor Volksgezondheid en Milieuhygiëne) (1998) Leefomgevingsbalans; Voorzet voor vorm en inhoud, RIVM, Bilthoven.

Roo, G. de (eds) (1996) *Milieuplanning in vierstromenland*, Samsom H.D. Tjeenk Willink, Alphen aan den Rijn.

Staatsblad 53 (1990) Besluit Genetisch Gemodificeerde Organismen, Den Haag.

StoWa (1996) Hinderonderzoek en bedrijfseffectentocts bij rioolwaterzuiveringsinrichtingen in Nederland, Utrecht.

Thibodeau, F.R., H.H. Field (eds) (1984) *Sustaining Tomorrow: A strategy for world conservation and development*, Tufts University/University Press of New England, Hanover.

TK (Tweede Kamer) (1988) Vierde nota over de ruimtelijke ordening, Vergaderjaar 1987-1988, 20490, nrs. 1-2, SDU uitgeverij, Den Haag.

TK (Tweede Kamer) (1989) Nationaal Milieubeleidsplan 1989-1993, Kiezen of verliezen, Vergaderjaar 1988-1989, 21137, nrs. 1-2, SDU-uitgeverij, Den Haag.

TK (Tweede Kamer) (1991) Milieukwaliteitsdoelstellingen bodem en water, Vergaderingsjaar 1990-1991, 21990, nrs. 1-2, SDU uitgeverij, Den Haag.

TK (Tweede Kamer) (1993) Nationaal Milieubeleidsplan 2; Milieu als maatstaf, Vergaderjaar 1993-1994, 23560, nrs. 1-2, SDU uitgeverij, Den Haag.

Udo de Haes, H.A. (1991) Milieukunde, begripsbepaling en afbakening, in J.J. Boersema, J.W. Copius Peereboom en W.T. de Groot (eds) *Basisboek milieukunde*, Boom, Meppel, pp. 21-34.

United Nations (1992) Agenda 21; Programme of Action for Sustainable Development, Rio Declaration on Environment and Development, Statement of Forest Principles, The final text of agreements negotiated by Governments at the United Nations Conference on Environment and Development (UNCED), 3-14 June 1992, Rio de Janeiro, Brazil, United Nations Publication, New York.

Velze, K. van, R.J.M. Maas (1991) Lokale milieuproblemen, in Rijksinstituut voor volksgezondheid en milieuhygiëne, Nationale Milieuverkenning 1990-2010, nr. 2, Samsom H.D. Tjeenk Willink, Alphen aan den Rijn, pp. 397-427.

Vlek, C.A.J. (1990) Beslissingen over risico-acceptatie, Publikatienummer A90/10, Gezondheidsraad, Den Haag.

VNG (1986) Bedrijven en milieuzonering, Groene reeks nr. 80, VNG Uitgeverij, Den Haag.

Voogd, H. (1985) Prescriptive analysis in planning, in *Environment and Planning B, Planning and Design*, 12, pp. 303-312.

Voogd, H. (1987) Ruimtelijke kwaliteit; ook voor toekomstige generaties, in *Stedebouw en Volkshuisvesting*, juni 1987, pp. 206-211.

Voogd, H. (1995) *Methodologie van ruimtelijke planning*, Coutinho, Bussum.

Voogd, H. (1996) *Facetten van de planologie*, Samsom H.D. Tjeenk Willink, Alphen aan den Rijn.

VROM (Ministerie van Volkshuisvesting, Ruimtelijke Ordening en Milieubeheer) (1984a) Meer dan de Som der Delen; Eerste nota over de planning van het milieubeleid, Tweede Kamer, 1983-1984, 18292, nr. 2, Den Haag.

VROM (Ministerie van Volkshuisvesting, Ruimtelijke Ordening en Milieubeheer) (1984b) Indicatief Meerjaren Programma Milieubeheer 1985-1989, Tweede Kamer, 1984-1985, 18602, nr. 2, Den Haag.

VROM (Ministerie van Volkshuisvesting, Ruimtelijke Ordening en Milieubeheer) (1989) Project KWS2000; Bestrijdingsstrategie voor de emissies van vluchtige organische stoffen, Projectgroep Koolwaterstoffen 2000, Directie Lucht, DGM, Den Haag.

VROM (Ministerie van Volkshuisvesting, Ruimtelijke Ordening en Milieubeheer) (1990) Ministriële handreiking voor een voorlopige systematiek voor de integrale milieuzonering, Integrale milieuzonering deel 6, Directie Geluid, DGM, Den Haag.

VROM (Ministerie van Volkshuisvesting, Ruimtelijke Ordening en Milieubeheer) (1994) Thema-document verstoring: naar een betere milieukwaliteit ten behoeve van een hoogwaardige leefomgeving, C.C.M. Gribling, J.A. Verspoor, Publikatiereeks verstoring, nr. 7, Directoraat-Generaal Milieubeheer, Den Haag.

VROM (Ministerie van Volkshuisvesting, Ruimtelijke Ordening en Milieubeheer) (1996) Thuis: op weg naar een integrale aanpak van het leefomgevingsbeleid, Interne VROM-notitie, Den Haag.

VROM (Ministerie van Volkshuisvesting, Ruimtelijke Ordening en Milieubeheer) (1998) Volkshuisvesting in cijfers, Directie Bestuursdienst, DGHV, Den Haag.

VROM (Ministerie van Volkshuisvesting, Ruimtelijke Ordening en Milieubeheer), minsterie van Economische Zaken, ministerie van Landbouw, Natuurbeheer en Visserij, ministerie van Verkeer en Waterstaat, ministerie van Financiën, ministerie van Buitenlandse Zaken (1998) Nationaal Milieubeleidsplan 3, VROM, Den Haag.

Wal, L. van der, P.P. Witsen (1995) De grenzen van de compacte stad; Duurzame eisen kunnen meer ruimtebeslag inhouden, *ROM Magazine*, nr. 3, pp. 3-7.

Wanink, J.H. (1998) *The pelagic cyprinid Rastrineobola argentea as a crucial link in the disrupted ecosystem of Lake Victoria: Dwarfs and Giants*, Ponsen & Looijen, Wageningen.

Werf, H. van der, H. Borst, G. de Roo, H. Voogd (1995) *Ruimte voor zoneren; Een onderzoek naar de planologische konsekwenties van Integrale Milieuzonering*, Deelonderzoek B, Faculteit der Ruimtelijke Wetenschappen, Rijksuniversiteit Groningen, Groningen.

Winsemius, P. (1986) *Gast in eigen huis; Beschouwingen over milieumanagement*, Samsom H.D. Tjeenk Willink, Alphen aan den Rijn.

World Commission on Environment and Development (1987) *Our Common Future*, Oxford University Press, Oxford.

Zeedijk, H. (1995) *Stof in de Houthavens, Centrum voor Milieutechnologie*, Technische Universiteit Eindhoven, Eindhoven.

Zon, H. van (1991) Milieugeschiedenis, in J.J. Boersema, J.W. Copius Peereboom en W.T. de Groot (eds) *Basisboek Milieukunde*, Boom, Meppel.

Chapter 3

Akkerman S.S. (1990) Tellen van woningen binnen zones 2. Handmatig (incl. Nieuwbouw), Integrale Milieuzonering, nr. 13, Directoraat Generaal Milieubeheer, Ministerie van VROM, Leidschendam.

Arts, E.J.M.M. (1992) Milieu-effectrapportage in de ruimtelijke ordening: waarom geen MER voor Vinex?, *Planologische Diskussiebijdragen 1992*, Delftse Uitgeverij Mij, Delft, pp. 245-254.

Ashworth, G.J., H. Voogd (1990) *Selling the City: Marketing Approaches in Public Sector Planning*, Belhaven, London/New York.

Bakker, H. (eds) (1992) Milieu en RO in conflict bij Dordtse kantorenbouw, *ROM Magazine* 10/1992, p. 46.

Bakker, H. (1997) Stad en milieu: milieubelangen lopen elkaar voor de voeten, *ROM Magazine*, nr. 4, pp. 4-8.

Bartelds, H.J., G. de Roo (1995) *Dilemma's van de compacte stad: uitdagingen voor het beleid*, VUGA Uitgeverij B.V., Den Haag.

Bartels, J.H.M., C.W.M. van Swieten (1990) Tellen van woningen binnen zones 1, Digitaal, Integrale Milieuzonering, nr. 12, Directoraat Generaal Milieubeheer, Ministerie van VROM, Leidschendam.

Beatley, T. (1995) Planning and Sustainability: The Elements of a New (Improved?) Paradigm, Comment, *Journal of Planning Literature*, Vol. 9, No. 4, pp. 383-395.

Beer, J. de (1997/45) Bevolkingsprognose 1996: minder bevolkingsgroei, meer vergrijzing, januari, *Sector Bevolking*, Voorburg/Heerlen, pp.6-12.

Beer, J. de, H. Roodenburg (1997/45) Drie scenario's van de bevolking, huishoudens, opleiding en arbeidsaanbod voor de komende 25 jaar, Maandstatistiek van de bevolking, Centraal Bureau voor de Statistiek, februari, Sector Bevolking, Voorburg/Heerlen, pp. 6-10.

Bever (1997) Beleidsvernieuwing Bodemsanering, Scope-document: voorstellen voor het bestuur, VROM, IPO, VNG, Den Haag.

Blanken, W. (1997) *Ruimte voor milieu; milieu/ruimte-conflicten in de stad; een analyse*, Faculteit der Ruimtelijke Wetenschappen, Groningen.

Borchert, J.G., G.J.J. Egbers, M. de Smidt (1983) *Ruimtelijk beleid van Nederland; Sociaal-geografische beschouwingen over regionale ontwikkeling en ruimtelijke ordening*, De Wereld in Stukken, Unieboek B.V., Bussum.

Borchert, J.G. (1983) De Randstad in het ruimtelijk beleid, in Borchert, J.G., G.J.J. Egbers, M. de Smidt, *Ruimtelijk beleid van Nederland; Sociaal-geografische beschouwingen over regionale ontwikkeling en ruimtelijke ordening*, De Wereld in Stukken, Unieboek B.V., Bussum.

Borst, H., G. de Roo, H. Voogd, H. van der Werf (1995) *Milieuzones in Beweging; Eisen, wensen, consequenties en alternatieven*, Samsom H.D. Tjeenk Willink, Alphen aan den Rijn.

Breheny, M. (1992) Towards Sustainable Urban Development, in A.M. Mannion, S. Bowlby, *Environmental Issues in the 1990s*, John Wiley & Sons, Chichester, UK, pp. 277-290.

Breheny, M., R. Rookwood (1993) Planning the sustainable city region, in A. Blowers (ed.) *Planning for a Sustainable Environment*, Earthscan, London, pp. 150-189.

Breheny, M. (1996) Centrists, Decentrists and Compromisers: Views on the Future of Urban Form, in M. Jenks, E. Burton, K. Williams, *The Compact City: A Sustainable Form?* E&FN Spon, London, pp. 13-35.

Brink, A.H. (1996) Efficiënt gebruik van ruimte op bestaande bedrijventerreinen; planologische regels geen beletsel voor doelmatig benutten, *ROM Magazine*, nr. 9, pp. 30-32.

Brouwer, J., P. Bijvoet, M. de Hoog (1997) Leve de stad!; Ontwerpverkenning binnen de stad, *ROM Magazine*, nr. 1/2, pp. 16-19.
Brussaard, W., A.R. Edwards (1988) De Vierde nota juridisch en bestuurlijk gezien, in Permanente Contactgroep Ruimtelijke Organisatie, *Commentaren op de Vierde nota over de ruimtelijke ordening*, Landbouwuniversiteit Wageningen, Wageningen, pp. 16-27.
Bueren, E. van (1998) Contouren in het restrictief beleid; verwachtingen van de bijdrage van een instrument aan ruimtelijke kwaliteit, *Planologische Discissiebijdragen 1998*, Thema 'Plannen met water', Stichting Planologische Discussiedagen, Delft, pp. 299-308.
Buijs, S.C. (1983) De compacte stad, in *Planologische Diskussiebijdragen 1983*, deel 1, Delftse Uitgevers Maatschappij b.v., Delft, pp. 133-142.
Camstra, R. (1995) Een mobiel partnerschap: mobiliteitskeuzen binnen geëmancipeerde huishoudens, Planologisch en Demografisch Instituut, Universiteit van Amsterdam, Amsterdam.
CBS (Centraal Bureau voor de Statistiek) (1996/44) Maandstatistiek van de bevolking, januari, Afdeling Bevolging, Voorburg/Heerlen.
CBS (Centraal Bureau voor de Statistiek) (1997/45) Maandstatistiek van de bevolking, februari, Afdeling Bevolking, Voorburg/Heerlen.
CEC (Commissie van de Europese Gemeenschappen) (1990) Groenboek over het Stadsmilieu, EU, Brussels.
Clercq, F. le, J.J.D. Hoogendoorn (1983) Werken aan de kompakte stad, in *Planologische Diskussiebijdragen 1983*, deel 1, Delftse Uitgevers Maatschappij b.v., Delft, pp. 155-166.
Dantzig, G., T. Saaty (1973) *Compact City: A Plan for a Liveable Urban Environment*, Freeman, San Francisco.
Elkin, T., D. McLaren, M. Hillman (1991) *Reviving the City: Towards Sustainable Urban Development*, Friends of the Earth, London.
Engelsdorp Gastelaars, R. van (1996) Steden in ontwikkeling; Stedelijke problemen en ruimtelijke verhoudingen binnen een toekomstige Randstad, *Stedebouw en Ruimtelijke Ordening*, nr. 5, pp. 19-23.
Engwicht, D. (1992) *Towards an Eco-City: Calming the Traffic*, Envirobook, Sydney.
EZ (Ministerie van Economische Zaken) (1994) Ruimte voor Economische Activiteit; Vestigingslokaties in de toekomst, een confrontatie van vraag en aanbod, Den Haag.
EZ (Ministerie van Economische Zaken) en Heidemij (Advies BV) (1996) Nieuwe kansen voor bestaande terreinen, een onderzoek naar de problemen en oplossingen voor verouderde bedrijventerreinen in Nederland, EZ, Den Haag.
Fikken, W., V. van Unen (1995) Naar een compactere invulling van Vinex-locaties; Mogelijkheden om in grotere dichtheid te verstedelijken, *ROM Magazine*, nr. 3, pp. 8-11.
Friedmann, J., J. Miller (1965) The Urban Field, *Journal of the American Institute of Planners*.
Frieling, D.H. (1995) Geen Stedenring Centraal Nederland maar een Hollandse Metropool, *Stedebouw en Volkshuisvesting*, nr. 5/6, pp. 6-12.
Geleuken, B. van, M.N. Boeve, C. Verdaas (1997) Ruimte, tijd en de betrokkenheid bij de Stad & Milieu-experimenten; Hoe milieuvriendelijk is het compacte-stadconcept, *ROM Magazine*, nr. 4, pp. 16-17.
Gemeente Apeldoorn (1996) Kanaalzone Stad & Milieu, Dienst Milieuhygiëne, Dienst Ruimtelijke Ordening en Volkshuisvesting en Stafeenheid Milieu-Educatie en -Voorlichting, Apeldoorn.
Gemeente Apeldoorn (1998) Raamnota Kanaalzone/Stad & Milieu, Apeldoorn.

Gemeente Arnhem (incl. Stichting Volkshuisvesting) (1997) Ontwikkelingsplan Malburgen, Concept, Arnhem.

Gemeente Delft (1998) Communicatieplan Stad & Milieu Zuidpoort, Bijlage 1; Projectbeschrijving Zuidpoort, Concept, Afdeling Milieu, Delft.

Gemeente Den Haag (1992) Stedebouwkundig ontwikkelingsplan Scheveningen haven, Dienst Ruimtelijke en Economische Ontwikkeling, Den Haag.

Gemeente Groningen (1997) CiBoGa; Concept-stedenbouwkundig plan en plankaarten, Samenvatting, Dienst RO/EZ, Groningen.

Gemeente Leiden (1998) Definitieve aanmelding voor Stad & Milieu van het project 'ontwikkeling Van Gend & Loosterrein e.o.', Leiden.

Gemeente Smallingerland (1997) A-voorstel, Ontwikkelingsvisie Drachtstervaart, Nr. 4, A. Muis, Sector Ontwikkeling, Beheer en Milieu, Drachten.

Gemeente Utrecht (1997) Verslag van inspraakavond over stedebouwkundige visie herontwikkeling Kromhout, Wijkbureau Oost, Gemeente Utrecht, Utrecht.

Gijsberts, P. (1995) Gebiedsgericht milieubeleid, in K. Bouwer en P. Leroy, *Milieu en Ruimte; Analyse en beleid*, Boom, Meppel, pp. 164-184.

Gordijn, H., H. Heida, H. den Otter (1983) *Primos, prognose-, informatie- en monitoringsysteem voor het volkshuisvestingsbeleid*, Planologisch studiecentrum TNO, voor het Directoraat-Generaal van de Volkshuisvesting, Ministerie van VROM, Den Haag.

Graaf, B. de, J. van den Heuvel, S.C. Mohr, D.A. Reitsma (1994) *Ruimte voor wonen 1995-2005*, Ministerie van Economische Zaken, Den Haag.

Hall, D. (1991) Altogether misguided and dangerous - a review of Newman and Kenworthy, *Town and Country Planning*, 60 (11/12), pp. 350-351.

Hall, D., M. Hebbert, H. Lusser (1993) The Planning Background, in A. Blowers (ed.) *Planning for a Sustainable Environment*, Earthscan, London, pp. 19-35.

Hauwert, P.C.M., R.W. Keulen (1990) Inventarisatie omvang zoneerbare milieubelasting met het oog op integrale milieuzonering; de situatie voor geheel Nederland, Integrale Milieuzonering, nr. 7. Ministerie van VROM, Directoraat Generaal Milieubeheer, Leidschendam.

Heide, H. ter (1992) Diagonal Planning: Potentials and Problems, *Planning Theory*, nrs. 7-8, pp. 116-134.

Heidemij (Advies BV) (1996) Herstructurering en efficiënt ruimtegebruik, Rijksplanologische Dienst, Den Haag.

Hemel, Z. (1996) Gevraagd: verstedelijkingsbeleid op een hoger niveau; Geen Randstad maar Stedenring, *Stedebouw en Ruimtelijke Ordening*, nr. 5, pp. 13-18.

Hofstra, H. (1996) Integrale aanpak bodemsanering, Milieudienst Groningen en Faculteit der Ruimtelijke Wetenschappen, Groningen.

Hollander, B. den, H. Kruythoff, R. Teule (1996) Woningbouw op Vinex-locaties: effect op het woon-werkverkeer in de Randstad, *Stedelijke en regionale verkenningen*, nr. 9, OTB, Delftse Universitaire Pers, Delft

Hoogland, J.S., F.J. Kolvoort (1993) Centraal Stadsgebied en IMZ: bestemmen & saneren, in G. de Roo (eds) *Kwaliteit van norm en zone; Planologische consequenties van (integrale) milieuzonering*, Geo Pers, Groningen.

Hooimeijer, P. (1989) Woningbehoefteprognoses, in F.M. Dieleman, R. van Kempen en J. van Weesep (eds) *Met nieuw elan, de herontdekking van het stedelijk wonen*, Volkshuisvesting in theorie en praktijk 23, Delftse Universitaire Pers, Delft, pp. 191-204.

Jacobs, J. (1961) *The Death and Life of Great American Cities*, Vintage Books (1992 edition), New York.

Jenks, M., E. Burton, K. Williams (1996) *The Compact City: A Sustainable Form?*, E&FN Spon, London.
Jong, D. de, M.A. Mentzel (1984) Waardering van woonmilieus in de compacte stad, in *Planologische Diskussiebijdragen 1984*, deel 2, Delftse Uitgevers Maatschappij b.v., Delft, pp. 463-484.
Kassenaar, B. (1997) Verstedelijken of herstedelijken?; Speuren naar middelen om ruimte intensiever te gebruiken, *ROM Magazine*, nr. 1/2, pp. 20-22.
Kempen, B.G.A. (1994) Wonen, wensen & mogelijkheden na 2000, Nationale Woningraad, Almere.
Koekebakker, M.O. (1997) Herstructureren van bestaande bedrijventerreinen, Kwaliteit op Locatie, nr. 9, Ministerie van Volkshuisvesting, Ruimtelijke Ordening en Milieubeheer, Den Haag, pp. 3-9.
Kok, J., F. van Wijk (1986) Haalbaarheid compacte stad: Verkenningen in de planologie, nr. 37, Planologisch en Demografisch Instituut, Universiteit van Amsterdam, Amsterdam.
Kolpron (1996a) *Inventarisatie van Vinex-bouwlocaties*, Kolpron Consultants, Rotterdam.
Kolpron (1996b) *Consequenties van een marktprogramma voor grondproductiekosten, grondopbrengsten en ruimtegebruik van Belstato-plannen*, Kolpron Consultants, Rotterdam.
Kolpron (1996c) *Herijking Belstato 1997; grondproduktiekosten, grondopbrengsten en ruimtegebruik in binnenstedelijk gebied*, Kolpron Consultants, Rotterdam.
Korthals Altes, W. (1995) *De Nederlandse planningdoctrine in het fin de siècle; Voorbereiding en doorwerking van de Viede nota over de ruimtelijke ordening (Extra)*, Van Gorcum, Assen.
Kreileman, M. (1996) *Vinex-locaties (on)bewoonbaar?; Een onderzoek naar milieu/ruimte-conflicten op Vinex-locaties*, Tauw Infra Consult/Faculteit der Ruimtelijke Wetenschappen, Deventer/Groningen.
Kreileman, M., G. de Roo (1996) Aanspraken op Vinex-uitleglocaties kunnen huizenbouw stevig frustreren; Bestaande milieudruk op woningbouwlocaties gewogen, *ROM Magazine*, nr. 12, pp. 20-22.
Kuiper Compagnons (1996) Drachtstervaart - Ontwikkelingsvisie, Rotterdam/Arnhem.
Kuijpers, C.B.F., Th.L.G.M. Aquarius (1998) Meer ruimte voor kwaliteit; Intensivering van het ruimtegebruik in stedelijk gebied, *Stedebouw en Ruimtelijke Ordening*, nr. 1, pp. 28-32.
Kunstler, J.H. (1993) *Geography of Nowhere*, Touchstone, New York.
Laan, D. van der (1992) Nieuwe kantoorlokatie: gelijke bereikbaarheid voor auto en OV, *ROM Magazine*, nr. 10, pp. 16-19.
Linders, B.E.M. (1995) Weerstand rond de uitbreiding van Schiphol, in B. van der Moolen en H. Voogd (eds), *Niet in mijn achtertuin, maar waar dan?; Het Nimby-verschijnsel in de ruimtelijke planning*, Samsom H.D. Tjeenk Willink, Alphen aan den Rijn, pp. 122-128.
Martens, K. (1996) *ABC-locatiebeleid in de praktijk; De rol van gemeenten, provincies en Inspecties Ruimtelijke Ordening in de doorwerking van het ABC-locatiebeleid in strategisch beleid en operationele beslissingen*, Vakgroep Planologie, Katholieke Universiteit Nijmegen, Nijmegen.
McLaren, D. (1992) Compact or dispersed? Dilution is no solution, *Built Environment*, 18 (4), pp. 268-284.
Needham, D.B. (1995) *De gronden van ons bestaan: Ruimtelijk beleid voor een klein dichtbevolkt land*, KUN, Faculteit Beleidswetenschappen, Vakgroep Planologie, Nijmegen.
Newby, H. (1990) Revitalizing the countryside: the opportunities and pitfalls of counter-urban trends, *Journal of the Royal Society of Arts*, CXXXVIII (5409), pp. 630-636.

Newman, P.W.G., J.R. Kenworthy (1989) *Cities and Automobile Dependency. An International Sourcebook*, Gower Technical, Aldershot, UK.

NS Railinfrabeheer (1996) Brief aan Ministerie van VROM, DGM Bestuurszaken betreffende Stad & Milieu, 11 oktober, Utrecht.

Oosterhaven, J., P.H. Pellenbarg (1994) Regionale spreiding van economische activiteiten en bedrijfsmobiliteit, *Maandschrift Economie*, jrg. 58, pp. 388-404.

OPCS (Office of Population Censuses and Surveys) (1992) 1991 Census - Preliminary Report for England and Wales, OPCS, London.

Owens, S. (1986) *Energy Planning and Urban Form*, Pion, London.

Poll, W. Van der (1996) De weersverwachting Vinex, *Stedebouw en Ruimtelijke Ordening*, nr. 5, pp. 32-38.

Projectgroep Raaks (1997) Stedenbouwkundig en Ruimtelijk Functioneel Programma van Eisen Raaks, Haarlem.

RARO (Raad van advies voor de ruimtelijke ordening) (1993) Advies over de Trendbrief Volkshuisvesting 1993, SDU uitgeverij, Den Haag

Riel, P. van, B. Hendriksen (1996) Proefproject Stad & Milieu Gemeente Arnhem; Revitalisering in de wijk Malburgen, Dienst Milieu en Openbare Werken, Gemeente Arnhem, Arnhem.

Rijksbegroting voor het jaar 1979 (1978) TK 15300, Staatsuitgeverij, Den Haag.

Rijksbegroting voor het jaar 1983 (1982) TK 17600, Staatsuitgeverij, Den Haag.

RIVM (Rijksinstituut voor Volksgezondheid en Milieuhygiëne) (1991) Nationale milieuverkenning 2; 1990-2010, Samsom H.D. Tjeenk Willink, Alphen aan den Rijn.

Roseland, M. (1992) *Toward Sustainable Communities, National Roundtable on the Environment and the Economy*, Ottawa.

RPD (Rijksplanologische Dienst) (1985) De compacte stad gewogen, Studierapporten RPD, nr. 27, VROM, Den Haag.

RPD (1996) Bestuurlijke samenwerking is sleutel tot succes; hoogleraar Pellenbarg in gesprek met Menger van de RPD, in VROM (Ministerie van Volkshuisvesting, Ruimtelijke Ordening en Milieubeheer), Op weg naar 2015: Berichten over de uitvoering van de Vierde Nota Ruimtelijke Ordening, Nieuwsbrief nr. 9, RPD, Den Haag, pp. 8-9.

SCMO-TNO (1993) Evaluatie van de bedrijfstypen aan zoneerbare milieubelastingen, bevolkingsaantallen blootgesteld aan niet-verwaarloosbare belastingen, TNO, Delft.

Scoffham, E., B. Vale (1996) How Compact is Sustainable - How Sustainable is Compact? in M. Jenks, E. Burton, K. Williams, *The Compact City: A Sustainable Form?*, E&FN Spon, London, pp. 66-73.

Sherlock, H. (1991) *Cities are Good for Us*, Transport 2000, London.

Smyth, H. (1996) Running the Gauntlet: A Compact City within a Doughnut of Decay, in M. Jenks, E. Burton, K. Williams, *The Compact City: A Sustainable Form?* E&FN Spon, London, pp. 101-113.

Stb (Staatsblad) (1994) 331, Wet bodembescherming.

Sudjic, D. (1992) *Urban Villages*, Urban Villages Group, London.

Swieten, C.W.M. van, R.W. Keulen (1992) *Omvang zoneerbare milieubelastingen 1991*, SCMO-TNO, VROM, IMZ-reeks deel 23, Leidschendam.

Thomas, L., W. Cousins (1996) The Compact City: A Successful, Desirable and Achievable Urban Form?, in M. Jenks, E. Burton, K. Williams, *The Compact City: A Sustainable Form?* E&FN Spon, London, pp. 53-65.

TK (Tweede Kamer) (1989) Regels over experimenten inzake zuinig en doelmatig ruimtegebruik en optimale leefkwaliteit in stedelijk gebied (Experimentenwet Stad en Milieu), Vergaderjaar 1997-1998, 25848, nrs. 1-2, Den Haag.

TK (Tweede Kamer) (1993) Nationaal Milieubeleidsplan 2; Milieu als maatstaf, Vergaderjaar 1993-1994, 23560, nr. 4, Den Haag.

United Nations (1992) Agenda 21; Programme of Action for Sustainable Development, Rio Declaration on Environment and Development, Statement of Forest Principles, The final text of agreements negotiated by Governments at the United Nations Conference on Environment and Development (UNCED), 3-14 June 1992, Rio de Janeiro, Brazil, United Nations Publication, New York.

Veeken, T. van der (1997) Arnhem-Malburgen: van milieuprobleem naar stedebouwkundige uitdaging, *ROM Magazine* 15 (4), pp. 14-15.

Vijver, O. van de (1998) Op naar de complete stad!, *Binnenlands Bestuur*, Nr. 38, 18 februari 1998, pp 32-33.

VRO (Ministerie van Volkshuisvesting en Ruimtelijke Ordening) (1976) Verstedelijkingsnota, beleidsvoornemens over de spreiding, verstedelijking en mobiliteit, deel 2a van de Derde nota over de ruimtelijke ordening, kamerstukken 13754, nrs 1-2, Den Haag, Staatsuitgeverij.

VRO (Ministerie van Volkshuisvesting en Ruimtelijke Ordening) (1979) Verstedelijkingsnota; Tekst van de na parlementaire behandeling vastgestelde pkb, deel 2e van de Derde nota over de ruimtelijke ordening, kamerstukken 13754, Den Haag, Staatsuitgeverij.

VROM (Ministerie van Volkshuisvesting, Ruimtelijke Ordening en Milieubeheer) (1983) Structuurschets voor de stedelijke gebieden, deel a: beleidsvoornemen, TK 18048, nrs 1-2, Staatsuitgeverij, Den Haag.

VROM (Ministerie van Volkshuisvesting, Ruimtelijke Ordening en Milieubeheer) (1988) Trendrapport Woningbehoefte 1988; woningmarktonderzoek, nr. 66, Directoraat-Generaal van de Volkshuisvesting, Den Haag.

VROM (Ministerie van Volkshuisvesting, Ruimtelijke Ordening en Milieubeheer) (1988b) Vierde nota over de ruimtelijke ordening, deel a: beleidsvoornemen, TK 20490, nrs. 1-2, Sdu, Den Haag.

VROM (Ministerie van Volkshuisvesting, Ruimtelijke Ordening en Milieubeheer) (1988c) Vierde nota over de ruimtelijke ordening, deel d: regeringsbeslissing, TK 20490, nrs. 9-10, Sdu, Den Haag.

VROM (Ministerie van Volkshuisvesting, Ruimtelijke Ordening en Milieubeheer) (1990a) Vierde nota over de ruimtelijke ordening Extra; deel 1: ontwerp-planologische kernbeslissing, Op weg naar 2015, TK 21879, nrs. 1-2, SDU-Uitgeverij, Den Haag.

VROM (Ministerie van Volkshuisvesting, Ruimtelijke Ordening en Milieubeheer) (1990b) Nota Volkshuisvesting in de jaren negentig; Van bouwen naar wonen, TK 1988-1989, 20691, nrs. 2-3, SDU uitgeverij, Den Haag.

VROM (Ministerie van Volkshuisvesting, Ruimtelijke Ordening en Milieubeheer) (1990c) Het juiste bedrijf op de juiste plaats: naar een locatiebeleid voor bedrijven en voorzieningen in het belang van bereikbaarheid en milieu, Den Haag.

VROM (Ministerie van Volkshuisvesting, Ruimtelijke Ordening en Milieubeheer) (1992) Trendrapport Volkshuisvesting 1992; vraag en aanbod op de woningmarkt, nr. 43, Directoraat-Generaal van de Volkshuisvesting, Den Haag.

VROM (Ministerie van Volkshuisvesting, Ruimtelijke Ordening en Milieubeheer) (1993) Vierde nota over de ruimtelijke ordening Extra; deel 4: planologische kernbeslissing nationaal ruimtelijk beleid, TK 21.879, nrs. 65-66, SDU-Uitgeverij, Den Haag.

VROM (Ministerie van Volkshuisvesting, Ruimtelijke Ordening en Milieubeheer) (1994) Thema-document Verstoring, Publikatiereeks Verstoring, nr. 7, Den Haag.

VROM (Ministerie van Volkshuisvesting, Ruimtelijke Ordening en Milieubeheer) (1995) Trendrapport Volkshuisvesting 1995; de woningmarkt: een verkenning, Directoraat-Generaal van de Volkshuisvesting, Den Haag.

VROM (Ministerie van Volkshuisvesting, Ruimtelijke Ordening en Milieubeheer) (1996a) Verstedelijking in Nederland 1995-2005; de Vinex-afspraken in beeld, Rijksplanologische Dienst / Interprovinciaal Overleg, Den Haag.

VROM (Ministerie van Volkshuisvesting, Ruimtelijke Ordening en Milieubeheer) (1996b) Op weg naar 2015: Berichten over de uitvoering van de Vierde Nota Ruimtelijke Ordening, Nieuwsbrief nr. 6, RPD, Den Haag.

VROM (Ministerie van Volkshuisvesting, Ruimtelijke Ordening en Milieubeheer) (1996c) Op weg naar 2015: Berichten over de uitvoering van de Vierde Nota Ruimtelijke Ordening, Nieuwsbrief nr. 8, RPD, Den Haag.

VROM (Ministerie van Volkshuisvesting, Ruimtelijke Ordening en Milieubeheer) (1996d) Actualsering Vierde nota ruimtelijke ordening Extra deel 1, Partiële herziening, PKB nationaal ruimtelijk beleid, Den Haag.

VROM (Ministerie van Volkshuisvesting, Ruimtelijke Ordening en Milieubeheer) (1996e) Verslag van de Groene Hart gesprekken, Den Haag.

VROM (Ministerie van Volkshuisvesting, Ruimtelijke Ordening en Milieubeheer) (1996f) Op weg naar 2015: Berichten over de uitvoering van de Vierde Nota Ruimtelijke Ordening, Nieuwsbrief nr. 5, RPD, Den Haag.

VROM (Ministerie van Volkshuisvesting, Ruimtelijke Ordening en Milieubeheer) (1996g) Binnen regels naar kwaliteit, Stad & Milieu, Rapportage deelproject, VROM, Den Haag.

VROM (Ministerie van Volkshuisvesting, Ruimtelijke Ordening en Milieubeheer) (1997) Op weg naar 2015: Berichten over de uitvoering van de Vierde Nota Ruimtelijke Ordening, Nieuwsbrief nr. 9, RPD, Den Haag.

VROM (Ministerie van Volkshuisvesting, Ruimtelijke Ordening en Milieubeheer) (1998) Naar een complete stad; kiezen voor milieu en kwaliteit in het bestaand stedelijk gebied, VROM, Den Haag.

VROM (Ministerie van Volkshuisvesting, Ruimtelijke Ordening en Milieubeheer) (1999) Reactie van de minister van VROM op de Startconferentie Ruimtelijke Ordening, in Startconferentie Ruimtelijke Ordening, 8 februari 1999, VROM, Den Haag, p. 14.

VROM (Ministerie van Volkshuisvesting, Ruimtelijke Ordening en Milieubeheer) (1999b) Primos Pprognose; De toekomstige ontwikkeling van bevolking, huishoudens en woningbehoefte, DGHV, VROM, Den Haag.

VROM (Ministerie van Volkshuisvesting, Ruimtelijke Ordening en Milieubeheer) (2000) Concept Vijfde Nota Ruimtelijke Ordening, RPD, Den Haag.

VROM (Ministerie van Volkshuisvesting, Ruimtelijke Ordening en Milieubeheer) (2001) Vijfde Nota Ruimtelijke Ordening, RPD, Den Haag.

VROM (Ministerie van Volkshuisvesting, Ruimtelijke Ordening en Milieubeheer), ministerie van Economische Zaken, ministerie van Landbouw, Natuurbeheer en Visserij, ministerie van Verkeer en Waterstaat, ministerie van Financiën, ministerie van Buitenlandse Zaken (1998) Nationaal Milieubeleidsplan 3, VROM, Den Haag.

VROM (Ministerie van Volkshuisvesting, Ruimtelijke Ordening en Milieubeheer), Ministerie van Economische Zaken, Ministerie van Landbouw, Natuurbeheer en Visserij, Ministerie van Verkeer en Waterstaat (1999) De ruimte van Nederland; Startnota ruimtelijke ordening 1999, VROM, Sdu Uitgevers, Den Haag.

VROM (Ministerie van Volkshuisvesting, Ruimtelijke Ordening en Milieubeheer), V&W (Ministerie van Verkeer en Waterstaat), EZ (Ministerie van Economische Zaken) (1990) Werkdocument Geleiding van de mobiliteit door locatiebeleid voor bedrijven en voorzieningen, RPD, Den Haag.

VROM (Ministerie van Volkshuisvesting, Ruimtelijke Ordening en Milieubeheer), IPO (Interprovinciaal Overleg), VNG (Vereniging Nederlandse Gemeenten) (1995) Streefbeeld bodem, Den Haag.

VROM-Raad (1998) Stedenland-plus, Advies over 'Nederland 2030 - Verkenning ruimtelijke perspectieven' en de 'Woonverkenningen 2030', Den Haag.

VROM-Raad (1999) Corridors in balans: van ongeplande corridorvorming naar geplande corridorontwikkeling, Den Haag.

V&W (Ministerie van Verkeer & Waterstaat) (2000) Van A naar Beter: Nationaal verkeers- en vervoersplan 2001-2020, Beleidsvoornemen, Deel A, V&W, Den Haag.

Wal, L. van der, P.P. Witsen (1995) De grenzen ven de compacte stad; Duurzame eisen kunnen meer ruimtebeslag inhouden, *ROM Magazine*, nr. 3, pp. 3-7.

Welbank, M. (1996) The Search for a Sustainable Urban Form, in M. Jenks, E. Burton, K. Williams, *The Compact City: A Sustainable Form?*, E&FN Spon, London, pp. 74-82.

World Commission on Environment and Development (1987) *Our Common Future*, Oxford University Press, Oxford.

WRR (Wetenschappelijke Raad voor het Regeringsbeleid) (1998) Ruimtelijke ontwikkelingspolitiek, Sdu Uitgevers, Den Haag.

Zoete, P.R. (1997) *Stedelijke knooppunten: virtueel beleid voor een virtuele werkelijkheid?; Een verkenning van de plaats van indicatief rijksbeleid in de wereld van gemeenten*, Thesis Publishers, Amsterdam.

Zonneveld, W.A.M. (1991) *Conceptvorming in de Ruimtelijke Planning; Patronen en processen*, Planologische Studies 9A, Planologisch en Demografisch Instituut, Universiteit van Amsterdam, Amsterdam.

Zonneveld, W.A.M. (1997) Recensie 'Stedelijke knooppunten', *Stedebouw en Ruimtelijke Ordening*, nr. 2, p. 49.

Zwanikken, T., W. Korthals Altes, B. Needham, A. Faludi (1995) Lessen voor de actualisatie van de Vinex; evaluatie van het Vinex-verstedelijkingsbeleid, *Stedebouw en Volkshuisvesting*, nr. 5/6, pp. 18-26.

Chapter 4

Ackoff, R.L. (1981) Beyond prediction and preparation, *S3 Papers*, University of Pennsylvania, pp. 82-106.

Albers, P., J. Bloem, W. Gooren (1994) Veranderende Sturingsconcepties en Beleidsevaluatie, *Beleidsanalyse*, Vol. 94-3, pp. 5-13.

Alexander, E.R. (1984) After Rationalism, What?, A review of responses to paradigm breakdown, *Journal of the American Planning Association*, nr. 1, pp. 62-69.

Alexander, E.R. (1986) *Approaches to Planning; Introducing current planning theories, concepts, and issues*, Gordon and Breach Science Publishers, New York.

Alexander, E.R., A. Faludi (1990) *Planning Doctrine; Its uses and implications, paper for the conference on planning theory; Prospects for the 1990s*, Vakgroep Planologie en Demografie, Universiteit van Amsterdam, Amsterdam.

Amdam, R. (1994) *The Planning Community; An example of a voluntary communal planning approach to strategic development in small communities in Norway*, Rapport 9404, Moreforsking Volda.

Arnstein, S.R. (1969) A Ladder of Citizen Participation, *Journal of the American Institute of Planners*, Vol. 35, pp. 216-244.

Ashby, W.R. (1956) *An Introduction in Cybernetics*, Chapman & Hall, New York.

Aufenanger, J. (1995) *Filosofie*, Prisma, Uitgeverij Het Spectrum B.V., Utrecht.

Bahlmann, J.P. (1996) De ondraaglijke lichtheid van complexe besluitvormingsprocessen, in P. Nijkamp, W. Begeer en J. Berting (eds), *Denken over complexe besluitvorming: een panorama*, SDU Uitgevers, Den Haag, pp. 87-100.

Bakker, H. (1997) Stad en milieu: milieubelangen lopen elkaar voor de voeten, *ROM Magazine*, pp. 4-6.

Barry, B., R. Hardin (eds) (1982) *Rational Man and Irrational Society?*, Sage Publications, Beverly Hills/London/New Delhi.

Berg, G.J. van den (1981) *Inleiding in de planologie; Voor ieder een plaats onder de zon?* , Samsom H.D. Tjeenk Willink, Alphen aan den Rijn.

Berry, D.E. (1974) The Transfer of Planning Theories to Health Planning Practice, *Policy Sciences*, Vol. 5, pp. 343-361.

Bertalanffy, L. Von (1968) *General System Theory: Foundations, Development and Applications*, Braziller, New York.

Berting, J. (1996) Over rationaliteit en complexiteit, in P. Nijkamp, W. Begeer en J. Berting (eds), *Denken over complexe besluitvorming: een panorama*, SDU Uitgevers, Den Haag, pp. 17-29.

Blanco, H. (1994) *How to Think about Social Problems: American Pragmatism and the Idea of Planning*, Greenwood Press, Westport, Connecticut.

Bohman, J. (1996) *Public Deliberation; Pluralism, Complexity and Demogracy*, The MIT Press, Cambridge, USA.

Borst, H., G. de Roo, H. Voogd, H. van der Werf (1995) *Milieuzones in beweging; Eisen, wensen, consequenties en alternatieven*, Samsom H.D. Tjeenk Willink, Alphen aan den Rijn.

Boulding, K.E. (1956) General system theory - the skeleton of science, *General Systems I*, Vol. 2, pp. 197-208.

Bouwer en Klaver (1995) Milieuzonering, in Bouwer, K. en P. Leroy (eds) *Milieu en Ruimte*, Boom, Meppel, pp. 207-235.

Braudel, F. (1992) *De Middellandse zee; Het landschap en de mens*, deel een, Uitgeverij Contact, Amsterdam/Antwerpen.

Braybrooke, D., C.E. Lindblom (1963) *A Strategy of Decision*, The Free Press, New York.

Bruijn, J.A. de, E.F. ten Heuvelhof (1991) *Sturingsinstrumenten voor de overheid; Over complexe netwerken en een tweede generatie sturingsinstrumenten*, Stenfert Kroese Uitgevers, Leiden.

Bruijn, J.A. de, E.F. ten Heuvelhof (1995) *Netwerkmanagement: Strategieën, instrumenten en normen*, Lemma Uigeverij B.V., Utrecht.

Bryson, J.M., P. Bromiley, Y.S. Jung (1990) Influences of Context and Process on Project Planning Success, *Journal of Planning Education and Research*, Vol. 9(3), pp. 183-195.

Bryson, J.M., A.L. Delbecq (1979) A Contingent Approach to Strategy and Tactics in Project Planning, *Journal of the American Planning Association*, April, pp. 167-179.

Buiks (1981) Institutie/institutionalisering, in Rademaker, L. (eds) *Sociologische grondbegrippen 1: Theorie en analyse*, Aula-boeken 685, Uitgeverij Het Spectrum, Utrecht / Antwerpen.

Cammen, H. van der (1979) *De binnenkant van de planologie*, Dick Coutinho, Muiderberg.

Casseres, J.M. de (1929), Grondslagen der planologie, *De Gids*, nr. 93, pp. 367-394.

Casti, J.L. (1995) *Complexification; Explaining a Paradoxical World Through the Science of Surprise*, Abacus, London.

Chadwick, G. (1971) *A Systems View of Planning: Towards a theory of the urban and regional planning process*, Pergamon Press, Oxford.

Chapin, F.S., E.J. Kaiser (1985) *Urban land use planning*, University of Illinois Press, Urbana.

Checkland, P.B. (1991) From Optimizing to Learning: A Development of Systems Thinking for the 1990s, in R.L. Flood and M.C. Jackson (eds), *Critical Systems Thinking*, John Wiley & Sons, Chichester, UK, pp. 59-75.

Cohen, M.D., J.G. March, J.P. Ohlsen (1972) A garbage can model of organizational choice, *Administrative Science Quarterly*, Vol. 17, pp. 1-25.

Cohen, J., I. Stewart (1994) *The Collapse of Chaos; Discovering Simplicity in a Complex World*, Penguin Books USA, New York.

Coveney, P., R. Highfield (1995) *Frontiers of Complexity; The Search for Order in a Chaotic World*, Ballantine Books, New York.

Dalton, L.C. (1986) Why the rational paradigm persists; the resistance of professional education and practice to alternative forms of planning, *Journal of Planning Education and Research*, Vol. 5, pp. 147-153.

Dekker, A., B. Needham (1989) De handelingsgerichte benadering van de ruimtelijke planning en ordening: een uiteenzetting, in N. Muller en B. Needham, *Ruimtelijk Handelen: Meewerken aan de ruimtelijke ontwikkeling*, Kerckebosch BV, Zeist, pp. 1-12.

Dror, Y. (1968) *Public Policymaking Reexamined*, Chandler, San Fransisco.

Dryzek, J. (1990) *Discursive Democracy; Politics, policy and political science*, Cambridge University Press, Cambridge.

Durkheim, E. (1927) *Les règles de la méthode sociologique*, Alcan, Paris

Elster, J. (1983) *Sour grapes; Studies in the subversion of rationality*, Cambridge University Press, Cambridge.

Emery, F.E., E.L. Trist (1965) The causal texture of organizational environments, in F.E. Emery (ed.) *Systems Thinking*, Penguin Books, Harmondsworth, pp. 241-257.

Etzioni, A. (1967) Mixed-scanning, a 'Third' approach to decision-making, in *Public Administration Review*, Vol. 27, pp. 385-392.

Etzioni, A. (1968) *The Active Society*, Collier-Macmillan, London.

Etzioni, A. (1986) Mixed Scanning Revisited, *Public Administration Review*, pp. 8-14.

Eve, R.A. (1997) Afterword: So where are we now? A final word, in Eve, R.A., S. Horsfall, E.M. Lee (1997) *Chaos, Complexity and Sociology; Myths, Models and Theories*, Sage Publications, Thousand Oaks, USA.

Eve, R.A., S. Horsfall, E.M. Lee (1997) *Chaos, Complexity and Sociology; Myths, Models and Theories*, Sage Publications, Thousand Oaks, USA.

Faludi, A. (1973) *Planning Theory*, Pergamon Press, Oxford.

Faludi, A. (1986) *Critical Rationalism and Planning Methodology*, Pion Limited, London.

Faludi, A. (1987) *A Decision-Centred View of Environmental Planning*, Pergamon Press, Oxford.

Faludi, A., A. van der Valk (1994) *Rule and Order; Dutch planning doctrine in the twentieth century*, Kluwer Academic Publishers, Dordrecht.

Feigenbaum, M. (1978) Quantitative Universality for a Class of Nonlinear Transformations, *Journal of Statistical Physics*, Vol. 19, pp. 25-52.

Flood, R.L. (1989) Six scenarios for the future of systems 'problem solving', *System Practice*, Vol 2, pp. 75-99.

Flood, R.L., M.C. Jackson (eds) (1991) *Critical Systems Thinking*, John Wiley & Sons, Chichester, UK.

Flood, R.L., M.C. Jackson (1991a) *Creative Problem Solving: Total Systems Intervention*, John Wiley & Sons, Chichester, UK.

Forester, J. (1989) *Planning in the Face of Power*, University of California Press, Berkeley, USA.

Forester, J. (1993) *Critical Theory, Public Policy and Planning Practice; Toward a Critical Pragmatism*, State University of New York Press, Albany.

Friedberg, E. (1993) *Le Pouvoir et la Règle; dynamiques de l'action organisée*, Editions du Seuil, Paris.
Friedmann, J. (1971) The future of comprehensive planning: a critique, *Public Administration Review*, nr. 31, pp. 315-326.
Friedmann, J. (1973) *Retracking America: A Theory of Transactive Planning*, Anchor Press/Doubleday, Garden City/New York.
Friedmann, J. (1987) *Planning in the Public Domain From Knowledge to Action*, Princeton University Press, Princeton.
Friedmann, J., C. Weaver (1979) *Territory and Function; The Evolution of Regional Planning*, Arnold, London.
Friend, J.K., N. Jessop (1969) *Local Government and Strategic Choice*, Pergamon, Oxford.
Friend, J.K., J.M. Power, C.J. Yewlett (1974) *Public Planning: The inter-corporate dimension*, Tavistock Publication, London.
Fuenmayor, R. (1991) Between Systems Thinking and Systems Practice, in R.L. Flood, M.C. Jackson (eds), *Critical Systems Thinking*, John Wiley & Sons, Chichester, UK, pp. 227-244.
Galbraith, J. (1973) *Designing Complex Organisations*, Addison Wesley, Reading, Mass.
Gell-Mann, M. (1994) *De quark en de jaquar*, Uitgeverij Contact, Amsterdam.
Giddens, A. (1984) *The Construction of Society; Outline of a theory of structuration*, Policy Press, Cambridge.
Giddens, A. (1984a) *The Constitution of Society*, University of California Press, Berkeley.
Gill, E., E. Lucchesi (1979) Citizen participation in planning, in F.S. So, I. Stollman, F. Beal and D.S. Arnold (eds) *The Practice of Local Government Planning*, International City Management Association, Washington DC. pp. 552-575.
Gillingwater, D. (1975) *Regional Planning and Social Change, A responsive approach*, Saxon House/Lexington Books, Westmead, UK.
Glasbergen, P. (eds) (1989) *Milieubeleid: Theorie en praktijk*, VUGA Uitgeverij B.V., Den Haag.
Gleick, J. (1987) *Chaos; De derde wetenschappelijke revolutie*, Uitgeverij Contact, Amsterdam
Goldberg, M.A. (1975) On the Inefficiency of Being Efficient, *Environment and Planning*, Vol. 7, pp. 921-939.
Goudappel, H.M. (1996) Het metafysisch domein van besluitvormingsprocessen: een verkenning, in P. Nijkamp, W. Begeer en J. Berting, *Denken over complexe besluitvorming: een panorama*, SDU Uitgevers, Den Haag, pp. 59-86.
Habermas, J. (1972) *Knowledge and Human Interests*, Heinemann, London.
Habermas, J. (1984) *The Theory of Communicative Action*, Beacon Press, Boston.
Habermas, J. (1987) *The Philosophical Discourse of Modernity*, Polity Press, Cambridge.
Hanf, K., F.W. Sharpf (eds) (1978) *Interorganizational Policy Making*, Sage, London.
Harper, Th.L., S.M. Stein (1995) Out of the Postmodern Abyss: Preserving the Rationale for Liberal Planning, *Journal of Planning Education and Research*, Vol. 14, pp. 233-244.
Healey, P. (1983) 'Rational method' as a mode of policy information and implementation in land-use policy, *Environment and Planning B: Planning and Design*, Vol. 10, pp. 19-39.
Healey, P. (1992) Planning through debate; The communicative turn in planning theory, *Town Planning Review*, Vol. 63(2), pp. 143-162.
Healey, P. (1996) The communicative turn in planning theory and its implications for spatial strategy formation, *Environment and Planning B, Planning and Design*, Vol. 23, pp. 217-234.
Healey, P. (1997) *Collaborative planning: shaping places in fragmented societies*, Macmillan, Basingstoke, UK.

Hemmens, G.C. (1980) New directions in planning theory, *Journal of the American Planning Association*, Vol. 46(3), pp. 259-260.

Herweijer, M., G.J.A. Hummels, C.W.W. van Lohuizen (1990) *Evaluatie van indicatieve planfiguren: handleiding en begrippen*, Studierapporten Rijksplanologische Dienst, nr. 50, VROM, Den Haag.

Hickling, A. (1974) *Managing Decisions: The strategic choice approach*, Mantec, Rugby.

Hickling, A. (1985) Evaluation is a five-finger exercise, in A. Faludi, H. Voogd, *Evaluation of complex policy problems*, Delfsche Uitgevers Maatschappij B.V., Delft.

Hirschman, A.O., C.E. Lindlblom (1962) Economic development, research and development, policy making: some converging views, in F.E. Emery (ed.) (1969) *Systems Thinking*, Penguin Books, Harmondsworth, UK, pp. 351-371.

Hofstee, W.K.B. (1996) Psychologische factoren bij besluitvormingsprocessen, in P. Nijkamp, W. Begeer en J. Berting (eds), *Denken over complexe besluitvorming: een panorama*, SDU Uitgevers, Den Haag, pp. 49-58.

Holland, C., P. Holdert (1997) Projecten genereren vaak nodeloos verzet, Het ruimte debat, De Volkskrant, 27-08-1997, p. 9.

Hoogerwerf, A. (1989) Beleid, beleidsprocessen en effecten, in A. Hoogerwerf (eds) *Overheidsbeleid*, Samsom H.D. Tjeenk Willink, Alphen aan den Rijn.

Horgan, J. (1996) *The End of Science; Facing the Limits of Knowledge in the Twilight of the Scientific Age*, Broadway Books, New York.

Houten, D.J. van (1974) *Toekomstplanning, planning als veranderingsstrategie in de welvaartsstaat*, Boom, Meppel.

Hufen, J.A.M., A.B. Ringeling (1990) *Beleidsnetwerken; Overheids-, semi-overheids- en particuliere organisaties in wisselwerking*, VUGA, Den Haag.

Innes, J.E. (1995) Planning Theory's Emerging Paradigm: Communicative Action and Interactive Practice, *Journal of Planning Education and Research*, Vol. 14(3), pp. 183-189.

Innes, J.E. (1996) Planning through Consensus Building; A New View of the Comprehensive Planning Ideal, *Journal of the American Planning Association*, no. 62(4), pp. 460-472.

Jackson, M.C. (1985) Systems inquiring competence and organisational analysis, in *Proceedings of the 1985 Meeting of the Society for General Systems Research*, pp. 522-530.

Jackson, M.C. (1987) New directions in management science, in M.C. Jackson, P. Keys, *New Directions in Management Science*, Gower, Aldershot.

Jackson, M.C. (1991) The origins and nature of critical systems thinking, *Syst. Pract. 4*, pp. 131-149.

Jackson, M.C., P. Keys (1991) Towards a System of Systems Methodologies, in R.L. Flood, M.C. Jackson (eds), *Critical Systems Thinking*, John Wiley & Sons, Chichester, UK, pp. 140-158.

Kaiser E.J., D.R. Godschalk, F.S. Chapin (1995) *Urban Land Use Planning*, University of Illinous Press, Chicago,

Kastelein, J. (1996) Inrichting en sturing van complexe besluitvorming, in P. Nijkamp, W. Begeer en J. Berting, *Denken over complexe besluitvorming: een panorama*, SDU Uitgevers, Den Haag, pp. 101-128.

Kauffman, S. (1995) *At Home in the Universe; The Search for Laws of Complexity*, Penguin Books Ltd, London.

Kickert, W.J.M. (1986) *Overheidsplanning, theorieën, technieken en beperkingen*, Van Gorcum, Assen/Maastricht.

Kickert, W.J.M. (eds) (1993) *Veranderingen in management en organisatie bij de rijks-overheid*, Samsom H.D. Tjeenk Willink, Alphen aan den Rijn.

Kleefmann, F. (1984) *Planning als zoekinstrument; Ruimtelijke planning als instrument bij het richtingzoeken*, VUGA, Den Haag.

Klijn, E.H. (1994) *Policy Networks: An Overview, Research programme policy and governance in complex networks*, Department of Public Administration, Erasmus University, Rotterdam.

Klijn, E.H. (1996) *Regels en sturing in netwerken; De invloed van netwerkregels op de herstructurering van naoorlogse wijken*, Eburon, Delft.

Knaap, P. van der (1997) *Lerende overheid, intelligent beleid; De lessen van beleidsevaluatie en beleidsadvisering voor de structuurfondsen van de Europese Unie*, Phaedrus, Den Haag.

Kooiman, J. (1996) Stapsgewijs omgaan met politiek-maatschappelijke problemen, in P. Nijkamp, W. Begeer en J. Berting (eds), *Denken over complexe besluitvorming: een panorama*, SDU Uitgevers, Den Haag, pp. 31-48.

Korsten, A.F.B. (1985) Uitvoeringsgericht ontwerpen van overheidsbeleid, *Bestuur*, nr. 8, pp. 12-19.

Korthals Altes, W. (1995) *De Nederlandse planningdoctrine in het fin de siecle: voorbereiding en doorwerking van de Vierde nota over de ruimtelijke ordening (Extra)*, Van Gorcum, Assen.

Kramer, N.J.T.A., J. de Smit (1991) *Systeemdenken*, Stenfert Kroese, Leiden.

Kreukels, A.M.J. (1980) *Planning en planningproces; Een verkenning van sociaal-wetenschappelijke theorievorming op basis van ruimtelijke planning*, VUGA bv, Den Haag.

Kuijpers, C.B.F. (1996) Integratie en het gebiedsgericht milieubeleid, in G. de Roo (eds), *Milieuplanning in vierstromenland*, Samsom H.D. Tjeenk Willink, Alphen aan den Rijn, pp. 52-67.

Kunstler, J.H. (1993) *The Geography of Nowhere; The rise and decline of America's man-made landscape*, Touchstone, New York.

Lange, M. de (1995) *Besluitvorming rond strategisch ruimtelijk beleid; Verkenning en toepassing van doorwerking als beleidswetenschappelijk begrip*, Thesis Publishers, Amsterdam.

Leeuw, A.C.J. de (1974) *Systeemleer en Organisatiekunde*, Stenfert Kroese, Leiden.

Lewin, R. (1997) *Complexity; Life on the Edge of Chaos*, Phoenix, London.

Lim, G.C. (1986) Toward a Synthesis of Contemporary Planning Theory, *Journal of Planning Education and Research*, Vol. 5(2), pp. 75-85.

Lindblom, C.E. (1959) The Science of Muddling Through, *Public Administrator Review*, nr. 19, pp. 78-88.

Lucy, W. (1988) *Close to Power*, Planners Press, Chicago.

Lui, C. (1996) Holism vs. Particularism, a Lesson from Classical and Quantum Physics, *Journal for General Philosophy of Science*, Vol. 27, pp. 267-279.

Maarse, J.A.M. (1991) Hoe valt de effectiviteit van beleid te verklaren? Deel 1: Empirisch onderzoek, in J.Th.A. Bressers, A. Hoogerwerf (eds), *Beleidsevaluatie, Serie Maatschappijbeelden*, Samsom H.D. Tjeenk Willink, Alphen aan den Rijn, pp. 122-135.

Mainzer, K. (1996) *Thinking in Complexity; The Complex Dynamics of Matter, Mind and Mankind*, Springer-Verlag, Berlin/Heidelberg.

Mandelbrot, B.B. (1982) *The Fractal Geometry of Nature*, Freeman, San Fransisco.

Mannheim, K. (1940) *Man and Society in an Age of Reconstruction*, Routledge & Kegan Paul, London.

Mannheim, K. (1949) *Ideology and Utopia*, Harcourt Brace, New York.

March, J.G., J.P. Olsen (1976) *Ambiguity and Choice in Organisations*, Universitetsforlaget, Bergen.

March, J., H. Simon (1958) *Organizations*, John Wiley & Sons, New York.

Mastop, J.M. (1987) *Besluitvorming, handelen en normeren; Een methodologische studie naar aanleiding van het streekplanwerk*, Planologisch en Demografisch Instituut, Amsterdam.
Mastop, J.M., A. Faludi (1993) Doorwerking van strategisch beleid in dagelijkse beleidsvoering, *Beleidswetenschap*, Vol. 1, pp. 71-90.
McLennan, G. (1995) *Pluralism*, Open University Press, Buckingham, UK.
Meyerson, M., E. Banfield (1955) *Politics, Planning and the Public Interest; The case of public housing in Chicago*, Free Press, New York.
Midgley, G. (1995) What is this thing called critical systems thinking?, in Ellis, K., A. Gregory, B.R. Mears-Young, G. Ragsdell, *Critical Issues in Systems Theory and Practice*, Plenum Press, New York, pp. 61-71.
Miller, D., G. de Roo (eds) (1997) *Urban Environmental Planning: Policies, instruments and methods in an international perspective* , Avebury, Aldershot, UK.
Milroy, M.B. (1991) Into Postmodernism Weightlessness, *Journal of Planning Education and Research*, Vol. 10(3), pp. 181-187.
Nelissen, N.J.M. (1992) Besturen binnen verschuivende grenzen, Inaugurele rede, Kerckebosch, Zeist.
Neufville, J.I. de, S.E. Barton (1987) Myths and the definition of policy problems, *Policy Sciences*, Vol. 20, pp. 181-206.
Nijkamp, P. (1996) De enge marges van het beleid en de brede missie van de beleidsanalyse, in P. Nijkamp, W. Begeer en J. Berting, *Denken over complexe besluitvorming: een panorama*, SDU Uitgevers, Den Haag, pp. 129-146.
Noordzij, G.P. (1977) *Systeem en beleid*, Boom, Meppel.
Ozbekhan, H. (1969) Toward a General Theory of Planning, in E. Jantsch (ed.), *Perspectives of Planning*; Proceedings of the OECD Working Symposium on Long-Range Forecasting and Planning, OECD, Paris.
Parsons, T. (1951) *The Social System*, Routledge & Kegan Paul, London.
Partidário, M., H. Voogd (1997) *An endeavour at integration in environmental analysis and planning*, Paper presented at the Second International Symposium on Urban Planning and Environment, International Urban Planning and Environment Association, University of Groningen, Groningen.
Piaget, J. (1980) *Six Psychological Studies*, The Harvester Press, Brighton.
Pirsig, R.M. (1991) *Lila, een onderzoek naar zeden*, Uitgeverij Bert Bakker, Amsterdam.
Popper, K.R. (1961) *The Poverty of Historicism*, Routledge & Kegan Paul, London.
Prigogine, I. (1996) *Het einde van de zekerheden; Tijd, chaos en de natuurwetten*, Lannoo, Tielt.
Prigogine, I., I. Stengers (1990) *Orde uit chaos; De nieuwe dialoog tussen de mens en de natuur*, Uitgeverij Bert Bakker, Amsterdam.
Ringeling, A.B. (1987) *De voortdurende discussie over het politiebestel*, Gouda Quint, Arnhem.
Rittel, H.W.J., M.M. Webber (1973) Dilemmas in a General Theory of Planning, *Policy Sciences*, nr. 4, pp. 155-169.
Rondinelli, D.A., J. Middleton, A.M. Verspoor (1989) Contingency Planning for Innovative Projects; Designing Education Reforms in Developing Countries, *Journal of the American Planning Association*, Winter, pp. 45-56.
Roo, G. de (1995) Gebiedsgericht milieubeleid in vierstromenland; Strategische keuzen en randvoorwaarden voor het gebiedenbeleid, *ROM Magazine*, nr. 7/8, pp. 16-20.
Roo, G. de (1996) Inleiding, in G. de Roo (eds) *Milieuplanning in vierstromenland*, Samsom H.D. Tjeenk Willink, Alphen aan den Rijn, pp. 11-18.
Rosenau, J. (1990) *Turbulence in World Politics: A theory of change and continuety*, Princeton University Press, Princeton.

Rosenblueth, A., N. Wiener (1945) The role of models in science, *Philosophy of Science*, Vol. 12.

Russell, B. (1995) *Geschiedenis van de westerse filosofie; in verband met politieke en sociale omstandigheden van de oudste tijd tot heden*, Servire Uitgevers bv, Cothen.

Sager, T. (1994) *Communicative Planning Theory*, Avebury, Aldershot.

Sagoff, M. (1988) *The Economy of the Earth: Philosophy, law and the environment*, Cambridge University Press, Cambridge.

Scharpf, F.W. (1978) Interorganizational policy studies: issues, concepts and perspectives, in K. Hanf, F.W. Scharpf (eds) *Interorganizational Policy Making; limits to coordination and central control*, Sage Publications, London, pp. 345-370.

Simon, H.A. (1960) *The New Science of Management Decision*, Harper & Row, New York.

Simon, H.A. (1967) *Models of Man, Social and Rational: Mathematical essays on rational human behavior in a social setting*, New York.

Simon, H.A. (1976) *Administrative Behavior: a Study of Decision-Making Processes in Administrative Organisations*, Free Press, New York.

Smith, H.E. (1963) Toward a clarification of the concept of social institution, in *Sociology and Social Research*, Vol. 48, 2, pp. 197-206.

Snellen, I.Th.M. (1987) *Boeiend en geboeid; ambivalenties en ambities in de bestuurskunde*, Samsom H.D. Tjeenk Willink, Alphen aan den Rijn.

Stacey, R.D. (1993) *Strategic Management and Organisational Dynamics*, Pitman Publishing, London.

Steigenga, (1964) *Moderne planologie*, Aula-boeken, Utrecht.

Störig, H.J. (1985) *Geschiedenis van de filosofie; deel 2*, Aula, Het Spectrum, Utrecht.

Tatenhove, J.P.M. van (1993) *Milieubeleid onder dak?; Beleidsvoeringsprocessen in het Nederlandse milieubeleid in de periode 1970-1990; nader uitgewerkt voor de Gelderse Vallei*, Wageningse sociologische studies, Universiteit van Wageningen, Wageningen.

Teisman, G.R. (1992) *Complexe besluitvorming; een pluricentrisch perspektief op besluitvorming over ruimtelijke investeringen*, VUGA, Den Haag.

Thompson, J.D. (1976) *Organizations in Action*, McGraw Hill, New York.

TK (Tweede Kamer) (1989) Nationaal Milieubeleidsplan 1989-1993, Kiezen of verliezen, Vergaderjaar 1988-1989, 21137, nrs. 1-2, SDU-uitgeverij, Den Haag.

Toffler, A. (1990) Wetenschap en verandering, voorwoord in I. Prigogine en I. Stengers, *Orde uit chaos; De nieuwe dialoog tussen de mens en de natuur*, Uitgeverij Bert Bakker, Amsterdam, pp. 9-24.

Verma, N. (1996) Pragmatic rationality and planning theory, *Journal of Planning Education and Research*, Vol. 16, pp. 5-14.

Vermuri, V. (1978) *Modeling of Complex Systems*, Academic Press, New York.

Vickers, G. (1968) *Value Systems and Social Process*, Tavistock Publications, London.

Voogd, H. (1986) Van denken tot doen, inaugurale rede, Rijksuniversiteit Groningen, Groningen.

Voogd, H. (ed.) (1994) *Issues in Environmental Planning*, Pion Ltd., London.

Voogd, H. (1995a) *Facetten van de Planologie*, Samsom H.D. Tjeenk Willink, Alphen aan den Rijn.

Voogd, H. (1995b) *Methodologie van ruimtelijke planning*, Couthino, Bussum.

Vries, G. de (1985) *De ontwikkeling van wetenschap: een inleiding in de wetenschapsfilosofie*, Wolters-Noordhoff, Groningen.

VROM (Ministerie van Volkshuisvesting, Ruimtelijke Ordening en Milieubeheer) (1995) Waar vele willen zijn is ook een weg, Stad&Milieu-rapportage, Den Haag.

Vroom, C.W. (1981) Organisatie, in Rademaker, L. (eds) *Sociologische grondbegrippen 1: Theorie en analyse*, Aula-boeken 685, Uitgeverij Het Spectrum, Utrecht / Antwerpen.

Waldrop, M.M. (1992) *Complexity: the emerging science at the edge of order and chaos*, Simon & Schuster, New York.

Waring, A. (1996) *Practical Systems Thinking*, International Thomson Business Press, Boston.

Weaver, C., J. Jessop, V. Das (1985) Rationality in the public interest: notes towards a new synthesis, in M. Breheny and A. Hooper (eds), *Rationality in Planning: Critical Essays on the Role of Rationality in Urban and Regional Planning*, Pion, London, pp. 145-165.

Webber, M.M. (1963) Comprehensive planning and social responsibility, *Journal of the American Institute of Planners*, Vol. 29, pp. 232-241.

Weick, K.E. (1969) *The Social Psychology of Organizing*, Addison-Wesley, Reading, UK.

Wells, A. (1970) *Social Institutions*, Heinemann, London.

Wiener, N. (1948) *Cybernetics*, John Wiley & Sons, New York.

Wissink, G.A. (1986) Handelen en ruimte, een beschouwing over de kern van de planologie, *Stedebouw en Volkshuisvesting*, pp. 192-194.

Wissink G.A. (1987) Nieuwe oriëntaties en werkterreinen voor de planologie, *Stedebouw en Volkshuisvesting*, pp. 197-205.

Woltjer, J. (1997) De keerzijde van het draagvlak; Ruimtelijke ordening niet altijd gebaat bij maatschappelijke discussie, *Stedebouw en Ruimtelijke Ordening*, nr. 4, pp. 47-52.

Zonneveld, W.A.M. (1991) *Conceptvorming in de ruimtelijke planning: Patronen en processen, Planologische studies 9A*, Planologisch Demografisch Instituut, Universiteit van Amsterdam, Amsterdam.

Chapter 5

Aart, Y.F., P.P.J. Driessen, P. Glasbergen (1993) Evaluatie van het ROM-gebiedenbeleid; deelstudie Rijnmond; Publikatiereeks gebiedsgericht milieubeleid, nr. 93/2, Ministerie van VROM, Den Haag.

Adviescommissie Geluidhinder door Vliegtuigen (Commissie Kosten) (1967) Geluidhinder door Vliegtuigen, Delft.

AGS (Ministers für Arbeid, Gesundheid und Soziales des Landes Nordrhein-Westfalen) (1974) Abstände zwischen Industrie - bzw. Gewerbegebieten und Wohngebieten in Rahmen der Bauleitplanung, III BI 8804 v. (25 juli), Düsseldorf.

Aiking, H., J. de Boer, V.M. Sol, P.E.M. Lammers, J.F. Feenstra (1990) Haalbaarheidsstudie Milieubelastingsindex, Reeks Integrale Milieuzonering nr. 8, Instituut voor Milieuvraagstukken, Directoraat-Generaal Milieubeheer, Ministerie van VROM, Den Haag.

Anderson, N., E. Hanhardt, I. Pasher (1997) From measurement to measures: landuse and environmental protection in Brooklyn, New York, in D. Miller and G. de Roo (eds) *Urban Environmental Planning*, Ashgate, Aldershot, UK, pp. 41-47.

ANWB (en Koninklijke Nederlandse Toeristenbond) (1989) Advies inzake het Nationaal Milieubeleidsplan, Brief aan de minister van VROM, d.d. 27 September 1989, Den Haag.

Ast, J.A. van, H. Geerlings (1995) *Milieukunde en milieubeleid*, Samsom, Alphen aan den Rijn.

Baaijen, A.J. (1997) Aanbiedingsbrief behorend bij de VROM publikatie 'Onderhandelen langs de zone', Directie Geluid en Verkeer, Directoraar-Generaal Milieubeheer, Den Haag.

Bakker, H. (1989) Stank is een hinderlijk milieuprobleem, maar er wordt aan gewerkt, *ROM Magazine*, nr. 7, pp. 3-6.

Bakker, H. (1994) 'Afstand' is net zo schaars als een primaire grondstof; ruimtelijke ordening en milieu in Amsterdam, *ROM Magazine*, nr. 10, pp. 4-8.

Bakker, H. (1997) Stad en milieu: milieubelangen lopen elkaar voor de voeten, *ROM Magazine*, pp. 4-6.

Bartelds, H.J. (1993) *De bodem beschermd? Gebiedsgerichte bodembescherming in bodembeschermingsgebieden en milieubeschermingsgebieden*, Faculteit der Ruimtelijke Wetenchappen, Rijksuniversiteit Groningen, Groningen.

Bartelds, H.J., G. de Roo (1995) *Dilemma's van de compacte stad; Uitdagingen voor het beleid*, VUGA, Den Haag.

Beerkens, H.J.J.G. (1998) *Gebiedsgericht milieubeleid & het besturingsvraagstuk; Posities en optreden van de overheid*, Faculteit der Ruimtelijke Wetenschappen, Rijksuniversiteit Groningen, Groningen.

Berg, B. van den (1993) *Milieubeleid in de Gemeente; Een praktische handleiding*, Stichting Burgerschapskunde, Nederlands Centrum voor Politieke Vorming, Leiden.

Bever/UPR (2000) Eindrapport Bever/UPR, Den Haag.

BEVER-werkgroep (1997) Scope-document: voorstellen voor het bestuur, Beleidsvernieuwing Bodemsanering, VROM, Den Haag.

BEVER-werkgroep (1998) Uitvoeringsprogramma Bever, VROM, Den Haag.

Biebracher, C.K., G. Nicolis, P. Schuster (1995) *Self-Organization in the Physico-Chemical and Life Sciences*, Rapport EUR 16546, Europese Commissie, Brussels.

Biekart, J.W. (1994) Uitvoeringsproblematiek milieubeleid, Discussienotitie, Stichting Natuur en Milieu, Utrecht.

Bierbooms, P.F.A. (1997) Bodemvervuiling en de verplichting tot schadevergoeding ex art. 75 Wbb, in P.F.A. Bierbooms, G.A. van der Veen en G. Betlem, *Aansprakelijkheden in de Wet bodembescherming*, Serie Aansprakelijkheidsrecht, Deel 4, Gouda Quint bv, Deventer.

Biezeveld, G.A. (1990) Uitspraken uit de discussie naar aanleiding van de 23ste ledenvergadering van de Vereniging voor Milieurecht, in Vereniging voor Milieurecht, Juridische en bestuurlijke consequenties van het Nationaal Milieubeleidsplan, nr. 2, W.E.J. Tjeenk Willink, Zwolle.

Biezeveld, G.A. (1992) Plaats van gebieden in het milieubeheer, in Chr. Backes, G.A. Biezeveld, J.C.M. de Bruijn, J.H. van der Put, H.F.M.W. van Rijswick en A. Wolters-Laansma, *Gebiedsgericht milieubeleid*, Rapport van de werkgroep Gebiedsgericht milieubeleid, Vereniging voor Milieurecht, nr. 1, pp. 1-6.

Blanco, H. (1999) Lessons from an adaption of the Dutch model for Integrated Environmental Zoning (IEZ) in Brooklyn, NYC, in D. Miller and G. de Roo (eds), *Integrating City Planning and Environmental Improvement*, Ashgate, Aldershot, UK, pp. 159-180.

Blanken, W., G. de Roo (1998) *Het project 'Stad & Milieu'*, Stichting Planologische Diskussiedagen, Delft, pp. 281-290.

Boei, P.J. (1993) Integrale milieuzonering op en rond het industrieterrein Arnhem-Noord, in G. de Roo (eds) *Kwaliteit van norm en zone; Planologische consequenties van (integrale) milieuzonering*, Geo Pers, Groningen, pp. 75-81.

Boer, W. de, A. van Bolhuis (1980) Milieunormen ruimtelijk vertaald, *Stedebouw en Volkshuisvesting*, nr. 11, pp. 588-592.

Boer, J. de, V.M. Sol, F.H. Oosterhuis, J.F. Feenstra, H. Verbruggen (1996) *De stadsstolpmethode; een afwegingskader voor de integratie van milieu, economie en ruimtelijke ordening bij stedelijke ontwikkeling*, Instituut voor Milieuvraagstukken, Vrije Universiteit, Amsterdam.

Booij, P.J. (1997) *Ernstige niet-urgente gevallen van bodemverontreiniging; saneren of accepteren*, Oranjewoud en Faculteit der Ruimtelijke Wetenschappen, Heerenveen en Groningen.

Borst, H. (1996) Integrale milieuzonering, in G. de Roo, *Milieuplanning in vierstromenland*, Samsom H.D. Tjeenk Willink, Alphen aan den Rijn, pp. 94-108.

Borst, H., G. de Roo (1993) Integrale milieuzonering en de ontketening van de ruimtelijke ordening, in *Planologische Diskussiebijdragen*, deel 1, Stichting Planologische Diskussiedagen, Delft, pp. 81-90.

Borst, H., G. de Roo, H. Voogd, H. van der Werf (1995) *Milieuzones in Beweging; Eisen, wensen, consequenties en alternatieven*, Samsom H.D. Tjeenk Willink, Alphen aan den Rijn.

Bos, E.C., C.W.L. de Bouter, T. Engelberts, G. Zandsteeg (1980) Naschrift Milieunormen ruimtelijk vertaald, *Stedebouw en Volkshuisvesting*, nr. 11, pp. 592-593.

Bouwer, K. (1996) Begrip van milieu en ruimte, in G. de Roo, *Milieuplanning in vierstromenland*, Samsom H.D. Tjeenk Willink, Alphen aan den Rijn, pp. 39-51.

Bouwer, K., B. van Geleuken (1994) Beleid op schaal - De zin van gebiedsgericht milieubeleid, *Bestuurskunde*, November, nr. 7, pp. 295-304.

Bouwer, K., J. Klaver, M. de Soet (1983) *Nederland stortplaats, Een milieukundige en geografische visie op het afvalprobleem*, Ekologische uitgeverij, Amsterdam.

Bouman, R. (1998) Brief stand van zaken nota MIG aan de leden van de klankbordgroepen MIG, mbg98006088, 20 January 1988, DGM, Directie Geluid en Verkeer, Projectburo MIG, Den Haag.

Braak, C.J. (1984) Knelpuntenonderzoek Wet geluidhinder: interimrapport vooronderzoek nr. 5, Interuniversitaire Interfaculteit Bedrijfskunde, Delft.

Breemen, A.J.G. van (1999) Projectplan Geluidsnota Amsterdam, Milieudienst, Gemeente Amsterdam, Amsterdam.

Brink, W.J. van den, et al. (1985) *Bodemverontreiniging*, Aula Paperback 127, Uitgeverij Het Spectrum, Utrecht.

Brussaard, W, G.H. Addink (1993) *Milieurecht*, W.E.J. Tjeenk Willink, Zwolle.

Buysman, J. (1997) *Provinciale Omgevingsplannen; Een analyse anno 1997*, Hogeschool IJsselland, Grontmij Groep, Deventer, De Bilt.

Carson, R. (1962) *Silent Spring*, Houghton Mifflin, Boston.

Cate, F. ten (1992) Teveel stank, roet en herrie voor woonwijken grenzend aan complexe industrieterreinen, *Binnenlands Bestuur*, 24 January 1992, pp. 16-18.

Cate, F. ten (1993) Consequenties van 'onwrikbare' milieuzones niet meer te accepteren, *Binnenlands Bestuur*, 9-4-1993, pp. 23-25.

CEA (Bureau voor communicatie en advies over energie en milieu B.V.) (1998) Koersdocument sturing bodemsaneringsbeleid, Rotterdam.

Cleij, J. (1997) Aanbiedingsbrief van het rapport 'De Stadsstolp Methode; Een afweginskader voor de integratie van milieu, economie en ruimtelijke ordening bij stedelijke ontwikkeling' door De Boer et al. 1996, 9200027/25, Milieudienst Gemeente Amsterdam, Amsterdam.

Colstee-Wieringa, F. (1988) Maastricht zet milieubelasting van de hele stad in geel, rood en oranje op de kaart; Inventarisatie geluid, luchtvervuiling, risico's, *ROM Magazine*, nr. 8-9, pp. 13-17.

Commissie Evaluatie Wet Geluidhinder (1985) Evaluatie van de werking van de Wet geluidhinder, Eindrapport, Distributiecentrum Overheidspublikaties, Den Haag.

Commissie Rey (1976) Veehouderij en Hinderwet, Landbouwschap, Dan Haag.

Commissie Ringeling (1993) Stappen verder ..., Eindrapport van de adviescommissie Evaluatie Ontwikkeling Gemeentelijk Milieubeleid, Den Haag.

Commoner, B. (1972) *The Closing Circle, Confronting the environmental crisis*, Jonathan Cape, London.

CRMH (Centrale Raad voor de Milieuhygiëne) (1989) Advies in hoofdlijnen over het Nationaal Milieubeleidsplan, nr. 14, Den Haag.
CRMH (Centrale Raad voor de Milieuhygiëne) (1991) Advies Plan van Aanpak Schiphol en omgeving, CRMH-reeks 91/17, Den Haag.
CRMH (Centrale Raad voor de Milieuhygiëne) (1992) Enkele opmerkingen vooruitlopend op het Milieuprogramma 1993-1996 en het Nationaal Milieubeleidsplan 2.
Dongen, J.E.F. van (1983) *De "wet van behoud van ellende" en de geluidhinder*, IMG-TNO, Delft.
Dönszelmann, E. (1993) Nota Stankbeleid, naar minder hinder?, *ROM Magazine*, nr. 3, pp. 23-25.
Driessen, P.P.J. (1996) Het ROM-gebiedenbeleid, in G. de Roo, *Milieuplanning in vierstromenland*, Samsom H.D. Tjeenk Willink, Alphen aan den Rijn, pp. 68-84.
dRO Amsterdam (dienst Ruimtelijke Ordening), Ingenieursbureau Amsterdam, Onderzoeksdienst voor milieu en grondmechanica Amsterdam (1995) Milieu effectrapport IJburg; Tweede fase, Amsterdam.
Drupsteen, T.G. (1990) Plannen en organiseren: Het Nationaal Milieubeleidsplan, in *Vereniging voor Milieurecht, Juridische en bestuurlijke consequenties van het Nationaal Milieubeleidsplan*, nr. 2, W.E.J. Tjeenk Willink, Zwolle, pp. 12-22.
EG (Europese Gemeenschap) (1973) Publicatieblad van de EG nr. C 112, Brussels.
Eikenaar, D., Th. de Muinck, G. de Roo, P.C.D. Vetkamp (2001) Modernisering van geluidbeleid; Ontwikkeling in de beleidspraktijk, Faculteit der Ruimtelijke Wetenschappen, Rijksuniversiteit Groningen, Groningen.
EK (Eerste Kamer) (1998) Regels inzake plannen op het terrein van het verkeer en vervoer (Planwet Verkeer en Vervoer). Gewijzigd voorstel van wet, vergaderjaar 1997-1998, 25337, nr. 258, Den Haag.
Evaluatiecommissie S&M (2000) Werkplan tussenevaluatie Stad & Milieu, Afd. Bestuurszaken, DGM, VROM, Den Haag.
Faludi, A., A. Van der Valk (1994) *Rule and Order; Dutch planning doctrine in the twentieth century*, Kluwer Academic Publishers, Dordrecht.
Flohr, A.P., H.C.J. Meijvis (1993) Milieuzonering in Hengelo, in G. de Roo, *Kwaliteit van norm en zone; Planologische consequenties van (integrale) milieuzonering*, Geo Pers, Groningen.
Friend, J.K., N. Jessop (1969) *Local Government and Strategic Choice*, Pergamon, Oxford.
Geest, H.J.A.M. van (1996) Marktwerking, deregulering en wetgevingskwaliteit: Forensen tussen macht en markt, *Milieu & Recht*, nr. 6, pp. 118-122.
Gemeente Amsterdam (1994) Ontwerp Beleidsnota Ruimtelijke Ordening en Milieu, dienst Ruimtelijke Ordening in samenwerking met de Milieudienst, Amsterdam.
Gemeente Amsterdam (1996) Beleidsnotitie Het Amsterdamse bodemsaneringsbeleid herzien, Gemeenteblad 1996, bijlage A, Amsterdam.
Gemeente Groningen (1992) Evaluatie IMZ-proefproject Noordoostflank, Groningen.
Gemeente Groningen (1997) Als de bodem maar goed genoeg is; nota over actief bodembeheer in de gemeente Groningen, Groningen.
Gemeente Utrecht (1997) Bandbreedte integraal afwegingskader; rapportage over de tweede fase, Utrecht.
Gezondheidsraad (1971) Geluidhinder, Rapport Gezondheidsraad Commissie Geluidhinder en Lawaaibestrijding, Den Haag.
Gezondheidsraad (1994) Geluid en gezondheid, nr. 15, Den Haag.
Gezondheidsraad (1995) Beoordeling van de IVM-milieubelastingsindex, Beraadsgroep Omgevingsfactoren en Gezondheid, publikatienummer 1995/09, Den Haag.
Gijsberts, P. (1995) Gebiedsgericht milieubeleid, in K. Bouwer en P. Leroy, *Milieu en ruimte; Analyse en beleid*, Boom, Meppel, pp. 164-184.

Gijsberts, G., B. van Geleuken (1996) De legitimiteit van gebiedsgericht milieubeleid, *Beleidsanalyse*, nr. 2, pp. 4-10.

Gilhuis, P.C. (1991) Wet algemene bepalingen milieuhygiëne, in W. Brussaard, T.G. Drupsteen, P.C. Gilhuis en N.S.J. Koeman, *Milieurecht*, W.E.J. Tjeenk Willink, Zwolle, pp. 59-103.

Glasbergen, P. (1989) Milieuproblemen als beleidsvraagstuk, in P. Glasbergen (eds) *Milieubeleid; theorie en praktijk*, VUGA, Den Haag, pp. 15-32.

Glasbergen, P. (ed.) (1998) *Co-operative Environmental Governance; Public-Private Agreements as a Policy Strategy*, Kluwer Academic Publishers, Dordrecht.

Glasbergen, P., C. Dieperink (1989) Het Nationaal Milieubeleidsplan, de weg naar duurzaamheid? Over de noodzaak van bestuurskundige consequentie-analyses, *Milieu en recht*, nr. 7-8, pp. 298-307.

Glasbergen, P., P.P.J. Driessen (1993) *Innovatie in het gebiedsgericht beleid; Analyse en beoordeling van het ROM-gebiedenbeleid*, SDU Uitgeverij, Den Haag.

Glasbergen, P., P.P.J. Driessen (1994) New strategies for environmental policy: Regional network management in the Netherlands, in M. Wintle and R. Reeve (eds) *Rhetoric and Reality in Environmental Policy*, Avebury, Aldershot, UK, pp. 25-40.

Grondsma, T. (1984) *Een geschiedenis van de Wet geluidhinder; Een vooronderzoek ten behoeve van de evaluatie van de Wet geluidhinder*, Technische Hogeschool Delft, Delft.

GS (Gedeputeerde Staten van de Provincie Groningen), BW (Burgemeesters en Wethouders van de gemeente Groningen) Groningen (1995) Brief aan de minister van VROM over het Plan van aanpak IMR-suikerindustrie Groningen, 11 January 1995, RVL94.36B, Groningen.

Gun, V. van der, G. de Roo (1994) An integrated environmental approach to land use zoning, in H. Voogd (ed.) *Issues in Environmental Planning*, Pion, London, pp. 58-66.

Hardin, G. (1968) Tragedy of the commons, *Science*, Vol. 162, pp. 1243-1248.

Healey, P. (1997) *Collaborative planning: shaping places in fragmented societies*, Macmillan, Basingstoke, UK.

Heuvelhof, E. ten, K. Termeer (1991) Gebiedsgericht beleid en het bereiken van win-win-situaties, *Bestuurswetenschappen*, nr. 4, pp. 301-315.

Hof, G.J.J. van der (1988) Verantwoordingssysteem voor ALARA-optimalisatie, Directie Stralenbescherming, VROM, Den Haag.

Hofstra, H. (1996) Integrale aanpak bodemsanering, Milieudienst Groningen en Faculteit der Ruimtelijke Wetenschappen, Groningen.

Hofstra, H., G. de Roo (1997) Bodemsanering en ruimtelijke ordening, in Veul, M. (eds) *Leidraad Bodemsanering*, B8 pp. 1-23, SDU Uitgeverij, Den Haag.

Humblet, A.G.M., G. de Roo (eds) (1995) *Afstemming door inzicht; een analyse van gebiedsgerichte milieubeoordelingsmethoden ten behoeve van planologische keuzes*, Geo Pers, Groningen.

Hunfeld, J., F.A.M. Schreiner (1981) Milieunormen in Hinderwet en bestemmingsplannen, *Stedebouw en Volkshuisvesting*, nr. 10, pp. 471-479.

Hutten Mansfeld, A.C.B., A.A. Zijderveld (1982) Een aanzet tot meer systematiek in de milieuzonering, *Stedebouw en Volkshuisvesting*, juli/augustus, pp. 400-404.

Inspectiewerkgroep Stankhinder (1983) Geurnormering; onderbouwing van een stankconcentratienorm, Publikatiereeks Lucht, nr. 11, ministerie van VROM, Den Haag.

IPO (Interprovinciaal Overleg) (1996) Reisgids ROMIO; een handreiking voor Ruimtelijke Ordening en Milieu bij Industrie en Omgeving, Concept, Projectgroep ROMIO, Den Haag.

IPO, PGBO, VNG en VROM (1996) 1e werkboek actief bodembeheer, VNG Uitgeverij, Den Haag.

IPO, VNG en VROM (1997) Bever Beleidsvernieuwing bodemsanering; Verslag van het Bever-proces, S. Ouboter en W. Kooper (eds), VROM, Den Haag.

Jongh, P.E. de (1989) Hoofdlijnen van het Nationaal Milieubeleidsplan, in E.C. van Ierland, A.P.J. Mol en W.A. Hafkamp, *Milieubeleid in Nederland; Reacties op het Nationaal Milieubeleidsplan*, Stenfert Kroese, Leiden, pp. 11-26.

Kabinetsstandpunt (1997) Over de vernieuwing van het bodemsaneringsbeleid, 19 June 1997, TK 25411, nr. 1, Den Haag.

Kaiser, E.J., D.R. Godschalk, F.S. Chapin (1995) *Urban Land Use Planning*, University of Illinois Press, Urbana and Chicago.

Kasteren, J. van (1985) 15 jaarmilieubeleid; Geluid, *Intermediair*, nr. 48, pp. 33-35.

Kasteren, J. van (1987) De gevaren van chemische industrieën uitgedrukt in cijfers en lijnen op de kaart, *ROM Magazine*, nr. 12, pp. 12-16.

Kasteren, J. van (1989) De weg naar een beter ecologisch draagvlak is geen rechte weg; Marius Enthoven over milieubeleidsplan, *ROM Magazine*, nr. 7; pp. 5-8.

Kickert, W.J.M. (1986) *Overheidsplanning, theorieën, technieken en beperkingen*, Van Gorcum, Assen/Maastricht.

Koeman, J.H. (1989) Hoofdlijnen van het Nationaal Milieubeleidsplan, in E.C. van Ierland, A.P.J. Mol en W.A. Hafkamp, *Milieubeleid in Nederland; Reacties op het Nationaal Milieubeleidsplan*, Stenfert Kroese, Leiden, pp. 47-55.

Koning, M.E.L. de, F. Elgersma (1990) *Het Nederlandse milieubeleid; Van Hinderwet tot Nationaal Milieubeleidsplan*, AO, nr. 2345, Stichting IVIO, Lelystad.

Koning, M.E.L. de (1994) *In dienst van het milieu; Enkele memoires van oud-directeur-generaal Milieubeheer prof. ir. W.C. Reij*, Samsom H.D. Tjeenk Willink, Alphen aan den Rijn.

Kooper, W. (eds) (1999) Van Trechter naar Zeef, Quintens Advies en Management, Sdu-Uitgevers, Den Haag.

Kuijpers, C.J. (eds) (1992) *Bedrijven en milieuzonering*, Vereniging Nederlandse Gemeenten, VNG uitgeverij, Den Haag.

Kuijpers, C.B.F. (1996) Integratie en het gebiedsgerichte milieubeleid, in G. de Roo, *Milieuplanning in vierstromenland*, Samsom H.D. Tjeenk Willink, Alphen aan den Rijn, pp. 52-67.

Kuijpers, C.B.F., Th.L.G.M. Aquarius (1998) Meer ruimte voor kwaliteit; Intensivering van het ruimtegebruik in stedelijk gebied, *Stedebouw & Ruimtelijke Ordening*, nr. 1, pp. 28-32.

Kuiper, G. (1995) Milieucompensatie in stedelijke gebieden, *ROM Magazine*, nr. 12, pp. 9-11.

Kusiak, L. (1989) In de Gelderse Vallei loopt de mest de spuigaten uit; Barsten in de ecologische structuur, *ROM Magazine*, nr. 11, pp. 16-19.

Lambers, C. (1989) De bodem, in W. Brussaard, Th.G. Drupsteen, P.C. Gilhuis en N.S.J. Koeman (eds), *Milieurecht*, W.E.J. Tjeenk Willink, Zwolle, pp. 167-192.

Lammers, P.E.M., V.M. Sol, J. de Boer, H. Aiking, J.F. Feenstra (1993) Een milieubelastingsindex voor toepassing in integrale milieuzonering, *Milieu*, nr. 3, pp. 81-86.

Leidraad Bodembescherming (1997) M.F.X.W. Veul (eds), 1983-..., in opdracht en onder redactie van VROM, DGM, directie Bodem, Water, Stoffen, SDU Uitgeverij, Den Haag.

Leroy, P. (1994) De ontwikkeling van het milieubeleid en de milieubeleidstheorie, in P. Glasbergen (eds), *Milieubeleid: een wetenschappelijke inleiding*, VUGA, Den Haag.

Los Angeles 2000 Committee (1988) LA 2000: A city for the future, Mayor's Office, Los Angeles.

LNV (Ministerie van Landbouw, Natuurbeheer en Visserij) (1992) Structuurschema Groene Ruimte; Het landelijk gebied de moeite waard, Ontwerp-planologische kernbeslissing, Den Haag.

LNV (Ministerie van Landbouw, Natuurbeheer en Visserij) (1995) Uitwerking compensatiebeginsel SGR, Directie Groene Ruimte en Recreatie, Den Haag.

LNV en VROM (Ministeries van Landbouw, Natuurbeheer en Visserij en van Volkshuisvesting, Ruimtelijke Ordening en Milieubeheer) (1992) Ontwerp-planologische kernbeslissing Structuurschema Groene Ruimte, Den Haag.

Lurvink, J.G.M. (1988) in *Integrale milieuzonering en de flexibiliteit van ruimtelijke planning*, NIROV-werkgroep milieubeleid, NIROV, Den Haag, pp. 63-70.

McDonald, G.T. (1996) Planning as Sustainable Development, *Journal of Planning Education and Research*, Vol. 15, pp. 225-236.

MDW-werkgroep (1996) Het geluid geordend; Een decentraal model met ruimer perspectief, Rapport MDW-werkgroep Wet geluidhinder, ES/PRO/MDW/GD122-96, Den Haag.

Meadows, D., D. Meadows, J. Randers, W. Behrens (1972) *Rapport van de Club van Rome, De grenzen aan de groei*, Uitgeverij Het Spectrum, Utrecht.

Meijburg, E. (1997) Towards an integrated district oriented policy: a policy for urban planning and the environment in Amsterdam, in D. Miller and G. de Roo (eds), *Urban Environmental Planning*, Avebury, Aldershot, UK, pp. 107-121.

Meijburg, E., M. de Knegt (1994) Een stolp over Amsterdam; milieu en ruimtelijke ordening verstrengeld, *ROM Magazine*, nr. 10, pp. 9-12.

Meijden, D. van der (1991) As low as reasonable achievable, *Milieu en Recht*, nr. 1, pp. 12-19.

Menninga, H. (1993) De hoofdstukken plannen en milieukwaliteitseisen van de Wet milieubeheer, *Milieu en Recht*, nr. 2, pp. 75-86.

Michiels, F.C.M.A. (1989) Wet inzake de luchtverontreiniging, in Brussaard et al. (eds), *Milieurecht*, W.E.J. Tjeenk Willink, Zwolle.

MIG (Modernisering Instrumentarium Geluidbeleid) (1998) Nota MIG, 3e concept, versie 29-01-1998/NM, DGM, Directie Geluid en Verkeer, Projectburo MIG, Den Haag.

Milieudienst Groningen (1993) Handleiding Milieubeoordelingsmethode Groningen, DSW Stadspark Groningen, Gemeente Groningen, Groningen.

Milieudienst Tilburg (1994) Pilotproject Stadsmilieu Tilburg, Methodiek voor Stedelijk Omgevingsbeleid, Bouwfonds Adviesgroep bv, Gemeenten Tilburg, Tilburg.

Milieuvoorschriften (1971-..) Vergunningen, heffingen en subsidies, M.V.C. Aalders et al., Delwel, Den Haag.

Miller, D., G. de Roo (eds) (1999) *Integrating City Planning and Environmental Improvement*, Ashgate, Aldershot, UK.

Ministers van VROM en BZ (2000) Instellingsbesluit Evaluatiecommissie Stad & Milieu, Mbb 2000043034, Dir. Bestuurszaken, DGM, VROM, Den Haag.

Moet, D. (1995) *Bouwen op verontreinigde grond; Een gebruiksspecifieke benadering*, Milieureeks VNG, VNG Uitgeverij, Den Haag.

Mol, A.P.J. (1989) Hoofdlijnen van het Nationaal Milieubeleidsplan, in E.C. van Ierland, A.P.J. Mol en W.A. Hafkamp, *Milieubeleid in Nederland; Reacties op het Nationaal Milieubeleidsplan*, Stenfert Kroese, Leiden, pp. 27-38.

Nelissen, N.J.M. (1988) *Het milieu: vertrouw maar weet wel wie je vertrouwt: een onderzoek naar verinnerlijking en verinnerlijkingsbeleid op het gebied van het milieu*, Kerckebosch, Zeist.

Nelissen, N.J.M., A.J.A. Godfroij, P.J.M. de Goede (eds) (1996) *Vernieuwing van bestuur; inspirerende visies*, Coutinho, Bussum.

Nelissen, N.J.M., T. Ikink, A.W. van der Ven (eds) (1996) *In staat van vernieuwing; Maatschappelijke vernieuwingsprocessen in veelvoud*, Coutinho, Bussum.

Nentjes, A. (1993) Milieu-economie, in J.J. Boersema, J.W. Copius Peereboom en W.T. de Groot, *Basisboek Milieukunde*, Boom, Meppel, pp. 272-293.

Neuerburg, E.N., P. Verfaille (1991) *Schets van het Nederlandse milieurecht*, Samsom H.D. Tjeenk Willink / VUGA, Alphen aan den Rijn / Den Haag.

Nieuwenhof, R. van den, H. Bakker (1989) Zones rond industrieën om milieuverstoringen in te perken; Planologische maatregelen op grond van milieunormen, *ROM Magazine*, nr. 10, pp. 6-12.

Nieuwenhof, R. van den, M. Groen (1988) Amerikaans milieubeleid is nauwelijks voorbeeld voor Europa, maar toch leerzaam, *ROM Magazine*, nr. 7, pp. 7-9.

Nijpels, E.H.T.M. (1988) Brief van 2 december 1988 aan de Tweede Kamer aangaande de sloop van Sluiskil-Oost als onderdeel van het proefproject integrale milieuzonering, Den Haag.

Nijpels, E.H.T.M. (1989) Voorwoord, in T.G. Tan en H. Waller (1989) *Wetgeving als mensenwerk; De totstandkoming van de Wet geluidhinder*, Samsom H.D. Tjeenk Willink, Alphen aan den Rijn.

OECD (Organisation for Economic Co-operation and Development) (1975) *The Polluter Pays Principle; Definition, analysis, implementation*, Paris.

Oosterhoff, H.A., G. de Roo. M.J.C. Schwartz, H. van der Wal (2001) Omgevingsplanning in Nederland; Een stand van zaken rond sectoroverschrijdend, geïntegreerd en gebiedsgericht milieubeleid voor de fysieke leefomgeving, Rijksplanologische Dienst, VROM, Den Haag.

Osleeb, J., D. Kass, H. Blanco, S.R. Zoloth, D. Sivin, A. Baimonte (1997) Baseline Aggregate Environmental Loadings (BAEL) Profile of Greenpoint/Williamsburg, Brooklyn, Hunter College, The New York City Department of Environmental Protection and the Watchperson's Office, New York.

Otten, F.P.J.M. (1980) *Ruimtelijke ordening en milieubeheer: de bescherming van het leefmilieu via het instrumentarium van de Wet op de ruimtelijke ordening*, Vuga-boekerij, Den Haag.

Otten, F.P.J.M. (1993) Geluidhinderwetgeving, in Brussaard, W. et al, *Milieurecht*, W.E.J. Tjeenk Willink, Zwolle, pp. 253-284.

Paauw, M.S., G. de Roo (1996) Het provinciale omgevingsplan nader uitgewerkt, in G. de Roo (eds) *Milieuplanning in vierstromenland*, Samsom H.D. Tjeenk Willink, Alphen aan den Rijn, pp. 202-214.

Peperstraten, J. van (1989) Schiphol: Motor voor economie, plaag voor omgeving, *ROM Magazine*, nr. 11, pp. 8-15.

Peeters, M. (1993) De markt en het milieu: het instrument van verhandelbare vervuilings-rechten; Enkele gedachten over een communautaire vergunningenmarkt, *Milieu en recht*, nr. 1, pp. 2-10.

Prigogine, I. (1996) *Het einde van de zekerheden; Tijd, chaos en de natuurwetten*, Lannoo, Tielt.

Project Mainport & Milieu Schiphol (1991) Integrale versie Plan van Aanpak Schiphol en Omgeving, redactie F.L. Bussink,Den Haag.

Project Organisers IMZA (1991) Integrale milieuzonering Arnhem-Noord: Deelrapport toepassing VROM-systematiek met het DSS-systeem van GEOPS Wageningen, Provincie Gelderland, Arnhem.

Provincie Gelderland (1987) provinciaal milieubeleidsplan 1987-1991, Arnhem.

Provincie Noord-Holland (1987) Streekplan Amsterdam-Noordzeekanaalgebied, Haarlem.

Provincie Zeeland (1987) Een gebiedsgerichte benadering van het milieu in de Kanaalzone, Middelburg.

Raa, B.D. te (1995) 'Stankbestrijding: prima, maar dit wordt te gek'; strengere geurnormen bedreigen nieuwbouwplannen, *ROM Magazine*, nr. 3, pp. 19-20.

RARO (Raad van advies voor de ruimtelijke ordening) (1989) Advies over het Nationaal Milieubeleidsplan, SDU Uitgeverij, Den Haag.

RARO (Raad van advies voor de ruimtelijke ordening) (1991) Advies over het actieplan gebiedsgericht milieubeleid, SDU Uitgeverij, Den Haag.

RARO (Raad van advies voor de ruimtelijke ordening) (1992) Advies over ruimtelijke ordening en milieubeleid, deel 4: Milieuzonering, SDU Uitgeverij, Den Haag.

RARO (Raad van advies voor de ruimtelijke ordening) (1994) Advies over het Tweede Nationale Milieubeleidsplan, SDU Uitgeverij, Den Haag.

RAVO (Raad voor de Volkshuisvesting) (1989) Advies inzake het Nationaal Milieubeleidsplan, nr. 166, SDU Uitgeverij, Den Haag.

RAWB (Raad van Advies voor het Wetenschapsbeleid) (1989) Advies inzake het Nationaal Milieubeleidsplan, Brief aan de minister van VROM d.d. 27 November 1989, Den Haag.

Rheinisch-Westfälischer TÜV (1974) Abstände zwischen Industrie - bwz. Gewerbegebieten und Wohngebieten im Rahmen der Bauleitplanung, Düsseldorf.

Ringeling, A.B. (1990) Plannen en organiseren: Het Nationaal Milieubeleidsplan, in *Vereniging voor Milieurecht, Juridische en bestuurlijke consequenties van het Nationaal Milieubeleidsplan*, nr. 2, W.E.J. Tjeenk Willink, Zwolle, pp. 3-11.

RIVM (Rijksinstituut voor Volksgezondheid en Milieuhygiëne) (1988) Zorgen voor morgen; Nationale milieuverkenning 1985-2010, onder redaktie van F. Langeweg, Samsom H.D. Tjeenk Willink, Alphen aan den Rijn.

Roeters, J.H. (1997) *Anticiperend bodemsaneringsbeleid bij provincies*, Faculteit der Ruimtelijke Wetenschappen, Rijksuniversiteit Groningen, Groningen.

Rondinelli, D.A., J. Middleton, A.M. Verspoor (1989) Contingency Planning for Innovative Projects; Designing Education Reforms in Developing Countries, *Journal of the American Planning Association*, Winter, pp. 45-56.

Roo, G. de (1992) Milieuzonering stuit op planologische obstakels; Of hele wijken afbreken of industrie sluiten, *ROM Magazine*, pp. 14-17.

Roo, G. de (eds) (1993) *Gaten in de stilte; Het beheer van het stiltegebied Waddenzee*, Geo Pers, Groningen.

Roo, G. de (eds) (1993b) *Kwaliteit van norm en zone; Planologische consequenties van (integrale) milieuzonering*, Geo Pers, Groningen.

Roo, G. de (1993c) Environmental Zoning: The Dutch Struggle Towards Integration, *European Planning Studies*, Vol. 1, nr. 3. pp. 367-377.

Roo, G. de (1995) Gebiedsgericht milieubeleid in vierstromenland; Strategische keuzen en randvoorwaarden voor het gebiedenbeleid, *ROM Magazine*, nr. 7/8, pp. 16-20.

Roo, G. de (1996) Contouren van het gebiedsgericht milieubeleid, in G. de Roo (eds) *Milieuplanning in vierstromenland*, Samsom H.D. Tjeenk Willink, Alphen aan den Rijn, pp. 19-38.

Roo, G. de (1996b) Inleiding, in G. de Roo (eds) *Milieuplanning in vierstromenland*, Samsom H.D. Tjeenk Willink, Alphen aan den Rijn, pp. 11-18.

Roo, G. de (1996c) Compensatie binnen de stedelijke context van milieu en ruimte - Een beschouwing, Lezingen Stad & Milieu-conferentie regio Noord, 26 september 1996, Groningen, *Stad & Milieu*, Ministerie van VROM, Den Haag, pp. 51-56.

Roo, G. de (1997) Kwaliteit, Verantwoordelijkheid en Compactheid; Synthese van milieu en ruimte in de compacte stad van de toekomst, in VROM, Ruimtelijk milieu of milieu-ordening, kansen voor vergaande samenwerking, 5 essays ten behoeve van de Nota Milieu & Ruimte, Ministerie van VROM, Den Haag, pp. 1-25.

Roo, G. de (1998) *Structurering en Normering van het Milieubeleid in de Jaren Zeventig en Tachtig*, Faculteit der Ruimtelijke Wetenschappen, Rijksuniversiteit Groningen, Groningen.

Roo, G. de, H.J. Bartelds (1996) Opkomst en ondergang van bodembeschermingsgebieden, in G. de Roo (eds) *Milieuplanning in vierstromenland*, Samsom H.D. Tjeenk Willink, Alphen aan den Rijn, pp. 135-143.

Roo, G. de, D. Miller (1997) Transitions in Dutch environmental planning: new solutions for integrating spatial and environmental policies, *Environment and Planning B: Planning and Design*, Vol. 24, pp. 427-436.

Roo, G. de, B. van der Moolen (eds) (1991) *De Voorlopige Systematiek voor Integrale Milieuzonering; Een doelgroepenbenadering in drie proefprojecten*, Geo Pers, Groningen.

Roo, G. de, M. Schwartz (2001) *Omgevingsplanning, een innovatief proces; Over integratie, participatie en de gebiedsgerichte aanpak*, Sdu Uitgevers, Den Haag.

Roo, G. de, M. Schwartz (2001b) De beleidspraktijk van omgevingsplanning, *Rooilijn*, nr. 1, pp. 4-9.

Rosdorff, S., L.K. Slager, V.M. Sol, K.F. van der Woerd (1993) *De stadsstolp: meer ruimte voor milieu èn economie*, Instituut voor Milieuvraagstukken, Vrije Universiteit, Amsterdam.

RPD (Rijksplanologische Dienst) (1993) Ruimtelijke Verkenningen, Ministerie van VROM, Den Haag.

RPD (Rijksplanologische dienst) (1993) Ruimtelijke Verkenningen, H. Hearn-Sukkel et al. (eds), Ministerie van VROM, SDU Uitgeverij, Den Haag.

Schoof, D.J.W. (1989) Het Nationaal Milieubeleidsplan: kiezen voor winst, *Milieu*, pp. 105-111.

SCMO-TNO (1992) Omvang zoneerbare milieubelastingen 1991, IMZ-reeks deel 23, Ministerie van VROM, Leidschendam.

Schwartz, M. (1998) Omgevingsplanning: trends en vraagstukken, *Rooilijn*, nr. 1, pp. 38-42.

SER (Sociaal-Economische Raad) (1989) Advies Nationaal Milieubeleidsplan, nr. 17, Den Haag.

SER (Sociaal-Economische Raad) (1994) Advies Nationaal Milieubeleidsplan 2, nr. 4, Den Haag.

Slocombe, D.S. (1993) Environmental planning, ecosystem science and ecosystem approaches for integrating environment and development, *Environmental Management*, Vol. 17, pp. 289-303.

Smit, C.Th. (1989) Wet verontreiniging oppervlaktewateren, in Brussaard et al. (eds), *Milieurecht*, W.E.J. Tjeenk Willink, Zwolle.

Stad & Milieu (1994) Projectplan Stad & Milieu Startnotitie april 1994, Bestuurszaken, DGM, VROM, Den Haag.

Stad & Milieu (1994b) Potje biljarten? Nee, liever voetballen; Strategie-notitie Project Stad & Milieu, 7 november 1994, Bestuurszaken, DGM, VROM, Den Haag.

Stad & Milieu (1995) Waar vele willen zijn, is ook een weg; Rapportage project Stad & Milieu, Bestuurszaken, DGM, VROM, Den Haag.

Stad & Milieu (1995b) Binnen regels naar kwaliteit, Stad & Milieu Rapportage, Deelproject Cobber, Bestuurszaken, DGM, VROM, Den Haag.

Stafbureau NER (192) Nederlandse Emissierichtlijnen-Lucht, Bilthoven.

Stb. (Staatsblad) (Bulletin of Acts, Orders and Decrees) 684 (1998) Wet van 26 november 1998, houdende regels over experimenten inzake zuinig en doelmatig ruimtegebruik en optimale leefkwaliteit in stedelijk gebied (Experimentenwet Stad en Milieu).

Streefkerk, N. (1992) *Handboek beoordelingsmethode milieu, inhoudelijke achtergronden en handleiding*, VNG-uitgeverij, Den Haag.

Stuurgroep IMZS-Drechtsteden (1991) Rapportage 1e fase IMZS-Drechtsteden: Inventarisatie milieubelasting, Provincie Zuid-Holland, Den Haag.

Stuurgroep Plan van Aanpak Schiphol en Omgeving (1990) Plan van Aanpak Schiphol en Omgeving, Den Haag.

Stuurgroep ROM-IJmeer (1996) Plan van Aanpak ROM-IJmeer, Amsterdam.

Stuurgroep ROM Rijnmond (1992) Ontwerp Plan van Aanpak en Beleidsconvenant ROM-project Rijnmond, Rotterdam.

Stuurgroep Tien-Jaren Scenario Bodemsanering (1989) Advies over een beleidsscenario, Den Haag.

Tan, T.G., H. Waller (1989) *Wetgeving als mensenwerk; De totstandkoming van de Wet geluidhinder*, Samsom H.D. Tjeenk Willink, Alphen aan den Rijn.

Tatenhove, J.P.M. van (1993) *Milieubeleid onder dak?; Beleidsvoeringsprocessen in het Nederlandse milieubeleid in de periode 1970-1990; nader uitgewerkt voor de Gelderse Vallei*, Wageningse sociologische studies, Universiteit van Wageningen, Wageningen.

TCB (Technische Commissie Bodemsanering) (1996) Jaarverslag, Den Haag.

Teunisse, P.B.W. (eds) (1995) *Planning, uitvoering en beheersing van gemeentelijk milieubeleid*, SDU-Uitgeverij, Den Haag.

Tjallingii, S.P. (1996) *Ecological conditions: stratgies and structures in environmental planning*, DLO Institute for Forestry and Nature Research, Wageningen.

TK (Tweede Kamer) (1975) Nota betreffende de relatie landbouw en natuur- en landschapsbehoud: gemeenschappelijke uitgangspunten voor het beleid inzake de uit en oogpunt van natuur- en landschapsbehoud waardevolle agrarische cultuurlandschappen, TK 1974-1975, 13285, nrs. 1-2, Staatsuitgeverij, Den Haag.

TK (Tweede Kamer) (1984) Indicatief Meerjaren Programma Lucht 1985-1989, Vergaderjaar 1984-1985, 18605, nr. 2, Den Haag.

TK (Tweede Kamer) (1985) Indicatief Meerjaren Programma Milieubeheer 1986-1990, Vergaderjaar 1985-1986, 19204, nr. 2, Den Haag.

TK (Tweede Kamer) (1989) Nationaal Milieubeleidsplan 1989-1993, Kiezen of verliezen, Vergaderjaar 1988-1989, 21137, nrs. 1-2, SDU-uitgeverij, Den Haag.

TK (Tweede Kamer) (1989) Notitie Omgaan met risico's: De risicobenadering in het milieubeleid, Nationaal Milieubeleidsplan 1989-1993, Vergaderjaar 1988-1989, 21137, nr. 5, SDU-uitgeverij, Den Haag.

TK (Tweede Kamer) (1990) Actieplan Gebiedsgericht milieubeleid, TK 1990-1991, 21896, nrs. 1-2, SDU uitgeverij, Den Haag.

TK (Tweede Kamer) (1990b) Regeringsbeslissing Natuurbeleidsplan, TK 1989-1990, 21149, nrs. 2-3, Ministerie van Landbouw, Natuurbeheer en Visserij, Den Haag.

TK (Tweede Kamer) (1990c) motie Swildens-Rozendaal/Van Noord, TK 1989-1990, 20490, nr. 47, Den Haag.

TK (Tweede Kamer) (1992) Nota Stankbeleid, Vergaderjaar 1991-1992, 22715, nr. 1, Den Haag.

TK (Tweede Kamer) (1993) Nationaal Milieubeleidsplan 2; Milieu als maatstaf, Vergaderjaar 1993-1994, 23560, nrs. 1-2, Den Haag.

TK (Tweede Kamer) (1993b) Brief van de minister van VROM inzake het stankbeleid, Vergaderjaar 1993-1994, 22666, nr. 4, Den Haag.

TK (Tweede Kamer) (1994) Plan van Aanpak Marktwerking, deregulering en wetgevingskwaliteit, Vergaderjaar 1994-1995, 24036, nr. 1, Den Haag.

TK (Tweede Kamer) (1996) Marktwerking, deregulering en wetgevingskwaliteit: Brief van de minister van Volkshuisvesting, Ruimtelijke Ordening en Milieubeheer, Vergaderjaar 1995-1996, 24036, nr. 26, Den Haag.

TK (Tweede Kamer) (1996b) Nieuwe regelen ter bescherming van natuur en landschap, Vergaderjaar 1996-1997, 23580, nrs. 10-11, Den Haag.

TK (Tweede Kamer) (1997) Wijziging van de Wet op de Ruimtelijke Ordening, Memorie van Toelichting, Vergaderjaar 1996-1997, 25311, nr. 3, Den Haag.

TK (Tweede Kamer) (1997b) Regels inzake plannen op het terrein van het verkeer en het vervoer (Planwet verkeer en vervoer), Vergaderjaar 1996-1997, 25337, nrs. 1-5, nr. 258, Den Haag.

TK (Tweede Kamer) (1998) Regels over experimenten inzake zuinig en doelmatig ruimtegebruik en optimale leefkwaliteit in stedelijk gebied (Experimenteerwet Stad en Milieu), Vergaderjaar 1997-1998, 25848, nrs. 1-2, Den Haag.

Twijnstra Gudde (1992) Tussentijdse evaluatie proefprojecten IMZ 2, IMZ-reeks deel 21, Ministerie van VROM, Leidschendam.

Twijnstra Gudde (1994) Derde evaluatie proefprojecten IMZ, IMZ-reeks deel 29, Ministerie van VROM, Den Haag.

Vereniging Natuurmonumenten en DHV (1996) Compensatie IJburg; toepassing van het compensatiebeginsel, Vereniging Natuurmonumenten, 's-Graveland.

Verschuren, J. (1990) *Bodemsanering van bedrijfsterreinen*, Dombosch, Raamsdonkveer.

Vlist, M. van der (1998) *Duurzaamheid als planningopgave: gebiedsgerichte afstemming tussen de ruimtelijke ordening, het milieubeleid en het waterhuishoudkundige beleid voor het landelijke gebied*, Landbouwuniversiteit Wageningen, Wageningen.

VM (Ministerie van Volksgezondheid en Milieuhygiëne) (1972) Urgentienota Milieuhygiëne, Tweede Kamer, 1971-1972, 11906, nr. 2, Den Haag.

VM (Ministerie van Volksgezondheid en Milieuhygiëne) (1974) Instrumentennota, Tweede Kamer, 1974-1975, 13100, Den Haag.

VM (Ministerie van Volksgezondheid en Milieuhygiëne) (1976) Nota Milieuhygiënische normen 1976, Tweede Kamer, 1976-1977, 14318, nr. 2, Den Haag.

VNG (Vereniging van Nederlandse Gemeenten) (1986) Bedrijven en milieuzonering, Groene reeks, nr. 80, VNG uitgeverij, Den Haag.

VNG (Vereniging van Nederlandse Gemeenten) (1991) Kaderplan van aanpak NMP voor gemeenten, i.s.m. VROM, VNG uitgeverij, Den Haag.

VNG (Vereniging van Nederlandse Gemeenten) (1992) Omgaan met bodemsanering, een gemeentelijke visie, Den Haag.

VNG (Vereniging van Nederlandse Gemeenten) (1993) Gids gebiedsgerichte milieu-aanpal; een handreiking voor gemeenten, C.J. Kuijpers (eds), VNG uitgeverij, Den Haag.

VNG (Vereniging van Nederlandse Gemeenten) (1995) Kaderplan Gemeentelijk Milieubeleid, VNG uitgeverij, Den Haag.

VNG (Vereniging van Nederlandse Gemeenten) (1996) Praktijkboek Lokale Agenda 21, Milieureeks nr. 5, B. Roes (eds) en SME MilieuAdviseurs, VNG uitgeverij, Den Haag.

VNG en VROM (Vereniging van Nederlandse Gemeenten en het Ministerie van Volkshuisvesting, Ruimtelijke Ordening en Milieubeheer) (1990) Praktijkboek gemeentelijk milieubeleid; Op weg naar een duurzame ontwikkeling, K. Plug (eds), Den Haag.

VNO/NCW (1991) Brief aan de minister van VROM betreffende integrale milieuzonering d.d. 24 april 1991, Bureau Milieu en Ruimtelijke Ordening, Den Haag.

Voerknecht, H. (1994) Experiences with environmental zoning: the case of the Drechtsteden region, H. Voogd (ed.) *Issues in Environmental Planning*, Pion, London, pp. 67-77.

Voogd, H. (1995) *Methodologie van ruimtelijke planning*, Couthino, Bussum.

Voogd, H. (1996) Provinciale omgevingsplannen; Een niet te forceren leerproces, in G. de Roo (eds) *Milieuplanning in vierstromenland*, Samsom H.D. Tjeenk Willink, Alphen aan den Rijn, pp. 194-201.

VROM (Ministerie van Volkshuisvesting, Ruimtelijke Ordening en Milieubeheer) (1983) Plan Integratie Milieubeleid, Tweede Kamer, 1982-1983, 17931, nr. 6, Den Haag.

VROM (Ministerie van Volkshuisvesting, Ruimtelijke Ordening en Milieubeheer) (1983b) Leidraad Bodemsanering, Directie Bodem, Water en Stoffen, Directoraat-Generaal Milieuhygiëne, Den Haag.

VROM (Ministerie van Volkshuisvesting, Ruimtelijke Ordening en Milieubeheer) (1984a) Meer dan de Som der Delen; Eerste nota over de planning van het milieubeleid, Tweede Kamer, 1983-1984, 18292, nr. 2, Den Haag.

VROM (Ministerie van Volkshuisvesting, Ruimtelijke Ordening en Milieubeheer) (1984b) Indicatief Meerjaren Programma Milieubeheer 1985-1989, Tweede Kamer, 1984-1985, 18602, nr. 2, Den Haag.

VROM (Ministerie van Volkshuisvesting, Ruimtelijke Ordening en Milieubeheer) (1988) Workshop verslag, Integrale milieuzonering deel 1, Directie Geluid, DGM, Den Haag.

VROM (Ministerie van Volkshuisvesting, Ruimtelijke Ordening en Milieubeheer) (1988b) Meerjarig uitvoeringsprogramma geluidhinderbestrijding (MUG) 1989-1993, VROM, Den Haag.

VROM (Ministerie van Volkshuisvesting, Ruimtelijke Ordening en Milieubeheer) (1988c) Vierde nota over de ruimtelijke ordening, deel a, beleidsvoornemens, TK 20490, nrs. 1-2, Sdu Uitgevers, Den Haag.

VROM (Ministerie van Volkshuisvesting, Ruimtelijke Ordening en Milieubeheer) (1989) Project KWS2000; Bestrijdingsstrategie voor de emissies van vluchtige organische stoffen, Projectgroep Koolwaterstoffen 2000, Directie Lucht, DGM, Den Haag.

VROM (Ministerie van Volkshuisvesting, Ruimtelijke Ordening en Milieubeheer) (1989b) Projectprogramma Cumulatie van bronnen en integrale milieuzonering, Integrale milieuzonering deel 2, Directie Geluid, DGM, Den Haag.

VROM (Ministerie van Volkshuisvesting, Ruimtelijke Ordening en Milieubeheer) (1989c) Reacties en adviezen naar aanleiding van het Nationaal Milieubeleidsplan; Hoe het NMP leeft in Nederland, Den Haag.

VROM (Ministerie van Volkshuisvesting, Ruimtelijke Ordening en Milieubeheer) (1989d) Workshop verslag, Integrale milieuzonering deel 1, Directie Geluid, DGM, Den Haag.

VROM (Ministerie van Volkshuisvesting, Ruimtelijke Ordening en Milieubeheer) (1990) Ministriële handreiking voor een voorlopige systematiek voor de integrale milieuzonering, Integrale milieuzonering deel 6, Directie Geluid, DGM, Den Haag.

VROM (Ministerie van Volkshuisvesting, Ruimtelijke Ordening en Milieubeheer) (1992) Externe integratie van milieubeleid: knelpunten, kansen en keuzen in het stedelijk gebied, Hoofdrapport, BRO Adviseurs en VNG, Publikatiereeks milieubeheer 1992/2A, DGM, Leidschendam.

VROM (Ministerie van Volkshuisvesting, Ruimtelijke Ordening en Milieubeheer) (1993) Vierde nota over de ruimtelijke ordening Extra, deel 4: Planologische Kernbeslissing Nationaal Ruimtelijk Beleid, Den Haag.

VROM (Ministerie van Volkshuisvesting, Ruimtelijke Ordening en Milieubeheer) (1994) Interventiewaarden bodemsanering, Circulaire 9 mei 1994, Den Haag.

VROM (Ministerie van Volkshuisvesting, Ruimtelijke Ordening en Milieubeheer) (1994b) Circulaire tweede fase inwerkingtreding saneringsregeling Wet bodembescherming, 22 december 1994, Den Haag.

VROM (Ministerie van Volkshuisvesting, Ruimtelijke Ordening en Milieubeheer) (1994c) De Gordiaanse knoop ontward: ROM-projecten in Nederland, M. Groen, Den Haag.

VROM (Ministerie van Volkshuisvesting, Ruimtelijke Ordening en Milieubeheer) (1995) Waar vele willen zijn, is ook een weg; Stad & Milieu Rapportage, Den Haag.

VROM (Ministerie van Volkshuisvesting, Ruimtelijke Ordening en Milieubeheer) (1995b) Brief van de minister van VROM aan de voorzitters van de vaste kamercommissies voor VROM, EZ en LNV inzake het stankbeleid d.d. 21 maart 1995, Den Haag.

VROM (Ministerie van Volkshuisvesting, Ruimtelijke Ordening en Milieubeheer) (1996) Verstedelijking in Nederland 1995-2005; De Vinex-afspraken in beeld, RPD en IPO, Den Haag.

VROM (Ministerie van Volkshuisvesting, Ruimtelijke Ordening en Milieubeheer) (1996b) Binnen regels naar kwaliteit, Stad & Milieu, Rapportage deelproject Cobber, Den Haag.

VROM (Ministerie van Volkshuisvesting, Ruimtelijke Ordening en Milieubeheer) (1996c) Actualisering omvang Nederlandse bodemverontreiniging, Den Haag.

VROM (Ministerie van Volkshuisvesting, Ruimtelijke Ordening en Milieubeheer) (1997) Onderhandelen langs de zone; Vijf jaar integrale milieuzonering in Nederland, Den Haag.

VROM (Ministerie van Volkshuisvesting, Ruimtelijke Ordening en Milieubeheer) (1997b) Gerede grond voor groei, Interdepartementaal beleidsonderzoek bodemsanering, Ronde 1996, IBO-rapport nr. 3, Den Haag.

VROM (Ministerie van Volkshuisvesting, Ruimtelijke Ordening en Milieubeheer) (1998) Nota Leefomgeving: borg voor samenhang, VROM Visie, Kadernieuwsbrief over VROM-beleid en -actualiteiten, 24 februari 1998, Den Haag.

VROM (Ministerie van Volkshuisvesting, Ruimtelijke Ordening en Milieubeheer) (1998b) De proef op de ROM: ervaringen met gebiedsgericht beleid in 10 ROM-gebieden, G.C. Naeff, A.F. van de Klundert, Den Haag.

VROM (Ministerie van Volkshuisvesting, Ruimtelijke Ordening en Milieubeheer) (2000) Milieu in het Investeringsbudget Stedelijke Vernieuwing, Handreiking, VROM, Den Haag.

VROM en EZ (Ministerie van Volkshuisvesting, Ruimtelijke Ordening en Milieubeheer en het Ministerie van Economische Zaken) (1983) Actieprogramma Deregulering Ruimtelijke Ordening en Milieubeheer, Tweede Kamer, 1982-1983, 17931, nr. 4, Den Haag.

VROM, IPO en VNG (Ministerie van Volkshuisvesting, Ruimtelijke Ordening en Milieubeheer, Interprovinciaal Overleg, Vereniging van Nederlandse Gemeenten) (1995) Streefbeeld Bodemsaneringsbeleid; Tussenstand in de discussies tussen VNG, IPO en VROM, Verslaggeving Alons en Partners bv, VROM, Den Haag.

VROM, IPO en VNG (Ministerie van Volkshuisvesting, Ruimtelijke Ordening en Milieubeheer, Interprovinciaal Overleg, Vereniging van Nederlandse Gemeenten) (1997) Bever-1: basisdocument met scenario's voor typologie, bandbreedte en beslisondersteuning, Verslaggeving Bever-1 door Witteveen+Bos, VROM, Den Haag.

VROM en LV (Ministeries van Volkshuisvesting, Ruimtelijke Ordening en Milieubeheer en van Landbouw en Visserij) (1985) Nota ruimtelijk kader randstadgroenstructuur, Den Haag.

VROM/Task Force DSM (1987) Integrale zonering DSM: beleidsuitgangspunten en onderzoeksresultaten, Leidschendam.

VW en VROM (Ministeries van Verkeer en Waterstaat en van Volkshuisvesting, Ruimtelijke Ordening en Milieubeheer) (1996) Aanwijzing geluidcontouren luchtvaartterein Schiphol, Ex. art. 27, jo. art. 24 Luchtvaartwet, Den Haag.

Walgemoet, A. (1995) *Compensatie en perceptie in het stedelijk beleid; Een onderzoek naar de mogelijkheden van compensatie bij het oplossen van milieu/ruimte-conflicten in het stedelijk gebied en de rol van perceptie daarin*, Faculteit der Ruimtelijke Wetenschappen, Provincie Noord-Holland, Groningen en Haarlem.

Weertman, J., F. Nauta (1992) 'Dit is natuurlijk onzin'; Proefproject integrale milieuzonering in Dordrecht, *Rooilijn*, nr. 3, pp. 70-75.

Welschen, R.W. (1996) Aanbiedingsbrief betreffende de evaluatie van de aanpak bodemverontreiniging in stedelijke knooppunten (Welschen-2), Bu, 7 maart 1996, Gemeente Eindhoven, Eindhoven.

Werkgroep bodemsanering (1993) Saneren zonder stagneren; Eindrapport van de Werkgroep bodemsanering, 'Welschen-1', Den Haag.
Werkgroep bodemsanering (1996) Bodemsanering: met gezond verstand goede afspraken maken; De toepassing van de aanbevelingen van de werkgroep bodemsanering geëvalueerd, 'Welschen-2', Den Haag.
Werkgroep Visie op Omgevingsbeleid (1996) Aanzet voor omgevingsbeleid, Directeuren Strategie BIZA, EZ, LNV, V&W, Den Haag.
Westerhof, A. (1989) De face-lift van de Wet geluidhinder, *Tijdschrift voor de Ruimtelijke Ordening en Milieubeheer*, nr. 10, pp. 16-17.
Wiersinga, W.A., W.L.H. Ronken, H. ten Holt (1996) *Compenseren tussen Milieu en Ruimte; Een verkenning naar het gebruik en de toepassingsmogelijkheden van het compensatie-beginsel*, Publikatiereeks Milieustrategie 1996/10, DGM, VROM, Den Haag.
Willems, W.P. (1987) Vliegveld Beek: overheid in de knoop met haar eigen milieuregels en -procedures, *ROM Magazine*, nr. 4, pp. 3-10.
Willems, W.P. (1988) De nacht-vliegers van Beek, *Geluid en omgeving*, September, pp. 102-104.
Windt, H.J. van der (1995) *En dan: wat is natuur nog in dit land?: Natuurbescherming in Nederland 1880-1990*, Boom, Amsterdam.
Winsemius, P. (1986) *Gast in eigen huis, Beschouwingen over milieumanagement*, Samsom H.D. Tjeenk Willink, Alphen aan den Rijn.
Wissink, B. (2000) Ontworpen en ontstaan; Een praktijktheoretische analyse van het debat over het provinciale omgevingsbeleid, Voorstudies en Achtergronden, V 108, Wetenschappelijke Raad voor het Regeringsbeleid, Den Haag.
Wissink, B., O. Lingbeek (1995) Provinciale planning in beweging; ontwikkeling in acht provincies, *Stedebouw en Volkshuisvesting*, Vol. 1/2, pp. 35-37.
Zundert, J.W. (1993) Het bestemmingsplan als zoneringsinstrument, in G. de Roo, *Kwaliteit van norm en zone; Planologische consequenties van (integrale) milieuzonering*, Geo Pers, Groningen, pp. 31-43.
Zwiers, J. (1998) *Integratie, decentralisatie en actief bodembeheer; Onderzoek naar de rol van integratie, decentralisatie en actief bodembeheer bij het tegengaan van stagnatie bij ruimtelijk functionele ontwikkeling als gevolg van bodemverontreiniging*, Faculteit der Ruimtelijke Wetenschappen, Groningen.

Chapter 6

Arts, E.J.M.M. (1998) *EIA Follow-up; On the Role of Ex-Post Evaluation in Environmental Impact Assessment*, Geo Press, Groningen.
Bartelds, H.J., G. de Roo (1995) *Dilemma's van de compacte stad; Uitdagingen voor het beleid*, VUGA, Den Haag.
Boei, P.J. (1993) Integrale milieuzonering op en rond het industrieterrein Arnhem-Noord, in G. de Roo (eds) *Kwaliteit van norm en zone; Planologische consequenties van (integrale) milieuzonering*, Geo Pers, Groningen, pp. 75-82.
Borst, H. (1994) Integrale milieuzonering en ruimtelijke ordening, Integrale milieuzonering nr. 28, Directoraat-Generaal Milieubeheer, Ministerie van VROM, Den Haag.
Borst, H. (1996) Integrale milieuzonering, in G. de Roo (eds) *Milieuzones in vierstromenland*, Samsom H.D. Tjeenk Willink, Alphen aan den Rijn, pp. 94-108.
Borst, H. (1997) Integrated environmental zoning in local land use plans: some Dutch experiences, in D. Miller and G. de Roo (eds) *Urban Environmental Zoning; Policies, instruments and methods in an international perspective*, Avebury, Aldershot, UK, pp. 289-297.

Borst, H., G. de Roo, H. Voogd, H. van der Werf (1995) *Milieuzones in beweging; Eisen, wensen, consequenties en alternatieven*, Samsom H.D. Tjeenk Willink, Alphen aan den Rijn.

Flohr, A.P., H.C.J. Meijvis (1993) Milieuzonering in Hengelo, in G. de Roo (eds) *Kwaliteit van norm en zone; Planologische consequenties van (integrale) milieuzonering*, Geo Pers, Groningen, pp. 97-103.

Gemeente Amersfoort (1992) Uitvoeringsschema Centraal Stadsgebied, Amersfoort.

Gemeente Groningen (1992) Besluit van het college van B&W inzake de resultaten en aanbevelingen proefproject IMZ Noordoostflank, 26 mei 1992, Groningen.

Gemeente Hengelo (1993) Uitvoeringsplan Integrale Milieuzonering Twentekanaal, Dienst Stadsontwikkeling, Hengelo.

Gemeente Maastricht (1987) Project Integratie Milieubeleid: Hoofdrapport, Maastricht.

Gemeente Maastricht (1993) Eindrapport PISA: Ruimtelijke keuzes en milieusaneringen, Maastricht.

Hermsen, C.H. (1991) *Integrale milieuzonering Arnhem-Noord (samenvatting) Studiedag Integrale Milieuzonering; Ontwikkeling van een nieuw beleidsinstrument*, Geoplan Amsterdam.

Hoogland, J.S., F.J. Kolvoort (1993) Centraal Stadsgebied en IMZ, in G. de Roo (eds) *Kwaliteit van norm en zone; Planologische consequenties van (integrale) milieuzonering*, Geo Pers, Groningen, pp. 65-73.

Humblet, A.G.M., G. de Roo (eds) (1995) *Afstemming door Inzicht; een analyse van gebiedsgerichte milieubeoordelingsmethoden ten behoeve van planologische keuzes*, Geo Pers, Groningen.

Kerngroep Drechtoevers (1994) Masterplan Drechtoevers; Een kwaliteitssprong, Projectbureau Drechtoevers, Dordrecht.

Kuijpers, C.B.F. (1996) Integratie en het gebiedsgericht milieubeleid, in G. de Roo (eds), *Milieuplanning in vierstromenland*, Samsom H.D. Tjeenk Willink, Alphen aan den Rijn, pp. 52-67.

Nijkamp, P. (1996) De enge marges van het beleid en de brede missie van de beleidsanalyse, in P. Nijkamp, W. Begeer en J. Berting (eds), *Denken over complexe besluitvorming: een panorama*, SDU Uitgevers, Den Haag, pp. 129-146.

Nijkamp, P., J. Vleugel, R. Maggi, I. Masser (1994) *Missing Networks in Europe*, Avebury, Aldershot.

Oliemulders Punter & Partners bv (1993) Leefsituatie Onderzoek in de IJmond, Provincie Noord-Holland, Haarlem.

Pool, J. (1990) *Sturing van strategische besluitvorming: mogelijkheden en grenzen*, VU Uitgeverij, Amsterdam.

Pool, J., P.L. Koopman (1990) Strategische besluitvorming in organisaties: Onderzoeksmodel en eerste bevindingen, M&O, *Tijdschrift voor Organisatiekunde en Sociaal Beleid*, nr. 44, pp. 516-531.

Projectbureau Drechtoevers (1994) Masterplan Drechtoevers; Een kwaliteitssprong, o.v.v. de Kerngroep Drechtoevers, Dordrecht.

Provincie Friesland (1991) Integrale Milieuzonering Burgum/Sumar, Tussenrapport, Hoofdgroep Waterstaat en Milieu, Bureau Geluid en Lucht, Leeuwarden.

Provincie Limburg (1992) Streekplanuitwerking Westelijke Mijnstreek: Startnotitie, Maastricht.

Provincie Noord-Holland (1993) Proefproject Integrale Milieuzonering IJmond: Project programma, Haarlem.

Roeters, J.H. (1997) *Anticiperend bodemsaneringsbeleid bij provincies: Onderzoek naar de inhoud en mogelijkheden van vormen van vernieuwend bodemsaneringsbeleid bij provincies en de aansluiting daarvan op het BEVER-proces*, Faculteit der Ruimtelijke Wetenschappen, Rijksuniversiteit Groningen, Groningen.

Roo, G. de (1992) Milieuzonering stuit op planologische obstakels, *ROM Magazine*, nr. 6, pp. 14-17.

Roo, G. de (1993) Epiloog: de positie van de milieunorm verschuift, in G. de Roo (eds) *Kwaliteit van norm en zone; Planologische consequenties van (intergrale) milieuzonering*, Geo Pers, Groningen.

S.A.B. adviseurs voor ruimtelijke ordening bv (1994) Voorontwerp Bestemmingsplan Twentekaneel-Zuid, Fabelenweg en Zeggershoek, Gemeente Hengelo, Hengelo.

Simon, H.A. (1996) *The Science of the Artificial*, The MIT Press, Cambridge, Massachusetts.

Stuurgroep IMZ Hengelo Twentekanaal (1991) Integrale Milieuzonering Hengelo Bedrijventerrein Twentekanaal: Plan van Aanpak, Hengelo.

Stuurgroep IMZS Drechtsteden (1991) Rapportage eerste fase: Inventarisatie milieubelasting, Den Haag.

Stuurgroep IMZS Drechtsteden (1994) Rapportage tweede fase: Milieuperspectief, Den Haag.

Stuurgroep IMZ Theodorushaven (1993) Eindrapport fase 1 + 2: Inventarisatie en confrontatie van de milieubelasting en de ruimtelijke ordening, Conclusies en aanbevelingen, Bergen op Zoom.

Tesink, J. (1988) Integrale zonering DSM, in *Geluid en omgeving*, juni, pp. 67-69.

TK (Tweede Kamer) (1998) Regels over experimenten inzake zuinig en doelmatig ruimtegebruik en optimale leefkwaliteit in stedelijk gebied (Experimentenwet Stad & Milieu), Vergaderjaar 1997-1998, 25848, nrs. 1-2, Den Haag.

Twijnstra Gudde (1992) Tussentijdse evaluatie proefprojecten IMZ 2, IMZ-reeks deel 21, Ministerie van VROM, Leidschendam.

Twijnstra Gudde (1994) Derde evaluatie proefprojecten IMZ, IMZ-reeks deel 29, Ministerie van VROM, Den Haag.

Twijnstra Gudde (1994b) Voorlopige integrale milieuzone Arnhem-Noord en uitvoeringsplan, concept, Amersfoort.

Voerknecht, H. (1993) Saneren en bestemmen in een regionale aanpak, in G. de Roo (eds) *Kwaliteit van norm en zone; Planologische consequenties van (integrale) milieuzonering*, Geo Pers, Groningen, pp. 83-95.

VROM (Ministerie van Volkshuisvesting, Ruimtelijke Ordening en Milieubeheer) (1990) Ministeriële handreiking t.b.v. de proefprojecten integrale milieuzonering, Reeks Integrale Milieuzonering, nr. 5, Den Haag.

Witteveen en Bos (1991) Integrale Milieuzonering Burgum/Sumar: Tussenrapport, Provincie Friesland, Leeuwarden.

Chapter 7

Boer, M. de (1995) Brief aan de Voorzitter van de Tweede Kamer der Staten-Generaal betreffende Stad & Milieu (DGM/B/MBI Mbb 95026926, 22 dec. 1995), Ministerie van VROM, Den Haag.

Borst, H., G. de Roo, H. Voogd, H. van der Werf (1995) *Milieuzones in Beweging; eisen, wensen, consequenties en alternatieven*, Samsom H.D. Tjeenk Willink, Alphen aan den Rijn.

Cleij, J. (1994) Aanbiedingsbrief Integrale Milieuvisie Amsterdam 1994-2015, 9400018/02, Milieudienst Amsterdam, Amsterdam.

Commissie van Advies voor Volkshuisvesting, Stadsvernieuwing, Ruimtelijke Ordening en Grondzaken (1995) Strategie in relatie tot Cargill, SO 93/1710, Amsterdam.

Gemeente Amsterdam (1990) Conceptnota van Uitgangspunten voor de IJ-oever, Amsterdam.

Gemeente Amsterdam (1991) Nota van Uitgangspunten voor de IJ-oevers; Amsterdam naar het IJ, Amsterdam.

Gemeente Amsterdam (1994) Ontwerp Beleidsnota Ruimtelijke Ordening en Milieu, dienst Ruimtelijke Ordening in samenwerking met de Milieudienst, Amsterdam.

Gemeente Amsterdam (1994b) Integrale Milieuvisie Amsterdam 1994-2015, Milieudienst Amsterdam, Amsterdam.

Gemeente Amsterdam (1994c) Milieumatrix Structuurplan; Toetsingskader voor de beoordeling van milieueffecten van ruimtelijke ordeningsvoorstellen op structuurplanniveau, dienst Ruimtelijke Ordening, Amsterdam.

Gemeente Amsterdam (1994d) Amsterdam Open Stad, Ontwerp Structuurplan 1994, Deel II De toelichting, dienst Ruimtelijke Ordening, Amsterdam.

Gemeente Amsterdam (1994e) Bestemmingsplan IJ-oevers, dienst Ruimtelijke Ordening, Amsterdam.

Gemeente Amsterdam (1995) Milieuverkenning Amsterdam, Milieudienst, Amsterdam.

Gemeente Amsterdam (1995b) Stedebouwkundig programma van eisen, Houthavens, Gemeentelijke projektgroep Houthavensgebied, Amsterdam.

Gemeente Amsterdam (1995c) Brief aan de Commissie van Advies voor Volkshuisvesting, Stadsvernieuwing, Ruimtelijke Ordening en Grondzaken betreffende de Strategie Houthavens in relatie tot Cargill, SO 93/1710, Amsterdam.

Gemeente Rotterdam (1994) Rapport strategie Maas-Rijnhaven: Een inventarisatie naar de mogelijkheden van woningbouw in relatie met milieu-aspecten, June 1994, Werkgroep diverse diensten, Rotterdam.

Hoogstraten, S., L. de Laat (1994) Milieu eerder in de planvorming betrokken, dRO publikaties, nr. 11, Gemeente Amsterdam, Amsterdam.

Humblet, A.G.M., G. de Roo (eds) (1995) *Afstemming door Inzicht; een analyse van gebiedsgerichte milieubeoordelingsmethoden ten behoeve van planologische keuzes*, Geo Pers, Groningen.

Kloeg, D. (eds), L. van den Hoek Ostende, M. Marbus, H. van Wieringen (1991) *Natuur en Milieu Encyclopedie*, Zomer & Keuning Boeken B.V., Ede.

Knegt, M. de, E. Meijburg (1994) De Amsterdamse stedebouw stelt het milieu voorop; De Beleidsnota Ruimtelijke Ordening en Milieu verschenen, dRO publikaties, nr. 5, Gemeente Amsterdam, Amsterdam.

Kuijpers, C.J. (1992) *Bedrijven en milieuzonering*, Vereniging Nederlandse Gemeenten, Den Haag.

Meijburg, E., M. de Knegt (1994) Een stolp over Amsterdam; milieu en ruimtelijke ordening verstrengeld, *ROM Magazine*, nr. 10, pp. 9-12.

Provincie Noord-Holland (1994) Sanering industrielawaai, saneringsvoorstel Westpoort, Achtersluispolder, Westerspoor-Zuid e.o., Dienst Milieu en Water, Haarlem.

Provincie Noord-Holland (1995) Houthavens; geluidruimte voor toekomstige ontwikkelingen, interne notitie, augustus 1995, Dienst Milieu en Water, Afd. LVG, Haarlem.

Provincie Noord-Holland (1995b) Heroverweging ingevolge de Algemene Wet bestuursrecht; Houthavens Amsterdam, Brief aan Burgemeester en Wethouders van de Gemeente Amsterdam, 95-510098, 17 januari 1995, Dienst Milieu en Water, Haarlem.

Provincie Noord-Holland (1995c) Heroverweging ingevolge de Algemene Wet bestuursrecht, Brief aan Directie van Cargill B.V. en I.G.M.A. B.V., 95-510098, 17 januari 1995, Dienst Milieu en Water, Haarlem.

Provincie Noord-Holland (1995d) Verweerschrift aangaande beroep Cargill inzake vaststelling hogere grenswaarden Wet geluidhinder voor Houthavens (OBP IJ-oevers), Brief aan de Voorzitter van de afdeling bestuursrechtspraak van de Raad van State, 95-515493, 5 september 1995, Dienst Milieu en Water, Haarlem.

Provincie Noord-Holland (1995e) Concept-tekst t.b.v. milieukader Cargill-soja t.a.v. de component stof, interne notitie 23 augustus 1995, Dienst Milieu en Water, Haarlem.

Rangelrooij, P., R.C. Spaans (1995) *Onderzoek naar de mogelijkheden voor woningbouw in de Houthavens te Amsterdam*, Rapport R.95.126.A, dgmr raadgevende ingenieurs bv, Den Haag.

Rosdorff, S., L.K. Slager, V.M. Sol, K.F. van der Woerd (1993) *De stadsstolp: meer ruimte voor milieu èn economie*, Instituut voor Milieuvraagstukken, Vrije Universiteit, Amsterdam.

Sandig, J., F.J.H. Vossen (1994) *Geuronderzoek Houthavens Amsterdam Eindrapportage*, Project Research Amsterdam BV, Amsterdam.

Stadig, D. (1995) Saldobenadering: vervanging of aanvulling van de norm?, in *Milieuzones in Beweging*, congresbundel, Rijksuniversiteit Groningen, Route IV, Groningen, Nijmegen.

Stafbureau NER (1992) Nederlandse Emissierichtlijnen-Lucht, Bilthoven.

Steiner, G.A. (1997) *Strategic Planning: What every manager must know*, Simon & Schuster, New York

Timár, E. (1996) Improving Environmental Performance of Local Land Use Plans, in Miller, D. and G. de Roo, *Urban Environmental Planning; Policies, instruments and methods in an international perspective*, Rijksuniversiteit Groningen, Groningen.

Vos, A.H.M.T. (1994) Gebiedsgericht beleid en Amsterdams structuurplan, *ROM Magazine*, nr. 10, pp. 13-16.

VROM (Ministerie van Volkshuisvesting, Ruimtelijke Ordening en Milieubeheer) (1992) Nota Stankbeleid, Tweede Kamer, vergaderjaar 1991-1992, 22715, nr. 1, Den Haag.

VROM (Ministerie van Volkshuisvesting, Ruimtelijke Ordening en Milieubeheer) (1994) Document Meten en rekenen geur, Publikatiereeks lucht & energie, nr. 115, DGM, Den Haag.

VROM (Ministerie van Volkshuisvesting, Ruimtelijke Ordening en Milieubeheer) (1995) Waar vele willen zijn, is ook een weg; Stad en Milieu Rapportage, Den Haag.

VROM (Ministerie van Volkshuisvesting, Ruimtelijke Ordening en Milieubeheer) (1996) Vierde Nota over de Ruimtelijke Ordening Extra, Den Haag.

Zeedijk, H. (1995) *Stof in de Houthavens*, Centrum voor MilieuTechnologie, Technische Universiteit Eindhoven, Eindhoven.

Interviews

Arents, F. (25 April 1996) Juridische Zaken, Bureau lokale planologie, Dienst Ruimte en Groen, Provincie Noord-Holland.

Bakker, S.B. (25 April 1996) Projectleider bureau geluidsanering industrielawaai, Dienst Water en Milieu, Provincie Noord-Holland.

Cleij, J. (3 April 1996) Directeur Milieudienst, Gemeente Amsterdam.

Meijburg, M.E. (3 April 1996 en 25 April 1996) Beleidsmedewerker Milieudienst, Gemeente Amsterdam.

Pijning, J. (25 April 1996) Beleidsmedewerker afdeling lucht, veiligheid en geluid, Dienst Milieu en Water, Provincie Noord-Holland.

Schoonebeek, C.A.M. (25 April 1996) Beleidsmedewerker afdeling lucht, veiligheid en geluid, Dienst Milieu en Water, Provincie Noord-Holland.

Letters

Arents, F. (18 November 1996) Juridische Zaken, Bureau lokale planologie, Dienst Ruimte en Groen, Provincie Noord-Holland; betreffende commentaar op concept-tekst.

Bakker, S.B. (7 July 1996) Projectleider bureau geluidsanering industrielawaai, Dienst Water en Milieu, Provincie Noord-Holland; betreffende correctie interviewtekst.

Pijning, J. (15 November 1996) Beleidsmedewerker afdeling lucht, veiligheid en geluid, Dienst Milieu en Water, Provincie Noord-Holland.; betreffende correctie interviewtekst.

Chapter 8

Christensen, K.S. (1985) Coping with Uncertainty in Planning, *Journal of the American Planning Association*, Winter, pp. 63-73.

Abbreviations

AMVB	=	governmental decree *(Algemene Maatregel van Bestuur)*
BEVER	=	new policy on soil remediation *(Beleidsvernieuwing Bodemsanering)*
BROM	=	Amsterdam Policy Document on Spatial and the Environment *(Beleidsnota Ruimtelijke Ordening en Milieu)*
BUGM	=	Subsidy Decree for Municipal Policy Implementation *(Bijdragebesluit Uitvoering Gemeentelijk Beleid)*
BZ	=	Ministry of Forreign Affairs *(Ministerie van Buitenlandse Zaken)*
CRMH	=	Dutch Advisory Council for the Environment *(Centrale Raad voor de Milieuhygiëne)*
EZ	=	Ministry of Economic Affairs *(Ministerie van Economische Zaken)*
FUN	=	NMP Implementation Budget *(Financiering Uitvoering NMP)*
IEZ	=	Integrated Environmental Zoning
IMP-M	=	Indicative Multi-year Programme on the Environment *(Indicatief Meerjarenprogramma Milieubeheer)*
IMZ	=	Integrated Environmental Zoning *(Integrale Milieuzonering)*
IPO	=	Interprovincial Platform *(Interprovinciaal Overleg)*
ISV	=	Investment Budget for Urban Renewal *(Investeringsbudget Stedelijke Vernieuwing)*
IVM	=	Institute of Environmental Studies of the Vrije Universiteit Amsterdam (VU)
KB	=	Royal Decree *(Koninklijk Besluit)*
LNV	=	Ministry of Agriculture, Nature Management & Fisheries *(Ministerie van Landbouw, Natuurbeheer & Visserij)*
MIG	=	Modernizing Instruments Noise Policy *(Modernisering Instrumentarium Geluidbeleid)*
NMP	=	National Environmental Policy Plan *(Nationaal Milieubeleidsplan)*
PDAES	=	1976 Policy Document on Ambient Environmental Standards *(Nota milieuhygiënische normen 1976)*
PIM	=	Environmental Policy Integration Project *(Project Integratie Milieubeleid)*

PGBO	=	Bureau for Integrated Soil Research *(Programmabureau Geïntegreerd Bodemonderzoek)*
RARO	=	Spatial Planning Council *(Raad voor de Ruimtelijke Ordening)*
RAWB	=	Advisory Council for Science Policy *(Raad voor Advies van het Wetenschapsbeleid)*
RIVM	=	National Institute of Public Health and the Environment *(Rijksinstituut voor Volksgezondheid en Milieu)*
RPD	=	National Spatial Planning Agency *(Rijksplanologische Dienst)*
ROM areas	=	designated Spatial Planning & Environment areas *(ROM-gebieden)*
ROMIO	=	Spatial Planning and the Environment in Industry and ??? *(Ruimtelijke Ordening en Milieu bij Industrie en Omgeving)*
SER	=	Socio-Economic Council *(Sociaal-Economische Raad)*
SCMO-TNO	=	TNO Study Centre for Environmental Research(TNO = Netherlands Organization for Applied Scientific Research)
TK	=	Lower House of Representatives (previously Second Chamber) *(Tweede Kamer)*
V&W	=	Ministry of Transport, Public Works and Water Management *(Ministerie van Verkeer en Waterstaat)*
VINEX	=	Fourth Policy Document on Physical Planning-Plus *(Vierde Nota Ruimtelijke Ordening Extra)*
VINO	=	Fourth Policy Document on Physical Planning *(Vierde Nota Ruimtelijke Ordening)*
VM	=	Ministry of Public Health and Environmental Hygiene *(Ministerie van Volksgezondheid en Milieuhygiëne)*
VNG	=	Association of Netherlands Municipalities *(Vereniging Nederlandse Gemeenten)*
VNO-NCW	=	Confederation of Netherlands Industries and Employers
VOGM	=	Follow-up Funding for the Development of Municipal Environmental Policy *(Vervolgbijdrageregeling Ontwikkeling Gemeentelijk Milieubeleid)*
VROM	=	Ministry of Housing, Spatial Planning and the Environment *(Ministerie van Volkshuisvesting, Ruimtelijke Ordening en Milieubeheer)*
VRO	=	Ministry of Housing and Spatial Planning, forerunner of VROM *(Ministerie van Volkshuisvesting en Ruimtelijke Ordening)*
VS-IMZ	=	Provisional System for Integrated Environmental Zoning *(Voorlopige Systematiek voor een Integrale Milieuzonering)*

Legislation

Air Pollution Act	=	*Wet inzake de luchtverontreiniging (Wet Luvo)*
Aviation Act	=	*Luchtvaartwet*
Buildings Decree	=	*Bouwbesluit*
Chemical Waste Act	=	*Wet chemische afvalstoffen*
Commodities Act	=	*Warenwet*
Environmental Protection Act	=	*Wet milieubeheer*
Environmental Protection (General Provisions) Act	=	*Wet algemene bepalingen milieuhygiëne (WABM)*
Environmental Retail Trade Decree	=	*Besluit Detailhandel Milieubeheer*
Indicative Multi-Year Programme	=	*Indicatief Meerjarenprogramma (IMP)*
Major Airports Noise-Nuisance Decree	=	*Besluit geluidbelasting grote luchtvaartterreinen*
Multi-Year Implementation	=	*Meerjarig uitvoeringsprogramma*
Programme for Noise Abatement	=	*Geluidhinderbestrijding*
Memorandum on Noxious Odours	=	*Nota Stankbeleid*
Noise Abatement Act	=	*Wet geluidhinder*
Nuclear Energy Act	=	*Kernenergiewet*
Nuisance Act	=	*Hinderwet*
Policy Document on Agriculture and Nature Conservation	=	*Relatienota*
Policy Document on Deregulation in Spatial Planning & Environmental Protection	=	*Nota Deregulatie ruimtelijke ordening en milieubeheer*
Pollution of Surface Waters Act	=	*Wet verontreiniging oppervlaktewateren (WVO)*
Prevention of Major Accidents Decree	=	*Besluit Risico's Zware Ongevallen (BRZO)*
Priority Policy Document on Pollution Control	=	*Urgentienota milieuhygiëne*
Rendering Act	=	*Destructiewet*
Soil Clean-up (Interim) Act	=	*Interimwet bodemsanering (IBS)*
Soil Protection Act	=	*Wet bodembescherming (WBB)*
Soil Protection Guidelines	=	*Leidraad Bodembescherming*
Spatial Planning Act	=	*Wet op de Ruimtelijke Ordening*
Spatial Planning Decree	=	*Besluit op de Ruimtelijke Ordening*
Urban and Rural Regeneration Act	=	*Wet op de stads- en dorpsvernieuwing*
Urban Renewal Act	=	*Wet stedelijke vernieuwing*
Waste Substances Act	=	*Afvalstoffenwet*
Working Conditions Act	=	*ARBO-wet*

Index

Printed and bound by CPI Group (UK) Ltd, Croydon, CR0 4YY
21/10/2024
01777082-0008